Computer Telephony on the Sun Platform

by

Patrick Kane, PhD.

A Flatiron Publishing, Inc. Book

Published by Flatiron Publishing, Inc.

Computer Telephony on the Sun Platform

ISBN 0-936648-85-6

Manufactured in the United States of America

First Edition, February 1996

Cover design by Mara Seinfeld

Printed at Bookcrafters, Chelsea, MI.

CONTENTS AT A GLANCE

CONTENTS

FOREWORD

WHY I LIKE SUN COMPUTERS
by Harry Newton

In many ways, Sun computers are the best platform to run computer telephony applications on. Go into telephone company central offices, you'll find Sun computers running the critical part of the phone network—Signaling System 7—the part that sets up calls and carries information, like who's calling and who is being called. That the phone industry chose Sun is a big compliment.

Sun computers are reliable, sturdy and very powerful. In recent years, the company has increasingly focused, with considerable success, on adding intelligence to telephony, not just at the central office, but throughout businesses. One of Sun's best-kept secrets is its XTL API, the Sun Solaris Teleservices Application Programming Interface.

According to Sun: "Sun's vision of the impact of widespread use of teleservices suggests that the computer workstation will become the new communications center, combining many existing communication media with new ones, while creating new paradigms for the expression and sharing of ideas. Information in the form of charts and pictures, schedules and plans, and audio and video will merge through application programs that provide a collaborative vehicle for decisions in the 1990's and beyond. The desktop will become the platform for a new set of productivity tools, seamlessly integrated into the critical business activities and methodology of today's companies, and providing a competitive edge for facing the global challenges of tomorrow. Individuals will gain new freedom in where they work and how they access information. And ideas will be communicated in more expedient and creative ways." Solaris Teleservices is the platform for workstation apps which leverage the benefits and capabilities of the telephone network. Teleservices applications, according to Sun, include:

- *Desktop Teleservices.* Workstation based telephone and answering machine applications allow users to efficiently plant, receive and manage telephone calls.

- *Remote Access.* Users can place calls for their workstation from any telephone and access applications and data through DTMF signaling, or through speech, using a workstation's speech recognition capabilities.
- *Wide Area Networking.* The ubiquity of telephone networks allow for the complete connectivity of all computers. Network links can be brought up or taken down on demand merely by placing or tearing down a telephone call.

The Solaris Teleservices Platform is called XTL, a multilayered software architecture based on client server computing model. XTL consists of four key components:

- *A client-side library.* Providing a high-level, object oriented application programming interface (API) to application programmers. Using the XTL API, an application can place and retrieve telephone calls. The API library consists of a collection of C++ objects which is linked to applications that wish to use the systems teleservices resources.
- *A server.* Providing multi-client and multi-device support, the server is the central point of contact for all teleservices, resource management and security are provided by the server. Communication between the XTL API and the server occurs through the XTL Server Protocol (XTLS).
- *One or more providers.* Manages each telecommunication device connected to the system. The Teleservices server communicates with an XTL provider using the XTL provider protocol (XTLP).
- *A data stream multiplexor* (Sun's spelling). The universal multiplexor (Umux) provides a uniform means for applications to access and share data channels associated with a telephone call.

The thrust of this book is towards Sun as a ubiquitous computer telephony tool. If you love stable apps and neat screens (I particularly like Sun's), this is the book for you.

Enjoy and profit.

Harry Newton
New York
March, 1996
harrynewton@mcimail.com

PREFACE

There is a megatrend underway which has dramatically changed the way we work, shop, communicate and spend our leisure. It has changed everything we know. It has assisted the downsizing, rightsizing and re-eingineering of businesses. It has resulted in telecommuting and the explosive growth of the SOHO market. It has expanded our reach globally without leaving the comforts of our home or office. It has put us in constant contact with the individuals, institutions and services that intersect our daily lives. Its tirelessly ubiquitous presence is near intrusive yet indispensable.

The explosive technological advancements in computer processing capabilities combined with innovation in communication network topology is the underpinning of this evolution. We have been freed from the constraints of yesterday's centralized technologies and are now capable of placing usable information in the hands of those who need it when they need it. It is a technology that assists; not constrains. It continually improves.

Late Summer 1995, Harry Newton, one of the industry's brightest stars, extended me the opportunity to detail the involvement of Sun Microsystems in this evolution in an easy to read resource, usable by students, practitioners and engineers alike, to assist their participation in, and understanding of this dynamically expanding megatrend.

Ensuing considerable research, conversation and thought I present you the maiden edition of this book. Any errors, omissions, inaccuracies, misconceptions or confusion caused by this work are unintentional yet part of the human element in preparing this treatise. To assist my future efforts in enhancing this work I request you take a few minutes to share your thoughts and suggestions with me.

Patrick Kane
New York City, January 1996
kanephd@aol.com

ACKNOWLEDGMENTS

There are many people, past and present, who influenced my authoring of this book. Those who have passed, and many still quite alive, will never know that they did, nor did I at the time. Special thanks go to George Dinsdale, Linkon's Mr. Sun, who back in the early 1990's introduced me to Sun Microsystems technology, and subsequently joined Linkon and brought our two organizations together. Without his assistance and contacts, this effort would have been much more painful than it was.

Thank you, Sun Microsystems, for being the industry leader that you are, and for the use of white papers and assorted documents in the preparaton of this work. Your assistance and input through the efforts of Doug Ehrenreich, Karen Richards, and their associates and contacts was invaluable. I am in your debt.

Many thanks to the Linkon team. Specific thanks go to Charles Castelli, who founded Linkon with me; and to Jim Linley, Linkon's Chief Architect—a man whose brilliance in hardware and software design has truly given the industry an open architecture "universal port" product, years before anyone really understood the benefits of this technology.

To Harry Newton, editor of *Teleconnect, Call Center, Computer Telephony* and *Imaging* magazines and author of *Newton's Telecom Dictionary*, whose unwavering evangelistic drive has inspired many individuals like myself to further move this industry into the mainstream: Harry, thanks for the opportunity of realizing a dream.

To Kenneth Gan Chin Leung for the illustrations and text design: Your talent and perspective is greatly appreciated.

In addition to maintaining my standard 70-plus-hour, work week, bodybuilding six days per week, and the obligations associated with daily life, I managed to author this book. This was only possible through the patience and resolve of my New York City family and friends. Peter Gabos, MD, my mind-double and training partner, whose friendship and inspiration kept the motor running when there truly was just fumes left in the tank. Janet Stampler and Travis Groff who regularly checked my progress and kept my eye on the ball.

Vincent Smith who has known me longer than most and seen me through some very difficult times. Dennis Kane with his endless enthusiasm and spirit, and last but certainly not least, Will Chafin, my spark plug, who sees true love and beauty in life, who found his way to keep me motivated while removing some of the jadedness from my New York personality. God love all of you.

|1|

What Is Computer Telephony?

Computer Telephony, the addition of mechanical intelligence to a telephone call, has had numerous names over the past decade. In earlier days it had no industry defined name at all. Pioneers of this industry have witnessed computer telephony's evolution: from simple barge entry dial-up services offered by AT&T in the early 1980's through telephone based weather, sports, time and other audiotex offerings in local markets by the RBOCs. The mid-1980's saw Bank-By-Phone getting its start while Chat Lines and 900 Pay-Per-Call services were clearly leaving their mark on society. Voice mail, voice messaging, fax, fax-on-demand, fax broadcasting and automated attendant grew from large corporate business tools to technology used by all corporations. Those tools are now available as a service for residential and small business use from some telephone carriers in local markets and small packaged offerings at the neighborhood computer store.

Computer Telephony has clearly transformed the way we work and live more dramatically than any other technology to date. The birth of this technology was a natural. Over 98% of all households and businesses had telephone service; the "computer" terminal for this new interaction. The model for the early uses of computer telephony integrated use was terminal access to a mainframe computer. The intelligence was located in the computer running the application at the other end of the network manipulated through the telephone keypad.

As is the case with computer evolution, more and more intelligence was added to this telephone "terminal". Technology improved. It became more robust and user friendly. The migration path proceeded closer and closer to the hands of the user. Both the computer and the telephone have achieved their ultimate union with their current availability on the workstation and desk-top. We are all the better for it.

The productivity gains realized through this unification are astounding. Pay back of capital outlays for this equipment is measured in months. The ability to unite the telephone network with the information needed throughout the day makes for a more complete process and a more productive individual.

Computer telephone integration began its successful journey with voice. It has evolved to include data and image. These technologies are now available simultaneously on the same telephone lines we use to converse daily. Who knows where the technological advancements of this rapidly expanding industry will take us next. The only limiting factor is our creative genius to determine just that.

Computer Telephony in Action

The following examples of Computer Telephony integrated products are provided to facilitate a better understanding of this technological evolution. Many of these systems you probably have experienced at one time or another. Experience will make you aware that all systems are not created equal. Some you will love; others you will hate. I apologize in advance for rekindling any bad memories.

Additional examples, with a greater amount of detail, can be found in Chapter 14.

Automated Attendants

Automated attendants simply provide a recorded message offering the caller a set of destinations, let the user make a choice by entering a number or extension, and route the call to that destination. Callers are typically required to respond from a phone which is capable of dual tone multi-frequency (DTMF) signaling. Automated attendants can route calls to an operator after a period of time with no response from the caller making the service accessible to those without DTMF service. Newer systems have speech recognition capabilities as well as DTMF. This reduces the number of calls requiring live operator intervention.

Voice Store and Forward

Messaging systems and voice mail are built upon voice store and forward technology. Many times after routing a call to a specific extension, via automated attendant, the called party not available to answer. In this case voice mail or voice messaging systems can prompt the caller to leave a message for the called party. A blinking light, warbling dial tone or an announcement over the intercom feature of the telephone handset can alert the called party that a message is waiting. Other features in commercial voice messaging systems include the ability to forward messages to another number, the use of distribution lists, alternate messages for extended absences and priority message designation.

Lesser developed countries, lacking infrastructure sufficient to provide the required telephone service, deploy "virtual telephone service". This service uses large voice messaging (voice mail) systems to provide mailboxes where subscribers can receive messages. They check for messages and return calls via public pay phones.

Voice Information Services and Interactive Voice Response

Voice information services, also known as audiotex services, allow users to obtain routine information such as stock quotes, product information, weather reports or sports scores. These are configured as simple prerecorded announcements and played on command by the caller. A variation of these are fax services that allow product data sheets, tax forms or other literature to be sent to callers on demand. This technology was most visibly employed by the pay-per-call industry.

More sophisticated applications allow the system to query a database and a synthesized voice message, using text-to-speech technology, can vocalize the information for the caller. Such systems are called interactive voice response

(IVR) which are often used by financial institutions to permit callers to check balances and transfer funds. IVR systems can replace live agents entirely or greatly reduce the headcount required to service customer inquiries.

Fax

Computer based fax technology has been widely embraced by many individuals and organizations. It certainly has substantially increased the volume of fax traffic on the telephone network. Following are a few examples of how it has been used.

When used for fax store and forward applications it works very similar to voice mail. Faxed documents are received either by a server or an independent computer and stored for later retrieval. It permit users ample storage and the ability to remotely retrieve documents. Some individuals use it in conjunc-

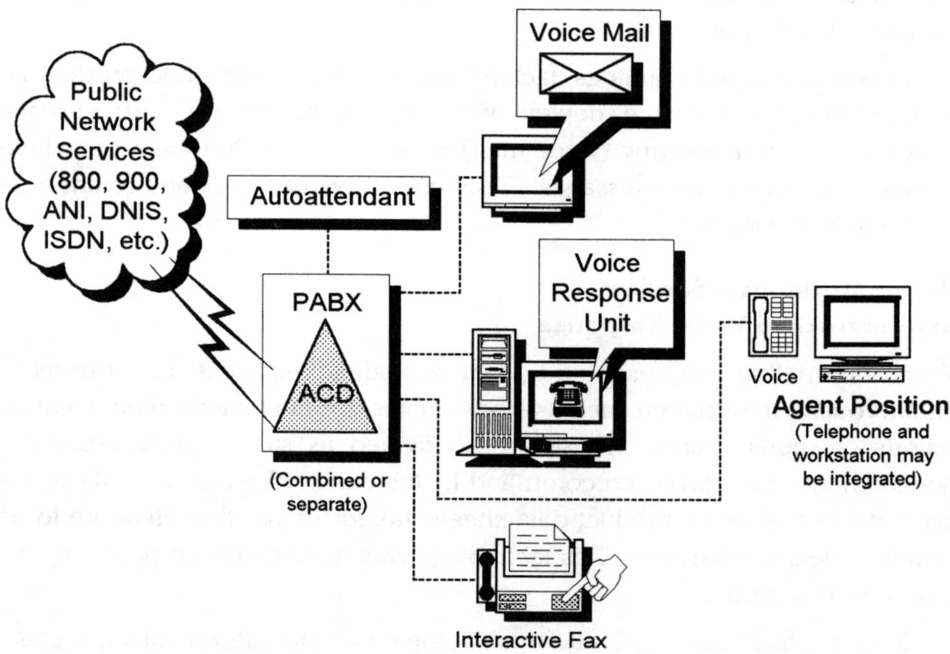

FIGURE 1.1 Voice Processing Systems are frequently configured as a set of options to a PBX/PABX system.

tion with their stand-alone fax machine. It provides the capability to handle overflow situations or receive faxes when their machine is in disrepair. Individuals not personally owning a fax machine can receive fax information via their "virtual fax machine" for later retrieval when they have access to a fax machine.

One and two call fax-on-demand systems is yet another widespread fax application. They provide a 24 hour day information source for customers and prospects to obtain product information or receive confirmation on an order they placed. Many companies have taken a substantial load off their sales support areas through deployment of fax-on-demand technology.

Fax broadcast system have substantially changed the direct mail industry. It has reduced their printing and mailing costs and cut their delivery times to minutes. Targeted marketing programs are more readily adjusted with increased response rate through to use of this application.

Figure 1.1 is a schematic of some of the above discussed technologies implemented in a network.

Voice Activated Dialing

Candice Bergman provided the most glamorous introduction to voice activated dialing known to this Computer Telephony technology. It is the first application with sufficient "sizzle" to warrant dedicated electronic media attention. Once the system is trained by the user the only ability he or she needs to use it is to speak. Voice Dialing provides interaction with the telephone network using natural language voice commands. This service introduces a new level of safety to mobile phone use and convenience when using a land line telephone.

Speaker dependent and speaker independent speech recognition make this service possible. It allows users to dial telephone numbers simply by speaking the number, name or for the person they would like to call. Voice activated dialing is a network based solution which permits the user to speak commands like "call mom" to connect with their mother's telephone. To use the service the user must train the system to their respective voice in the environment the service will be used. For example, to use the service with a car phone the caller must first record the trigger or command phrases to be used (i.e. call home, call office, call Harry), while in the car. Once trained the system should respond with approximately 95+% accuracy to the trained name or phrase. Voice command technology is now sparsely seen in the market. It will become commonplace in the near future.

Unified Messaging Systems

Unified messaging is the combination of voice mail, fax mail and E-mail in the same mail box for convenient retrieval by the box owner. Most of us receive a barrage of voice, fax and E-mail messages daily. Until recently we needed to check three separate sources to retrieve all messages sent to our attention. Unified messaging creates a central repository with documentation as to what was received, when and the degree of urgency. Most systems offering this service utilize terminal notification and retrieval capabilities to make the most productive use of these combined information sources. When used at the desktop and workstation level the user is presented an index listing of all calls received. The user can then scan the list for the most urgent messages an attend to them first. Less important communications can be retrieved at a more appropriate time. With Unified Messaging remote retrieval is possible over a standard telephone through the utilization of voice store and forward technology for voice mail messages and text-to-speech for fax and E-mail messages. Critical faxes and E-mail messages can be routed throughout the system just like standard voice mail or faxed or modemed to the location the user is currently at. Unified messaging is the ultimate convenience service for anyone out of their office for any reason.

Single Number Service

Single Number Service is used by frequent travelers, field service industry personnel and individuals who like to stay in touch. Callers dial a single phone number which is routed to the subscriber wherever he or she might be. The service is usually based on the subscribers predetermined schedule. More advanced services use a call screening function which records the caller's name and relays this information to the subscriber, who may elect to accept or decline the call. If a caller is unavailable or not taking calls the call is routed to a mailbox where it is stored for later retrieval.

Video-Conferencing

Through the utilization of digital ISDN circuits video-conferencing has become a reality. It provides the capability of having a face-to-face meeting with someone physically located on the other side of the globe. This real time communication can include document sharing, the viewing of power point slides and any other relative stationary communication form. To accomplish such a feat both systems used for this highly interactive technology must adhere to the same communications standard. Most video-conferencing systems available are based on a H.320/T.120 multiplexing protocol standard. Compliance permits dissimilar video-conferencing systems to communicate. Video-conferencing systems are based on computers fitted with a microphone and camera mounted above the monitor screen.

ACD (Automatic Call Distributor)

ACD technology exhibits the highest degree of Computer Telephony integration to date. This technology, used by corporate call centers and telemarketing firms, has dramatically increased the productivity of customer telephone contact. The majority of these productivity gains have been realized in amortization of capital expenses for equipment and manpower as well as the dramatic increase in quality information available to the agent and caller at the time of the call. Through ACD customer contact is like an ongoing dialogue due to database access capabilities integrated with screen pop technology. The customers and companies have both benefited substantially from its use.

ACD technology evolved from the automated attendant applications discussed earlier. It uses Automatic Number Identification (ANI) or voice prompted digits to identify the caller. This information is interfaced with a customer databases containing the documented relationship history with that organization. Information such as purchasing history, payment history, demographic profile of the caller and their family assists the agent in personalizing the call; building and sustaining the relationship. Through the use of ACD the call is routed to the next available agent or "relationship manager" agent assigned to that customer/client. When the call arrives at the agents' workstation it is preceded by the customer information with a customized script for the agent to read in the form of a screen pop. The information provided to the agent could have its origin in several databases. The customer data could

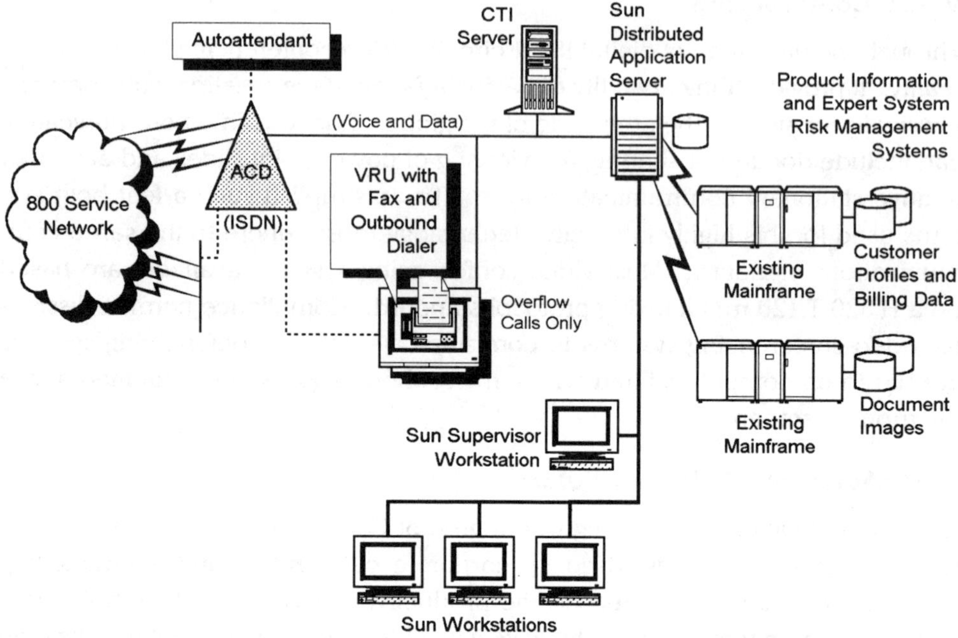

FIGURE 1.2 A Credit Card Customer Service Solution is one example of a generic customer service call center application.

originate in a customer data base where it is merged with accounting database information for the financial history. Customer purchase information may require the ACD to access an order entry and inventory data base. All this appears in the screen pop nanoseconds prior to the agent saying "Good morning Mr. Phelps".

Telemarketing organizations back-ending direct response and customer service related campaigns for several organizations utilize Dialed Number Identification Service (DNIS) to trigger the appropriate program at the agent workstation. This technology identifies the called number, combines it with ANI information detailed above and passes the call to the next available agent or "relationship manager" agent. ANI information with no prior customer history can be interfaced with reverse directory data to provide information to help personalize the call. This ANI information can also be interfaced with credit history databases to pre-establish potential parameters when handling the purchase of orders.

Remote and on-site supervisor and administrative control are features of most ACDs. These include call monitoring, inventory reporting, up-to-the-minute tabulation of call results, number of calls handled in total and by agent segmented by time of day etc. ACD technology is quite advanced and continues to progress.

Figure 1.2 is a schematic depicting a credit card customer service solution which utilizes much of the technology detailed above.

All the above examples of the technology rely on computer power for processing combined with the telephone network for transport of the communications. To get a better understanding of how Computer Telephony actually works we must first establish a basic understanding of the telephone network. The intention of the next chapter is to explain the telephone network.

|2|

Understanding the *Telephone System*

Background

Before the US phone system was deregulated, AT&T provided nearly every aspect of the domestic telephone service. This included local dial tones and long-distance service. The most significant aspect of the 1984, breakup of AT&T was the spinning off of local service to the seven Regional Bell Operating Companies (RBOCs or LECs). These LECs provide local dial tone in geographically defined markets.

These LECs have enjoyed a near exclusive right to provide dial tone in one or more local access and transport areas (LATAs), which usually correspond to an area served by a single area code. In addition to providing dial tone within its LATA, the LEC is also responsible for the completion and billing of calls that originate and terminate inside each of its LATAs. A variety of central offices (COs) or exchanges, which are identified by their prefixes, make up the LATA.

To complete calls that cross LATA boundaries, interexchange carriers (IXCs) become involved. Connections from a LEC to an IXC occur through an IXC point of presence (POP) within the LATA as seen in Figure 2.1. The most renowned of these IXCs are AT&T, MCI and Sprint. Each of these carriers owns

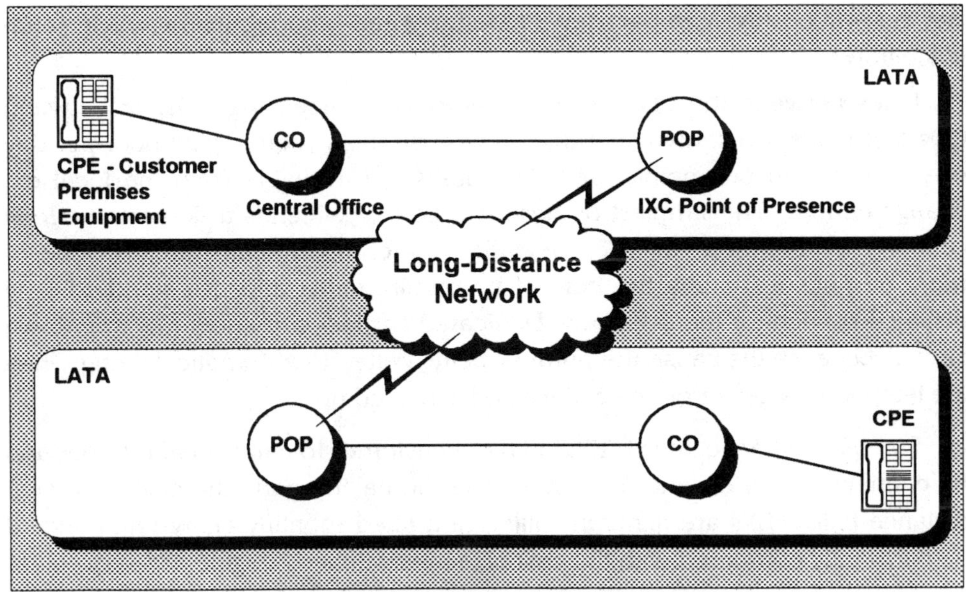

FIGURE 2.1 Commercial Phone System Topology. Long-distance service involves the cooperation of both local exchange carriers (LEC) and interexchange carriers (IXC) connecting at an IXC point of presence.

and operates its own network, which represent a huge capital investment. These networks, for the most part, are national and international fiber optic backbones with hundreds of POPs. Other IXC vendors include resellers and aggregators, both of whom simply resell long-distance and billing services under agreements with the big three.

The standard telephone hardware, which we are all grew up with, consists of a dialer (rotary or tone) and handset. This is traditionally referred to as a station set or telephone. Its basic connection is known as a plain old telephone service line (POTS) or central office (CO) line. Most homes still use analog connections for the local loop between the phone company and the home. Phones that serve these analog circuits are known as 2500 sets. Phone systems for business and new residential service often use phone sets that can communicate digitally. These digital connections allow for more sophisticated communication between the phone and switch. As the LECs continue to modernize their networks and become more digital oriented in their service delivery, this will positively impact service delivery and broaden the available product features.

Lines between the CO and the customer, between the COs, or between COs and POPs are known as trunks. For customers requiring 24 hour per day service between two points, dedicated lines are available from the local exchange carrier. The simplest of these is a tie line which is a dedicated circuit capable of carrying a single conversation. A company can purchase one or more tie lines to handle the bulk of its requirements and use standard long distance service for the overflow. Dedicated tie lines are known as ring-down circuits because the cause the number being called to automatically ring when the telephone receiver on the caller's end is picked up.

Foreign exchange lines (FEXs) allow a customer to call a local number and be connected to a distant location, saving the calling party the cost of a long-distance call. FEXs are generally billed at a fixed monthly charge and should be constantly compared with the cost of using switched 800 and 888 service.

Other special kinds of dedicated lines include the all digital T-1 and ISDN lines, which handle data and voice communications. T-1 lines have an aggregate throughput of 1.54 megabits per second (Mbps). Each T-1 line consists of 24, 64- Kbps channels. Each channel can be used separately for telephone or data communications sessions or can be combined to obtain higher data rates.

ISDN

The integrated service digital network (ISDN) refers to a specific set of services available through standardized sets of interfaces. ISDN is an architecture for digital communications that can provide a number of services currently requiring separate networks such as voice, circuit-switched and packet-switched data, video conferencing and fax. ISDN is available in Primary Rate Interface (PRI), Basic Rate Interface (BRI) and Broadband ISDN. Detailed information on ISDN circuits can be found in Chapter 6.

Although long-distance carriers have embraced ISDN, until recently the LECs took a less aggressive posture. Most of the LECs currently are aggressively marketing this service. In Europe, where no alternative for international data service exists, ISDN has been strongly adopted.

800, 888 and 900 Number Services

The use of toll-free 800 and 888 numbers by businesses and consumers has seen explosive growth since their introduction. These services are offered by the LECs and the IXCs. Such services from the LECs are limited to the LATA, making them of little use to organizations conducting national business. Switched 800 and 888 service is more useful because it can be directed to a specific CO line by the long-distance carrier. With the advent of switched 800 and 888 service both inbound 800 and 888 and standard outbound and inbound service can be handled by the same POTS or CO line. This makes the service more flexible and quicker to set up. Now with 800 and 888 portability, the customer has the ability to use the same 800 and 888 numbers with any long-distance IXC. Now the customer is no longer tied to a specific carrier for service. This has resulted in more competitively priced switched 800 and 888 service. Figure 2.2 is a schematic of this type of service.

Other kinds of 800 and 888 service include dedicated inbound CO lines and

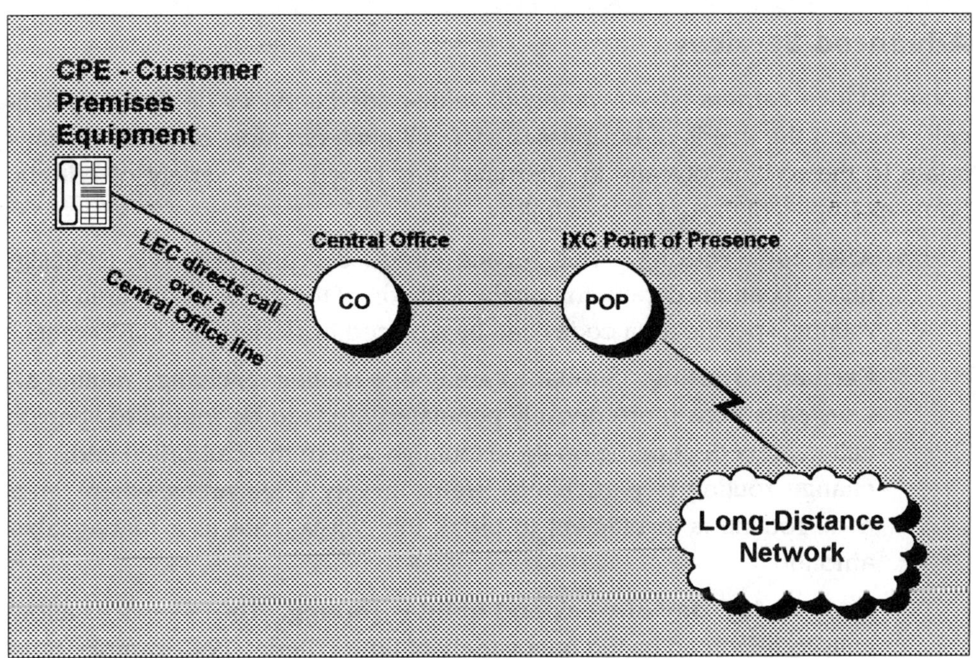

FIGURE 2.2 Switched 800 Number Service is gaining popularity because it uses existing lines and can be easily and quickly established.

special access lines, often using T-1 service. The use of these rather than switched 800 and 888 service is chiefly a matter of line usage and cost. On average, the cost of using switched 800 and 888 service on 10 analog circuits is the same as running the service on a T-1 circuit, which is 24 digital lines.

Alternative access providers such as MFS Communications Network (MFS) permit the customer to bypass the LECs and connect directly to the IXC through the alternative access provider's network. The LECs charge substantial fees to the IXCs for connection to their local network. Savings from the alternative access providers through this bypass are immediately realized by the customer.

With 900 number services the called party charges the caller both for the call and the price of special services obtained during the call. Billing for these services is provided by the RBOC under contract to the long-distance carrier. These services spawned a new pay-per-call industry in the late 1980's and have substantially decreased in use since then. They still represent an excellent way for companies to offer self-funding information and customer support services.

Advanced Services

Long-distance carriers offer a variety of value-added services designed specifically to meet the needs of businesses and call centers. Most of these services relate to the routing of calls to alternate sites based on a variety of criteria. Some examples of these services follow.

⇒ **Area code routing.** This service directs 800 and 888 calls to designated locations depending on the location of the caller. With area code routing, specific area codes can be serviced by designated call centers.

⇒ **Exchange routing.** Similar to area code routing, exchange routing allows calls from more densely populated areas to be routed to specific call centers depending on the exchange where the call originated. Exchange routing is particularly useful in very large areas where area code routing is insufficient to break the call volume to a manageable amount.

⇒ **Time of day, day of week and holiday routing.** These services allow calls to be routed based on a variety of time criteria. For example, time of day routing could allow calls to be directed to a particular call center of the company during normal business hours, but to another after 5

P.M. Day of week and holiday routing can direct calls based on predefined calendars. Using this service, companies can better schedule and utilize personnel.

⇒ **Emergency routing.** If a disaster or other act of God forces a call center out of operation, emergency routing allows calls to be directed by the long-distance carrier to an alternate site.

⇒ **All trunks are busy (ATB).** During peak periods or ensuing a commercial running on electronic media, all lines into a specific call center may be busy. ATB routing can redirect calls to alternate locations rather providing a busy signal thus requiring the caller to call back..

⇒ **Allocation routing.** Allocation routing allows calls to be split among designated sites on a percentage basis. This allocation scheme can be used to balance loads at any time.

Many carriers are now placing the control of call routing services in the hands of their larger customers. This is accomplished by the installation of a terminal on-site at the customer location.

Switches and Automatic Call Distribution

Telephone service connections have two basic components; lines and the switches that service them. Digital software controlled switching represents some of the most advanced technology in phone systems today. It incorporates millions of lines of code, advanced predictive algorithms and extensive fail-safe mechanisms. Switching systems are located in point of presence (POP), central offices (COs) and businesses. Carrier switches are an integral component of their advance intelligent network (see Chapter 15). Businesses use PBXs or Centrex service from the RBOC.

Central Office Switches

In the past central office switches had a basic function of connection between an incoming call and an outgoing line, based on a dialed number. Central office switches are rapidly being converted to digital technology, allowing the introduction of advanced services. These services include emergency 911, call waiting, call forwarding, *69 service, call conferencing and voice messaging to name a few.

One of the most widely known CO-based service is Centrex. Centrex provides such basic services as call transfer, call forwarding and conferencing without requiring equipment at the customer's site. It uses the CO switch. Cheaper than expensive PBX systems, Centrex is appropriate for temporary use by large corporations and for small offices. Because Centrex is a CO based service, its security and reliability are excellent.

Private Branch Exchanges

Essentially a scaled-down version of a CO switch, a private branch exchange (PBX), is a switch located at the customer site. Most PBXs are digital and can provide features such as call forwarding, conferencing and call waiting. The recent move toward open application interfaces (OAI) for switches is significant. In the mid-1980's, NEC America and InterCom were among the first to unbundle their switch interfaces. Many others have followed suit but PBXs are historically built on proprietary technology. Most of these companies are experiencing substantial fall-off in sales in this explosive Computer Telephony marketplace.

PBXs are actually sophisticated computers. Their switching capabilities can be enhanced by pairing them with phones, allowing the use of more complex protocols, programmable buttons and specialized phone displays.

Due to their proprietary nature, most computer-to-switch links are often serial interfaces. Major telephone manufacturers including Rolm, Northern Telecom and AT&T offer tool kits to help programmers to create switch applications.

More recently the advancements made in personal computers and workstations have spawned the development of dumb switches. As their name suggests, dumb switches are just simple switching systems that rely on external computer systems to control them. The advantage to this approach is that the PBX is not limited or hampered by the lack of software upgrades from the manufacturer. Instead the open interface encouraged third-party developments that result in lower prices and a larger array of choices.

Board Level Switches

With the continued miniaturization efforts, switch technology has come inside general purpose workstations and computers as one or more add-in boards.

Usually appropriate for small companies, board level switch products can support up to 100 lines. These systems provide the maximum amount of control and present some interesting low-cost integration opportunities.

Telephone Numbers

The North American Number Plan administered by Bellcore established the digit sequences by which all CO switches and PBXs manage call traffic in the United States. With approximately 300 countries in existence today, worldwide agreement on a country code system whereby international call completion can take place is also critical. Without this process in place communication, via telephone, as we know it today would be virtually impossible.

The standardized telephone numbering sequence for all calls, national and international has four components:

1. local number
2. area code
3. country code
4. access code

Local Numbers

To signal the network as to what type of phone call you intend to make requires specific numerical sequences which depend on where you dial from. Generally, to reach a number located in the same region or area you need only dial the local number. The local number is the smallest number of digits that can be dialed over the public network to reach another local phone. In the US, that is a seven-digit number. The first three digits, referred to as the NXX code, specify the CO switch to which the line is attached. The last four digits delineate the actual line. Each CO switch has several NXXs and lines attached to them. There are certain NXXs that have been reserved for special use by the LECs. For example NXXs ending with X11 have been reserved for emergency numbers. The 976 NXX is the LECs version of the IXCs 900 pay per call services. The business and residential population density of the area which the CO switch is located is the major determining factor for the number of lines and switches located at a particular CO.

Area Code

Calls to numbers outside the region or area require the inclusion an area code in the dialing sequence. In the United States area codes are three digits long. The first number can only be a digit from 2 through 9, the second digit can be any digit zero through 9 and the third digit can be any digit zero through 9 also. However, the area code can never have 00 or 11 as the second and third digit. In Europe and some other countries the area code can be variable length between two and three digits.

There are three area codes that indicate special services rather than geographically defined locations. The US services that fall into this category are:

⇒ Toll-free 800 and 888 service is a national WATS service that the called party pays for all charges.

⇒ AT&T's Easy Reach special service operates on the 700 area code. It is a transportable number where an individual can retain the number where ever the are located.

⇒ The 900 pay-per-call service requires the caller to pay premium rates for information provided by the called party. This premium service is available nationally. The content of these calls is monitored by the IXC providing the transport and billing. The calling party's LEC does all billing for these services.

Country Code

The country code is one- to three-digit number. It signifies to the worldwide telephone network the country of origin. For example the United States and Canada share the same country code; "1". The UK is 44, Brazil 55, Guatemala 502, Ireland 353, Singapore 65, Hong Kong 82, Russia 7 and France 33.

Access Codes

In the United States and Canada, when dialing a number requiring an area code the calling party is required to dial "1" which is an access code. This signals the network that an out of LATA call will be placed. There will be a ten-digit string; area code (3 digits) + NXX (3 digits) + CO line number (4 digits).

Other access codes used in the US include:

⇒ **0 + area code + number** ... This signals the network that an operator-assisted long distance call is being placed. The caller can either enter a phone credit card number or the local number and wait for the operator to come on line. If the operator comes on-line instructions can be given for billing purposes. These can include collect calls and third party billing.

⇒ **011 + country code + area code + number** ... This signals the network that a direct dial international call is being placed.

⇒ **01 + country code + area code + number** ... This is the network signal for operator assisted international call.

⇒ **10*xxx* + 1 + area code + number** ... The *xxx* is the equal access code which specifies which long distance (IXC) the caller has selected for the call completion. Each IXC has a three-digit equal access code which can be reached from any phone in the network. Since the 1984 break-up of AT&T residential and business phone owners select a IXC provider as their default carrier. This means every long-distance call placed from that CO line will be routed through that selected carrier. The most common select IXC carriers are AT&T, MCI and Sprint. When the 10*xxx* code is dialed the default IXC will be replaced by the selected carrier.

For example when calling a telephone number in the UK from my Manhattan office I would dial as follows

011 44 1 223 846 177

Using our earlier explanation:

011...signals the network that I am placing a direct dial international call

44 ...delineates UK as the called country

1 ...is an access code to signal that an area code and local number follow

223...is the area code for Cambridge

846...is the NXX for Great Shelford

177...is the CO line number on the switch

|3|

Why Computer Telephony on Sun

Those familiar with the power and bus speed available on Sun SPARCworkstations and SPARCservers have trouble comprehending why anyone would use another computer system for anything. Their loyalty to the Sun "pizza box" architecture is unparalleled. PC in their vocabulary translates to Plane Crash. They have good reason for their feeling.

The best way to explain why Sun makes an excellent choice for a computer telephony platform is to take a quick trip through their history and get the true flavor of Sun. Their demonstrated strength in processing coupled with their expertise in network management and distributed computing easily explains "why Sun."

A Historical View

Sun Microsystems Inc., took the computer industry by storm in 1982, with the introduction of their first computer system, built from off-the-shelf parts, in an oversized garage in Santa Clara, California. By using standard components, with a streamlined design and a UNIX operating system resulted in their bringing a superior product quickly to market at a cheap price. This product was the Sun-1.

The Sun-1 and subsequent Sun-2 were instant successes. The company was profitable within its first 6 months of operations and concluded their first year with $8 million in revenue.

Sun was founded by a group of energetic guys in their mid-twenties who were bright, shared a dream and realized it. Their drive and free-spirited entrepreneurial style was first laughed at by more traditional computer companies. This laughter soon turned into fear, loathing and envy as Sun ate their lunch. It was this approach that led to their dominance of the university market and subsequent movement into the commercial market as well. Their early success was uncanny. While researching Sun, I uncovered an example that encapsulates Sun's spirit.

Just 17 months after Sun's incorporation, the company was aggressively pursuing a large contract to sell workstations to Computer Vision, a major CAD (Computer Aided Design) supplier. Computer Vision had decided to abandon its' proprietary hardware and chose a new platform for its products. The winning company would make millions of dollars supplying Computer Visions with systems for resale. What happened next has become part of Sun folklore. Despite spirited arguments and an enticing deal from Sun, Computer Visions picked Apollo (subsequently acquired by Hewlett-Packard). Nevertheless, Sun personnel hurriedly drafted a new proposal and overnighted it to top CV executives. In addition, a Sun hit team took the redeye East in order to be in the CV lobby the next morning. During the hours spent in the lobby making phone calls to CV executives, the Sun team was told the deal was closed and that Apollo remained the winner. But the Sun group only agreed to leave after the CV sales chief said he would call later to explain the choice. During this call and a subsequent clandestine meeting, the Sun team made another offer, which was accepted. Of course, this put Sun on the map in the mainstream technical market.

The company's aggressive sales and marketing tactics, plus relentless product development, yielded a technologically superior product that resulted in the company's meteoric rise.

Sun has consistently led the industry with several innovative developments. From the beginning, Sun developers and management were cognizant of the need for an open system architecture in the computer industry. It was their belief that closed architectures limit expedient and efficient market expansion, thus limiting innovation and product improvement to the capabilities of very few developers.

Open systems, a recent posture of the computer telephony community and others, has been the mantra of Sun since 1984. This philosophy was demonstrated early out by their release of NFS (network file sharing). It was the first Sun product to be broadly licensed. It went on to become the industry standard.

In 1988, Sun, and the original UNIX creator, AT&T, forged a partnership under which the two would design a new version of UNIX that would unite many of the existing flavors of the operating system. The result was the industry standard UNIX System V Release 4, which is still widely used today.

April 1989, Sun released SPARCstation 1, the first high power RISC based workstation to appear on the market. At the time of its introduction Sun's competitors felt the need of this much power at the workstation level was unnecessary. Luck would have it that this new technology went on to dominate the computer industry's major growth segment at that time.

To expand their market and build critical mass, Sun began licensing its RISC based SPARC technology. When they began this licensing effort some skeptics argued that SPARC clones would hurt Sun, just as PC clones cost IBM market share. Fortunately Sun's management disagreed. They understood the difference between planning a cloning strategy and dealing with their unwanted existence after clones appear. After all, when you think about it, IBM grew a $7 billion PC business in spite of the clones. Was Sun wrong? I think not.

Sun's magic continued. By the end of 1990, they were a strong number one in the workstation market with a 39 percent share. Their closest competitor, Hewlett-Packard was trailing at a distant 20 percent.

By 1991, Sun financially outperformed the entire computer industry at a time when many of the earlier industry stars were rapidly sinking, becoming a distant glimmer of times passed. Sun continued its move forward. To maintain their fighting weight and keep the company as nimble as possible, Sun management divided the company into smaller subsidiaries. In effect, they created their own solar system of several smaller, fiercely competitive companies with fast decision-making focus. They were like several younger Suns, each addressing a core business function. Sun long ago learned how to run lean in order to offer the outstanding price/performance ratio that helped it outpace its rivals.

In 1992, several important milestones were achieved. The company replaced Wang on the prestigious Standard and Poor's 500 list. Fortune magazine named it the second largest exporter as a percentage of sales at 49.3%. Sun also made the Fortune International 500, ranked at number 425, which was the highest first-time ranking for a computer company.

While looking for new markets (such as SunSoft's porting of the Sun Solaris operating environment to the Intel X86 platform), the Sun operating companies continued to focus on proliferation of the basic Sun technologies of SPARC and Solaris. Making a major commitment to multiprocessing, SMCC (Sun Microsystems Computer Company) rolled out a series of important products in 1992. These products included a 20-way mainframe replacement server, the world's first multiprocessor desktop and the SPARCstation 10 line. These were accompanied by a new version of Solaris based on UNIX SVR4 with built-in symmetric multiprocessing support.

Although Sun was in the gun sights of every UNIX vendor, a bigger threat was already emerging. The PC was being reborn into a new system with some of the capabilities that UNIX workstations had all along. And the de facto PC monopoly, Microsoft and Intel, fully intended to control this PC replacement too. The Microsoft marketing machine was trying to convince the world that its NT operating environment was the solution for the 1990's.

To Sun, this issue was more than a contest of technologies. At stake was a fundamental business philosophy that would affect the quality and value of what was available to computer users. Microsoft and Intel advocate a closed world, in which a single company controls the technologies, rates of innovation, prices and supplies of a critical computer component. This contrasts greatly with Sun's open systems philosophy, in which the use of open interfaces creates broad, multi-vendor selections. More significantly, a competitive environment spurs vendors to offer better products and value.

The open systems philosophy is also seen in Wabi, a technology from SunSelect that has been widely adopted. It allows UNIX systems to run the most popular Windows applications and is available on Solaris and other UNIX operating systems.

Taking their open systems a final step, Sun recently introduced the Java™ language (see Chapter 18) which provides hardware and software independence. Java™, an object oriented computer language, possesses the unique capability of operating on any computer, having sufficient memory, running any operating system. The idea of Java™ is "build once run anywhere." Now, *that* is *open.*

Networking

"The Network is the Computer," the ever-present tag line of Sun, appears to have been years ahead of its time by today's activities. The RBOCs have begun building Advanced Intelligent Networks (AINs), and AT&T and others are incorporating groupware to their networks all in an effort at adding value. The Internet now receives high visibility. It appears that network computing and communications are now in.

Sun was the early pioneer of network computing. Client-server computing, a bandwagon most major computer manufacturers subsequently jumped on, was started by Sun in the late 1980's. It delivers effective organization-wide computing better than any other solution. To many it is a way of describing network computing where "client" workstations, PCs or terminals are linked to larger "server" computers that store, process and move data. To others, "client-server" is synonymous with distributed computing, in which resources such as computers and printers are distributed across a network. These descriptions are correct, but Sun actually took "client-server" technology much further.

Through Sun's open standard technology the resources of the entire organization can be no farther than a simple keystroke or click of the mouse. Using a Sun network, a user can access information located anywhere on the network, on a mainframe, or on a workstation from a different vendor, and whether the machine is located across the hall or across the globe.

Sun built an entire client-server environment rather than assuming a single application need be accessed by a group of users. With this foundation, a corporation need not anticipate their future needs today. Needs change rapidly. Future computing needs defined by today's standards might take you through 6 months. Beyond that, your luck would be better in Atlantic City. With Sun's client-server technology, network additions and deletions are easily made without affecting any other component.

Client-server computing has an appealing advantage in today's business environment. It maximizes the existing system technology and permits the integration of new technology with the existing infrastructure. For example, a traditional business client-server installation can easily be upgraded with computer telephony capabilities. This additional technology might be a fax server, an IVR providing customers access to information, or a total unified messaging capability seamlessly integrated into the existing architecture of an organization. This capability makes it quite easy to develop highly usable solutions which access existing information, route messages and provide a total CT server or desktop solution; whatever is needed.

Sun in the Telcos

The telcos have successfully integrated Sun's client-server technology into their infrastructure to provide billing services, front office support and supply the distributed call center technology used for their directory assistance and live operator services. The distributed nature of Sun's processing architecture parallels the overall telco network of central offices (COs), whereby calling areas are composed of several points of presence (POPs). Without distributed management, tracking and assistance these functions would need be centralized. Not quite an appropriate solution.

Sun in the Internet

You are probably thinking this architecture sounds very similar to the Internet. Well, it should. The Internet is a WAN using client-server technology based on a TCP/IP protocol. It should not surprise you that 56 percent of all Internet servers are Sun systems.

Sun as a Computer Telephony Solution

Sun's entry into the computer telephony market is well planned. The implementation of their Customer Management Solutions™ program (see Chapter 4) is critical to their strategy. By adding value to the powerful distributed architecture that their customer base has in place, they increase the value of their investment, and further expand the utility of their network. Sun's customers trust Sun's capabilities in designing tailored network solutions for their business. Computer Telephony is just one more addition to that solution.

|4|

Customer Management Solutions

The Implementation of Sun's Computer Telephony Program

Customer Management Solutions™ are the combination of computer technology with telephone access to improve communications between a company and its customers. This computer telephony integration (CTI) is the technological basis in a cross-industry drive to greater customer satisfaction for all forms of service delivery.

Contact with a company's customers is often concentrated in one or more customer service call centers (CCSCs) which have a high level of inbound and outbound telephone activity. A CSCC may handle any of the following tasks:

⇒ *Obtain information about the firm, its products, or services by phone, mail or fax.*

⇒ *Purchase goods or initiate a service.*

⇒ *Report a problem or learn the status of a repair.*

⇒ *Get help or advice regarding assembly or use of a product.*

⇒ *Track the status of shipments.*

⇒ *Check balance information for a bank account or benefits program.*

⇒ *Retrieve scheduling information for travel, entertainment, education, etc.*

⇒ *Sell products or services through telemarketing.*

⇒ *Conduct customer satisfaction surveys and other polling.*

⇒ *Contact customers to arrange bill payments.*

This is only a partial list. Use of the telephone to conduct interactions and transactions is becoming routine among nearly all companies. Every day, new applications are added to those carried out by CSCCs.

Call centers are generally a key contact point between a company and its community of prospects, customers, vendors and employees. On behalf of their customers, agents complete telephone based interactions and transactions that often require access to many computational resources, electronic archives and databases in many locations. As a result, the CSCCs are prime sites for the most aggressive development of client-server based, distributed processing applications in several of Sun's targeted industries.

Call center automation customarily has two objectives, whether it occurs during an initial effort or while enhancing existing systems:

⇒ *To improve customer satisfaction through the delivery of timely, accurate, complete information.*

⇒ *To provide selective automated voice response for ever-increasing call volumes, thereby reducing costly interaction.*

Although Sun is often seen solely as a workstation and server vendor, there are historically vast opportunities for Sun to deliver equipment, software, and services that comprise a more comprehensive solution. Chapter 14 details an assortment of these Sun CTI solutions.

You will notice when reading Chapter 14 that the primary focus of these solutions is CSCCs. Sun has selected CSCC applications as their first area of concentration for computer telephony solutions. This is due to their substantial installed base of customers and the tremendous growth within that market segment.

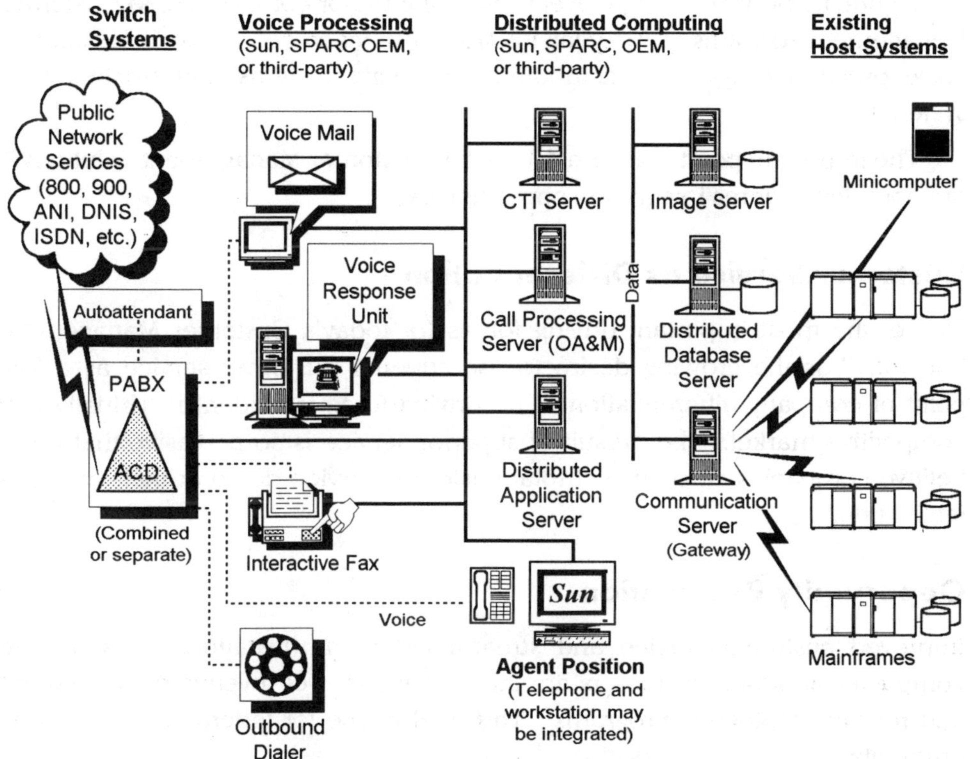

FIGURE 4.1 Customer Management Solutions Components

Sun's CSCC approach considers the concept of total customer service—which, by the way, is a company-wide philosophy. Total customer service requires every individual in an organization to understand and support the need to contribute to the highest level of customer satisfaction. Sun management felt that their organization certainly was not the only company that could benefit from this approach; therefore it became their initial computer telephony focus.

CSCCs are often placed as the hub of the movement towards total customer service. In competitive markets, the quality of customer support services (such as help desks) has become an important factor in revenue generation, image building, customer retention, cost reduction, morale building and strategic re-systemization programs.

Figure 4.1 provides a high-level view of the major components in Customer Management Solutions™. They fall into four distinct categories: switch systems, voice processing systems, distributed computing systems and existing host systems.

The important point of Figure 4.1, is that Customer Management Solutions™ are complex systems integration opportunities.

Customer Service as Differentiation

One of the most important driving forces for today's Customer Management Solutions™ is the growing desire to use superior customer service as a key point of company differentiation. The driving force is evident in virtually any competitive market. The pursuit of superior service is so pervasive that competitive companies who neglect total customer service do so at their own potential peril.

Opportunity Recognition

Improved customer service and streamlined communications links among companies and their customers are the cornerstones of revenue enhancement and revenue protection programs. This is also true for federal and local governments.

With the introduction of higher-speed network interfaces and more robust applications platforms, call centers can now deliver required information and messages in just about any form or format. Fueled by the desire to improve employee efficiency while enhancing customer service, call center specialists are driven to integrate more capabilities and functionality into customer service installations.

General Solutions

CSCCs are destined to be one of the early sites for companies to capitalize on implementing such advanced technologies as expert systems, image processing, video teleconferencing, voice/fax integration and unified messaging. This will have a profound impact on CSCC roles and functions. It will spur demand for high-end workstations: a higher level of computer power to process calls, to monitor and manage agent efficiency, to serve information and applications for a variety of systems, and to respond quickly to requests from customers and employees.

Integration of these technologies only enhances the revenue potential associated with the provision of CSCC systems and services. The trends that will have the greatest impact on this area are the following:

Intelligent Network Features. Long distance and international carriers have been quick to install new digital circuits and signaling links that transform the public network into a large computer network which can carry voice, data, images and moving video. CSCCs equipped with CTI hardware and software are among the first to take advantage of the improved, standard interfaces among central office (CO) switches, customer premises equipment (CPE), and external database servers which supply information crucial to call processing. Automatic number identification and answer supervision are examples.

Expert Systems. The agent workstation provides agents with prompts or scripts developed by the best employees to bolster the efficiency of new agents. Firms in the telecommunications industry have used artificial intelligence subsystems to march agents through the process of fault detection, fault isolation and correction without having to dispatch expensive on-site technicians.

Multimedia. Material in a variety of formats can support customer service agents and their customers. In the past, agents faced individual screens with forms requiring detailed input of customers' responses. Any required reference material was generally in print form. More recently, agents are being given multimedia workstations to display input from interactive voice response (IVR), archived data, reference manuals and even full-motion video to support their transactions.

Voice Response. A growing percentage of customer service inquiries and order entry functions are handled without human intervention, although the best systems give callers the option to transfer to a live agent at any time. In health care applications, for example, routine scheduling of appointments (for prescribed tests or reordering prescription medicine) can be done as an IVR database query, much as you would use a telephone-based teller for routine inquiries to a bank.

Voice Recognition. As the vocabulary of voice recognition equipment grows, so does its ability to respond to telephone-based queries without requiring callers to use the telephone keypad to spell things or to enter menu items. This speeds up transactions and also makes applications easier to use in a hands-free mode. A good example is voice dialing technology utilized in the automobile.

Interactive Facsimile. Fax machines are becoming nearly ubiquitous and many call center applications regard them as remote printers. In the context of CSCC automation projects, an agent may transmit printed product information by fax to the caller. Callers may also request fax responses through dual tone multi-frequency (DTMF) or voice input.

Imaging. The ability for agent workstations to display high-quality graphics should accelerate the incorporation of image processing and image systems into CSCC scenarios. In a financial services setting, for example, agents could retrieve digitized images of transaction confirmations or of personal checks, customarily photographed and destroyed.

Full-Motion-Video. Once the infrastructure is in place for high-speed communications services, enhanced switching, and high-resolution workstations, then there will be a drive to incorporate such TV-quality video images as teleconferencing tools to support long-distance learning and make high-quality reproductions of moving images. One of the most frequently cited beneficiaries of such a system is the health care industry, which according to a 1992 study by A.D. Little, could use video teleconferencing to reduce operating costs by an estimated $200 million per year.

Sun Customer Management Solutions™ Strategy

Sun's strategy for the Customer Management Solutions™ market introduces a complete technology suite wherein Sun hardware and software combine with communications processing hardware and software from such third party vendors as Linkon Corporation, for implementation by system integrators. Sun's initial targets are companies within the financial services, government, health care, pharmaceutical, telecommunications and transportation industry segments. The Sun platforms that the customer management solutions market requires include workstations, application servers, communications servers, and database servers in standard and high-availability configurations.

Why Sun's Customer Management Solutions™ Focus

A company that handles its customer contacts well will find that the superior service helps retain existing clients and gain new customers. Mishandling these contacts, however will decrease the company's reputation and drive customers to competitors who will serve them better.

Customer service and the company's image are very closely linked. For example, customers generally take their phone system or governmental services for granted until they need assistance or something goes wrong. Their only contact with those enterprises may be one telephone call to the customer service center. The response to that call may be the only opportunity to influence the customer's perception. Delayed answering, long periods on hold, repeated transfers and inadequate response can leave the customer with a negative impression of the whole company. However, if the customer can get a simple question answered with an easy-to-use, interactive voice response system without being put on hold, or be immediately connected with an agent who can answer complex inquiries without transferring the customer to someone else, then the company's image is either maintained or improved.

Employment conditions are another aspect of a company's image. Properly designed systems that improve working conditions for all employees and that enable the hiring of physically challenged workers will promote goodwill. Even if development of systems for the physically challenged is costly, the reputation for being a good place to work will affect not only employee retention and recruiting costs but will also attract new customers who are concerned about these issues themselves.

Retain Existing Customers

Customer retention should be a priority, even over attracting new customers. Recent studies reveal that 65 percent of the average company's business comes from present satisfied customers. This percentage is probably higher in the services sector. According to Terry G. Vavra in his book "Aftermarketing," it costs five times more to acquire a new customer as it costs to service an existing customer. In "Customer Service Operations: The Complete Guide," Warren Blanding calculated that over a ten year period, assuming a ten percent growth rate in accounts, a ten percent increase in customer retention would generate a 50 percent increase in total revenues. And a 30 percent increase in customer retention would generate an increase of 295 percent in total revenues.

All companies need focus on the following key factors to retain their customers.

Improve Accessibility. Repeated busy signals block even initial contact with a company. Once connected, customers do not want to wait on hold, whether calling to check on the status of a delivery by a freight company or to discuss unusual transactions on a bank statement. Customers want to direct and prompt access to their services.

Increase Customer Satisfaction Through Efficiency and Accuracy. Nothing frustrates customers more than when after finally being connected with a service agent and having a detailed discussion, they are then transferred to another party to repeat the same discussion. Efficiency occurs when a request is understood, appropriate action is taken, and the action is on time, accurate and of expected quality. When customers receive information that is incomplete, inaccurate, or different from information received from another department, they lose faith in the company as a whole. Customers want accurate information the first time, from any employee.

Facilitate Personal Service. Customers want the personal touch whether they are in contact with an actual agent or the company's voice response system. When the contact is personalized, the customer has greater confidence in the information given in the promise made, in the company itself. Lack of recognition can be damaging as well. When a customer has made repeated calls to obtain information or to have service repaired and there is no acknowledgment of the previous contacts, the customer is more likely to lose patience or to be unsatisfied, regardless of the outcome.

Offer Premier Service. To maintain existing customers is also paramount in mature industries such as the U.S. telecommunications wireline segment. In particular, the loss of a major account is disastrous to company revenues. Large business telecommunications customers, while representing only a small percentage of the total accounts, contribute a majority of a local exchange company's revenue. These customers are very aware of their value to their suppliers' bottom line. This knowledge, combined with their increased sophistication, makes them more demanding and more expensive to service. As competition becomes even fiercer in such industries as telecommunications and transportation, diligent and exceptional care for these valuable customers becomes paramount.

Attract New Customers

Attracting new customers is essential for the maintenance of market share. A growing market share mandates vigorous programs, such as Customer Management Solutions™, to create new customers. The following elements can have a significant effect on new customer growth rates:

Favorable Word of Mouth. Word-of-mouth can attract new customers or cause sales to go to competition. Studies have shown that a bad experience with a company may be retold to up to 55 other consumers, while a good experience will be retold to only a few. Customer services solutions must be crafted to minimize the opportunity for service failures.

Increase in Reference Accounts. As an indirect sales tool, customer service can also generate new customers through positive references. Most large customers will request references prior to buying a product. Purchasers will be much more confident about their decision to buy if there are significant positive references. The quality of customer service provided after a sale will affect the reference's rating of your company. This is particularly true of large business customers and government agencies.

Differentiation. When companies offer the same basic product or service as their competitors, they often strive to appear different by adding something that the customers want. A differentiator is a special touch or added value that cannot be overlooked. Travelers, when offered matching fares on competing airlines will make their choice based upon other factors such as frequent flyer plans or on-time record. Even hospitals are copying the hospitality industry, calling their patients *guests*, and decorating their rooms and halls to resemble quality hotels. Better customer service can be a prime differentiator.

Market Intelligence. Capturing market intelligence is a marketing imperative. However, the typical company throws away valuable intelligence with every customer contact. Generally, the customer services agent asks many questions during the contact with the customer. The customers' answers are used for the sole purpose of handling that one request, with no provisions for viewing the information as a reusable asset. Whether the customer is in contact with a person or an interactive voice response system, the system should be designed to capture and collate the customers' responses, enabling these to be input to marketing analysis applications for access during later contacts with the customer. Market intelligence, based on demographics, should also be purchased for use by the company's inbound and outbound telemarketing centers. This intelligence should be combined on an integrated expert system platform with applicable information stored by the company's various disparate information systems.

Increase Revenue Per Call or Opportunity

Increasing the rate of sales per customer call is another business issue for call center managers. This would also be applicable to collections, whether for a corporate or government agency responsible for the collection of taxes, licenses, fines and fees. To put it in perspective, calculate the value of a minimum sale to just 1 out of 100 customers, above the normal sales or collection rate, for service centers with large call volumes.

A typical telephone company, for example, may handle 150,000 calls per day. The increased revenue for selling one more service with a monthly value of $5 to 1 out of 100 customers calling the customer service center is more than $21.6 million annually.

Maximize Sales Per Call. Viewing every customer contact as an opportunity to sell should be the goal of every customer service center. Combining improved tools such as expert sales tools with information systems to facilitate personalized call handling will enhance the opportunity to maximize sales per customer contact. In addition, deployment of inbound and outbound telemarketing centers that use expert sales systems and sophisticated telemarketing software will maximize the company's sales results.

Increase Number of Opportunities. Improved call routing through integrated voice response systems, together with better integration of information systems and use of expert sales tools, will increase the number of calls that can be handled, thus increasing the number of opportunities to sell. An important additional element in this area is eliminating calls abandoned due to slow answer.

Create Sales Expertise. The proper use of expert sales and collections tools, plus the collation of existing customer profile information, allows each agent to emulate the skills of an experienced, successful sales agent. Use of scripts will ensure that the most effective sales messages are delivered consistently. The expert sales tools should be available to all customer service workers, even to those not specifically assigned to the sales offices.

Improve Postsales Analysis. Tools that improve analysis of the call center's sales results can pinpoint weak areas hourly or daily, facilitating immediate response and improvement. Sales agents can see on a call-by-call basis their sales results. This gives them immediate opportunities to find ways to increase their commissions and redouble their efforts, instead of a month later when traditional systems would produce sales reports.

Reduce Cost of Business Operations

The following aspects of the cost of call center operations are also functions of customer service, particularly affecting both customer and employee retention rates. However, it is important to quantify the cost savings associated with each of these to support the cost of retooling the customer service centers. Note that this objective is particularly applicable to cash-strapped government agencies that must deal with scheduling such diverse items as community college classes, jury duty and building inspections.

Reduce Call Handling Costs. Frequent and routine calls are expensive when handled by a live operator. Customer call centers should analyze their call statistics to determine what percentage fall into the routine category. These routine, standard procedure calls should be considered as candidates for interactive voice processing. "Inbound/Outbound" magazine states that every two-minute call handled by a voice response unit saves about $.80 over the cost of a live agent handling the call. Additionally, using automatic number identification to look up a caller's profile, and to deliver that information to the customer service agent's screen before the call even reaches the agent's telephone, can save a minimum of 20 to 30 seconds of time per call. And should the call then need to be transferred to another agent, the data is transferred along with the call for a savings of another 20 to 30 seconds.

Some estimates show that as many as 25 percent of the inbound calls that the call center receives each day are unnecessary. Improving the quality of the information that a customer obtains during the first call may prevent unnecessary, repeat calls to get correct or complete information, as well as improve the overall customer satisfaction rating.

No one likes to be put on hold, especially when they have not been told how long the wait will be. Putting customers on hold ties up the company's call processing capacity and blocks additional calls from getting through. If customers are calling on toll-free numbers, the cost to the company is even greater. Adequate staffing is probably more cost-effective than tying up expensive communications resources and blocking potential sales calls. However, reducing the number of calls that must be answered by live operators and reducing the length of the contact allows for an increase in the number of calls handled each day without increasing staff. Imagine the savings to a hospital if concerned family members could check a patient's status through a voice response unit without tying up nurses' limited time without being put on hold.

Calls sent to the wrong person, or multiple inquiries that cannot be answered by a single person, inappropriately expend the time and resources of personnel. This increases labor and call processing costs because the phone system is backed up with customers still waiting on hold. Each company should analyze its call processing statistics to find out how much time customers spend on hold, and how many calls a single person answers and handles completely.

Automate Data Collection Efforts. Call one agent and get one answer. Call another and get a different answer. They are not using the same script, they are not accessing the same systems or they are interpreting the data differently. Again, this will result in poorer customer satisfaction results and the cost of handling the same call over and over. Integrated platforms that give customer information about their accounts automatically, or that locate and collate all appropriate information for the customer services agent, will reduce the time spent logging on to the system and give the agent more time to focus on the customer.

Decrease Training Costs. The use of expert systems, combined with automated information collection from existing systems, may help relatively untrained personnel approach the effectiveness of traditionally trained workers.

This can enable the company to use employees more effectively and to shift resources where needed without significant training expenses formerly associated with departmental or work group transfers. When one call center is busy, instead of putting customers on hold, another can handle the overflow calls even if the calls are of a different nature. This reduces costs and improves customer satisfaction.

Do It Right the First Time. When the customer receives erroneous information from an agent because the correct information was not at hand, the company loses money because the customer has to call back (or worse, chooses not to). Customers who call back tie up time and resources. Integrated customer service platforms with expert systems and database integration, combined with intuitive, consistent user interfaces, can improve the opportunity to give complete and accurate information in a personalized and customized way.

A good example of wasting resources is the case of a customer calling a service center to report a service or repair problem. The agent taking the repair call is often just recording information to be passed along for someone else to deal with. The agent may have little or no information about the customer and little information about the type of problem being reported. This results in incorrect or incomplete information being passed to the next person, who wastes time gathering data. Additionally, someone with a higher labor cost may have to be dispatched to solve a problem that could have been taken care of while the customer was still on the line.

The cost to dispatch a telephone repair technician at one of the RBOCs is reported to be $132 per call. Assume a customer base of 10 million lines, an average 3 percent repair report rate, and dispatch of a technician on 40 percent of the calls. Just 5 percent reduction in those trips lessens expenses by $9.5 million. Since the U.S. local exchange carriers are struggling to remove 50 percent of costs out of their business, this approach is vital. It requires integrated, interactive, and intelligent customer service systems that combine voice response systems with existing customer systems and the addition of expert systems, providing customer service workers with consistent and easy-to-use interfaces. Such interfaces mask the complexity of the underlying systems and facilitate the delivery of personalized and appropriate one-stop shopping for every customer.

Improve Employee Morale

Customer service workers are frustrated by their inability to consistently provide personalized, flexible, timely, complete and accurate service to their customers. The lack of integrated systems, consistent friendly system interfaces, expert systems, and one-stop shopping capability contributes to job frustration, which is reflected in employee turnover rates. The costs of high turnover rates include recruitment and training costs, reduced productivity and lower morale among the remaining employees who have to work harder. With high turnover, the number of calls per employee declines, customers' hold time increases, length of the call increases, customers receive more incorrect or incomplete information and customer service declines overall.

Empower Employees. Empower employees by giving them the tools to perform work normally relegated to only the very experienced. Give them expert systems that allow them to make complex decisions without relying on supervision for assistance. The improved systems put the worker in control of the environment and reduce the stress associated with job dissatisfaction and high turnover rates.

Improve Work Environment. Installing voice response systems to handle the routine calls, allowing the customer service workers to handle more complex, satisfying problems, will improve morale. Providing workers with intelligent platforms with expert systems will put them in control of the contact and give them the opportunity to provide personalized, special service. Eliminate the "incompletion" factor by making systems finish paperwork so that when the service agent ends the call, the work is complete. These improvements will have a positive impact on reducing turnover.

Employ the Physically Challenged. New platforms enable the physically challenged to perform complex work at higher wages. Telecommunication devices for the deaf and voice recognition systems for the blind and those with limited mobility will enable these special workers to enjoy the same work opportunities as other employees.

In Summary

For all the reasons indicated above Sun feels that their Customer Management Solutions™ implementation will have the most meaningful effect on the computer telephony integrated solutions in the market.

Computer telephony is a self-financing technology that will assist in customer retention, win new customers and generate revenue while reducing costs. It is truly a win-win situation for all parties involved.

|5|

The Internet, Intranet, Netra™, & HotJava™

The Internet

Hardly a day goes by without something about the Internet appearing in print, on the tube or surfacing in casual conversation. Young and old, it seems to be discussed by everyone. It probably is the most talked about network of all times. The catalyst for its new found visibility must have been the Clinton-Gore "Information Superhighway" initiative.

The Internet with its purported 30 million users is rumored to be adding users to the tune of 1 million per month. Now, that's *significant.* Fact or fiction or somewhere in between no one wants to be left behind in this new way of doing business, communicating, information delivery, advertising, retailing or whatever else composes these dreams.

The Internet is the network of networks. The largest WAN (wide area network) on earth. It is hardly new. This historic communications medium, used by universities and scientists for a couple of decades, has recently been adopted by business as the latest way of participating in the information-age economy. Virtually overnight it has become the latest competitive tool. The Internet's capabilities include:

⇒ **Electronic Mail.** Private one-to-one and one-to-many global communication.

⇒ **Collaboration.** On-line conferences, private discussion groups, the ability to access files or log into remote computers.

⇒ **Information Gathering.** Financial information, airline schedules, on-line newspapers, government and regulatory databases, product and service information from businesses worldwide, and special collectors of all sorts.

⇒ **Information and Dissemination.** The ability to post information such as on-line catalogs, product and service information, and support systems.

Access to the Internet early-on required the user to be a proficient UNIX based computer operator and navigate via text based commands through the sundry of information available world wide. As knowledge of the Internet became more "main stream", dial-up access became available through service providers; many of whom were currently providing on-line services via local BBSes or more commercialized services such as America Online, CompuServe and Prodigy. This activity has spawned a new breed of service provider specializing in just Internet access. Companies like PSI offer Internet access, and Netscape provides a browser which streamlines Internet usage through icons, symbols, and the interactive graphics to which we have become accustomed. AT&T, MCI and the Regional Bell Operating Companies, not to miss this opportunity either currently offer access or soon will develop on-ramps to this recently recognized resource.

Sun Microsystems technology dominates the Internet. For more than a decade Sun has been delivering Internet solutions. Here are some interesting facts:

⇒ *According to the Internet Society 56 percent of all servers on the Internet are Sun systems.*

⇒ *80 percent of Internet applications were developed on Sun systems, including the most popular search and retrieval tools Mosaic™, Netscape and WAIS.*

⇒ *According to Dataquest, Sun has a 34 percent share of the installed operating system base among Internet service providers.*

⇒ *Sun SITE servers are the most popular WEB servers on the Internet. Over 100,000 file transfers occur on these systems daily.*

⇒ *Over 1 million internal and external Sun messages are processed daily.*

Internet access is in demand by businesses of all sizes, educational institutions, government bodies and individuals. Sun's Netra™ Internet Server provides a way for PC and Macintosh users on local area networks (LANs) to gain Internet access as a group, so that any user on the LAN can access the wealth of information on-line without dealing with the esoteric underlying mechanisms involved with Internet connection. The Netra™ Internet Server comes equipped to act as a World Wide Web (WWW) server, so that a company can publish information on the Internet.

The Internet as the Intranet

As this panacea of communication expands companies are adopting the Internet as the vehicle to tie their numerous locations together; the Intranet. Corporate Internet committees are adjudicating standards as to what communications are sufficiently secure or innocuous enough for Internet traffic. For the most part, accounting, personnel and strategic planning information will be restricted to the more pedestrian forms of transport, but just about everything else will move towards the new corporate Intranet (WAN).

Later in this chapter is a discussion about how huge savings in corporate telecommunications can be realized by utilizing the Internet/Intranet for voice traffic and messaging, as well as E-mail and fax.

Sun's Netra™

The Netra™ Internet Server is a simple, integrated secure solution that provides Internet and WWW access from PC and Macintosh LANs. It consists of a Sun Netra™ server configured with the hardware and software necessary to interface a LAN to the Internet through a connection from an Internet service provider. The Netra™ Internet Server allows businesses and users to access other WWW sites and to set up their own WWW Home Page to publish information. The Netra™ Internet Server enables an easy and transparent access to the Internet from LAN desktop computers through such key client applications as Mosaic™, Netscape or Gopher. These client applications are not bundled with Netra™ and must be either downloaded from the Internet, if publicly available, or purchased from third parties.

The Netra Internet Server includes WWW server software; anonymous file transfer protocol (FTP) server software; and E-mail, FTP and Telnet capabilities. Base-level server security is included in the product. The optional FireWall software package provides security for the entire LAN. The Netra Internet Server comes equipped to function as a primary or secondary DNS server.

The Netra Internet Server also comes with ease-of-use administration tools. Its software includes a graphical HTML (Hyper Text Markup Language) setup and administration toll that makes the Netra™ Internet Server easy for LAN administrators and resellers to install and manage. The product is designed for customers familiar with PC network elements, but not necessarily with UNIX operating systems and administration. This makes it easy for businesses to tap into the vast resources of the Internet.

A dedicated Internet server allows users on the LAN to access Internet E-mail or supported applications such as Mosaic™ or Netscape whenever they wish, just as they would access any other application on their LAN. The Netra™ Server makes the connection process completely invisible to the LAN user. Users can be notified of Internet E-mail as it arrives, not just when they dial up their Internet connection. Users can also surf the Internet using the most popular browsing tools and download information directly to their desktop. The Netra™ Internet Server offers a direct high-bandwidth connection to the Internet that is scaleable for the workgroup or the entire enterprise. It also enables businesses to present their products and services on the Internet through a cost-effective, global medium; a WWW server.

The Netra™ Internet Server offers support for heterogeneous LANs, that is, both TCP/IP and NetWare LANs. This feature, available as an option, enables NetWare clients to access the Internet without having to install TCP/IP, lowering their cost and maximizing ease of administration.

Netra's™ Target Markets

There are three primary target markets for Netra™:

Businesses and commercial users in many areas such as law firms, health care, publishing, entertainment and sports, financial institutions and brokerages, insurance, manufacturing and retail, customer service and scientific research. These organizations use the Internet to get fast, cost-effective access to critical information, to provide up-to-date information to serve their customers, and to deliver services to their customers and prospects. The commercial market for Internet is growing at a rate of 92 percent per year; faster than any other segment of the user base.

Community service providers, local and state government agencies, and institutions such as museums and libraries find in the Internet a medium for collaboration and information access, and a means to supply information to the citizens they serve.

Educational institutions from elementary schools through colleges, which need a relatively low cost way to access the information and research services of the Internet and have only PC and Macintosh expertise.

Netra™ System Overview

The Netra™ Internet Server manages the interface between a LAN and an Internet service provider. The Netra™ Internet Server consists of a Sun Netra™ server configured with Solaris 2.4 operating system optimized for Internet access; World Wide Web server software; anonymous FTP server software; Versions 2 and 3 of the post-office-protocol (POP3) mail server and IMAP mail server to support PC and Macintosh clients; and a set of tools to simplify installation, configuration and administration of the server. It connects to the LAN through its built-in Ethernet connection or through a variety of other options, such as token ring or 100 Mbps Fast Ethernet.

Clients on the LAN must have either a TCP/IP or IPX/SPX stack installed, as well as POP3-compatible E-mail, FTP, Telnet, Mosaic or Netscape, if those services are desired. Client software is not included. A number of third-party solutions are tested and referenced by but not supported by Sun. SunSoft's PC-NFS Pro and PC-NFS/SelectMAIL options include most of the needed features and are supported by Sun.

The actual Internet connection is supplied by the Internet service provider, which administers the domain name service (DNS) server, locates other hosts on the Internet, supplies the DNS name information and IP address for the Netra server and provides full Internet connectivity. The service provider typically furnishes the physical components for Internet access: a leased or high-speed dial-up communication line and the appropriate interface. It is usually a DSU (data service unit) and router for a leased line or a modem in the case of a dial-up line. A leased line is a high-speed, high-bandwidth line, such as a T-1 line, connected directly, point to point, from the customer to the service provider. A modem connected to a dial-up line allows a customer to determine the destination by the number they dial. Dial-up lines are slower than leased lines and

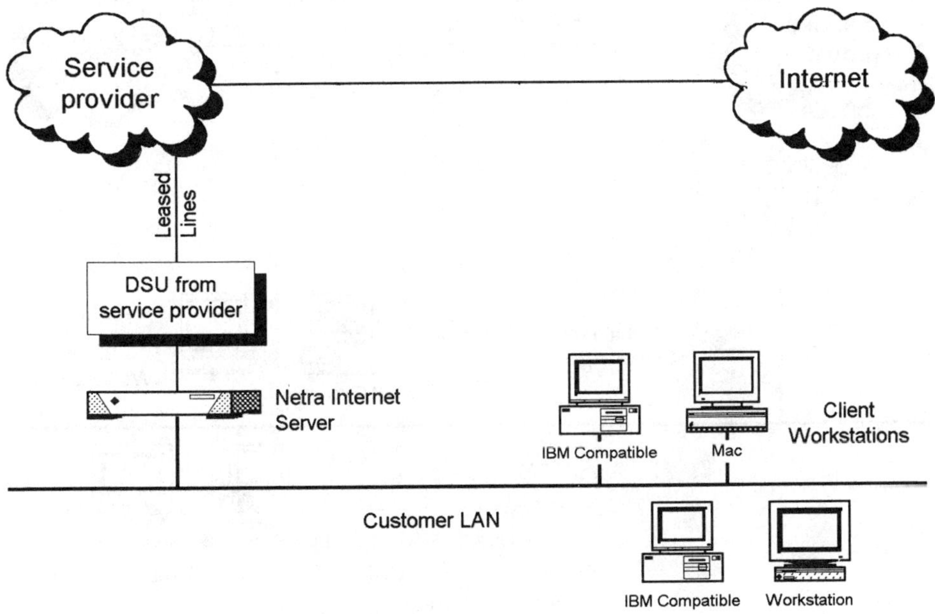

FIGURE 5.1 Leased Line Connection to the Internet

cannot carry as much information.

Figure 5.1 shows the relationship between the components in the leased-line configuration using the Netra™ Internet Server. Note that several band-width options are available. The proposed usage type and anticipated number of simultaneous users will determent the specific choice. In this scenario, the Netra™ Internet Server has an Ethernet Sbus card and acts as a router.

Figure 5.2 shows the Netra™ Internet Server in a dial-up scenario. The connection to the Internet is accomplished through a modem, which is directly connected to the Netra™ Internet Server. This is an intermittent connection designed for limited use.

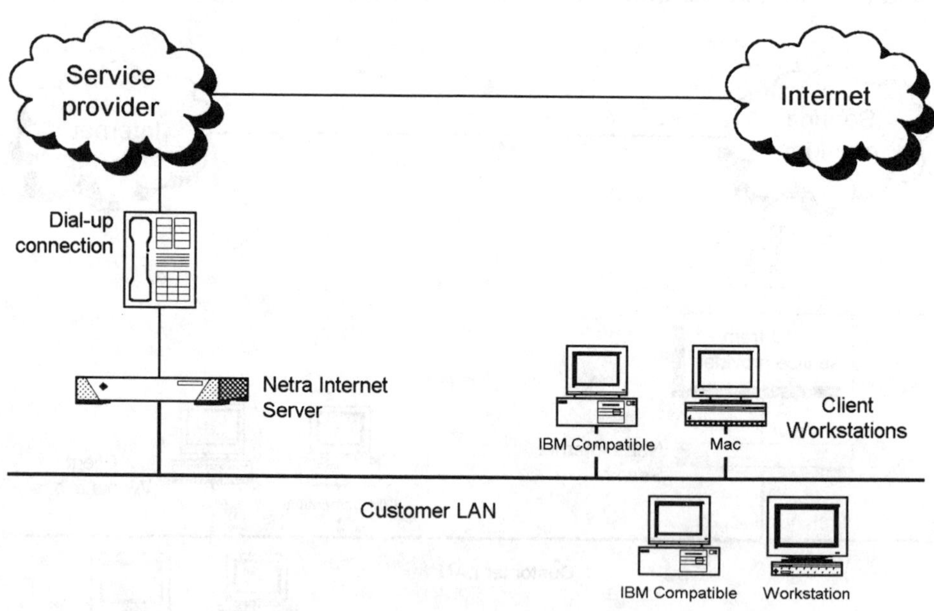

FIGURE **5.2** **Leased Line Connection to the Internet**

TABLE 5.1 The Netra™ Internet Server System at-a-Glance

	Netra™ i4	Netra™ i5	Netra™ i20
Packaging	Desktop	Desktop	Desktop
Processor	One 85-Mhz MicroSPARC-II	One-110-MHz MicroSPARC-II	One 75-MHz SuperSPARC with SuperCache
Multiprocessing	No	No	Yes
Cache Size	24 KB on chip	24 KB on chip	36 KB per CPU 1 MB external
Base Memory	16 MB	32 MB	32 MB
Maximum Memory	160 MB	256 MB	512 MB
Internal Disk Capacity	1.05 GB–4.2 GB	1.05 GB–4.2 GB	1.05 GB–4.2 GB
External Disk Capacity	28 GB	58 GB	73 GB
Sbus I/O slots	One 32-bit	Three 32-bit	Four 64-bit
Operating System	Solaris 2.4	Solaris 2.4	Solaris 2.4
Internal Backup / Distribution	3.5 in. floppy and CD-ROM standard	3.5 in. floppy and CD-ROM standard	3.5 in. floppy and CD-ROM standard
Typical Number of LAN Users	10–50	10–50	25–250
Installed Software	Web server, Internet server, Instal + Adm	Web server, Internet server, Instal + Adm	Web server, Internet server, Instal + Adm

Technical Server Details

Expansion to advanced networking

⇒ *Twisted pair and AUI Ethernet are integrated into all Netra™ configurations.*

⇒ *Sun Fast Ethernet adapters and networking technologies such as QuadEthernet, ISDN, HSI and Token ring can be integrated using the Netra™ Internet Server's high-performance Sbus.*

Upgradeability

⇒ *Netra™ i4 is upgradeable to Netra™ i5 or Netra™ i20 configurations, and the Netra™ i5 configuration is upgradeable to the Netra™ I20.*

⇒ *Identical package is used for all three configuration types with same internal components.*

⇒ *Upgrading requires only a simple board swap.*

Software Details

Figures 5.3 and 5.4, and Table 5.2 provide details about the availability and organization of the software.

Internet Browsers

The Internet is a vast sea of data represented in many formats and stored in many ways on an assortment of hosts. Much of the Internet is organized as the World Wide Web (WWW) which uses hypertext to make navigation easier than traditional ways such as anonymous FTP and Telnet. WWW browsers are used to navigate through the data found on the net.

Browsers allow people to treat the data spread across this vast network as a cohesive whole. Web browsers integrate the function of fetching the data with figuring out what it is and displaying it for the user. One of the most important file types that browsers understand is the hypertext markup language (HTML). HTML allows text data objects to imbed simple formatting information and references to other objects.

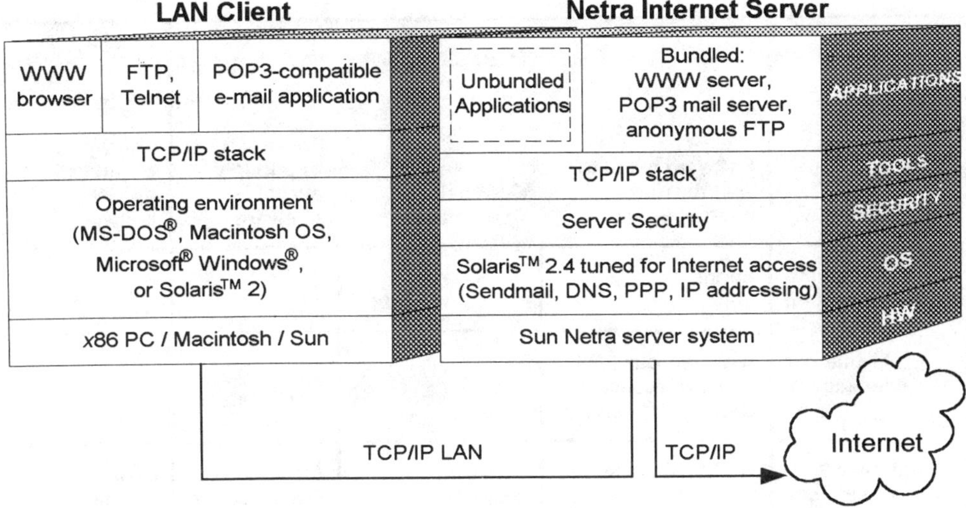

FIGURE 5.3 **Server and Client software for a TCP/IP LAN**

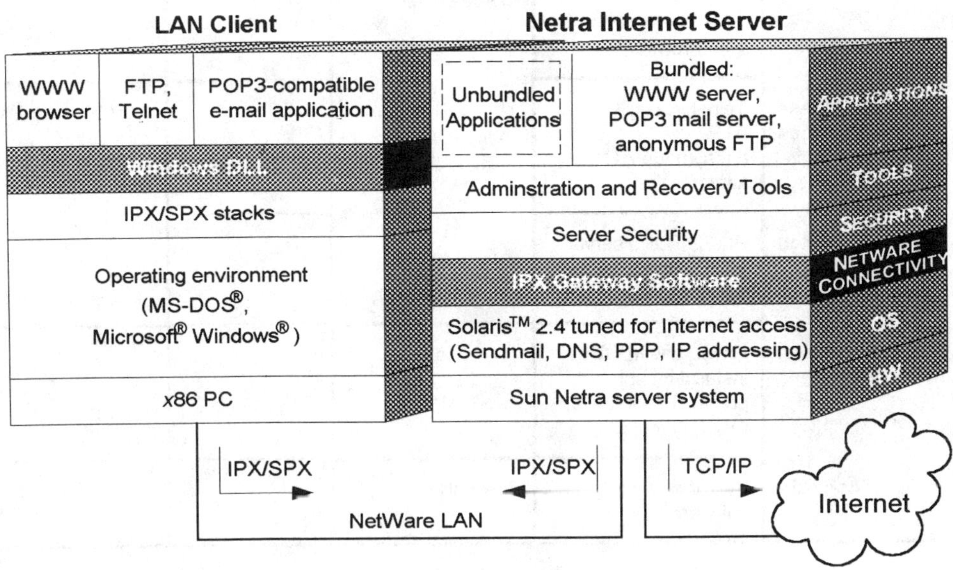

FIGURE 5.4 **Server and Client software for a NetWare LAN**

TABLE 5.2 Summary of Netra Internet Server and Related Software

	Server Software			**LAN Client Software**
	Software bundled with the Netra Internet Server	**Optional Sun Software**	**Third-party or publicly available software**	**Third-party or publicly available software**
Operating system	Solaris 2.4			
Internet connectivity	Asynchronous PPP, Primary/Secondary/ Caching DNS server			
World Wide Web services	NCSA WWW server, Netscape Communications server*, Netscape browser for Solaris**, TTY browser	Netscape News server, Netscape Proxy server, Netscape Com-merce server	Archie server, Gopher server	Mosaic or Netscape browser, authoring tools
Remote access services	Telnet, FTP, anonymous FTP server			TCP/IP stacks, with Telnet, FTP, and POP3 client
Mail	POP2/POP3 server, IMAP server, Sendmail V8			
Administration tools	HTML-based GUI, installation and administration tools, recovery tool			
Security	Server security software	FireWall-1		

Sun's HotJava™

HotJava™ is a Web browser that makes the Internet "come alive." HotJava™ builds on the Internet browsing techniques established by Mosaic™ and expands them by implementing the capability to add arbitrary behavior, which transforms static data into applications. The data viewed in other browsers is limited to text, illustrations, low-quality sounds and video.

Using HotJava™, you can add applications that range from interactive science experiments in educational material, to games and specialized shopping applications. You can implement interactive advertising and customized newspapers; the possibilities are endless.

In addition, HotJava™ provides a way for users to access these applications in a new way. Software transparently migrates across the network. There is no such thing as "installing" software. It just comes when you need it. Of course, some of it you must pay for. Content developers for the World Wide Web do not have to worry about whether or not some special software is installed in a user's system; it just gets there automatically. This transparent acquisition of applications frees developers from the boundaries of the fixed media types like images and text. It lets them do whatever they would like.

HotJava™ possesses these dynamic capabilities because it is written in a new software language called Java™. Keeping the explanation simple, Java™ is a simplified, safe and portable version of C++. It has an architecture-neutral distribution format, meaning that compiled Java™ code runs on any CPU architecture.

What Makes HotJava™ Special?

The central difference between HotJava™ and other browsers is that while other browsers have a lot of detailed, hard-wired knowledge about the many different data types, protocols and behaviors that are necessary to navigate the Web, HotJava™ need understand virtually none of them. This essential difference results in greater flexibility and allows new capabilities to be easily added.

Interactive Content

One of the most visible uses of HotJava's™ ability to dynamically add to its capabilities is called interactive content. For example, someone could write a Java™ program to the HotJava™ API and implement an interactive chemistry simulation. People using the HotJava™ browser on the Web could easily get this simulation and interact with it, rather than having a static picture with some text. They can do this and not worry over whether the code that brings their chemistry experiment to life might be contaminated with malicious, system-threatening code. Neither malicious code nor erroneous code can breach the walls placed around it by the security and robustness features of Java™.

Dynamic Types

HotJava's™ dynamic behavior is also used for understanding different types of objects. For example, most Web browsers can understand a small set of image formats; typically GIF, X11 pixmap and X11 bitmap. If they see some other type, they have no way to deal with it directly. HotJava™, on the other hand, can dynamically link the Java™ code from the host that has the image, allowing it to display the new format. So if someone invents a new compression algorithm, the inventor just has to ensure that a copy of the Java code is installed on the server that contains the image the inventor wants to publish. All other browsers currently available will have to be upgraded to take advantage of the new algorithm. HotJava™ upgrades itself on the fly when it sees this new type.

Dynamic Protocols

The protocols that the Internet hosts use to communicate among themselves are key components of the net. For, the World Wide Web, the hypertext transmission protocol (HTTP) is the most important of these communication protocols. In documents on the Web a reference to a document is called a uniform resource locator (URL). The URL contains the name of a protocol, HTTP for example, that is used to find that document. Current Web browsers have the knowledge of HTTP built-in. HotJava™, rather than having built-in protocol handlers, uses the protocol name to link in the appropriate handler. This allows new protocols to be incorporated dynamically.

The dynamic incorporation of protocols has special significance to how business is done on the Internet. Many vendors are providing new Web browsers and servers with added capabilities such as billing and security. These capabilities most often take the form of new protocols. Each vendor implements something unique, a new style of security for example, and sells a server and browser that speak this new protocol. If a user wants to access data on multiple servers each having a proprietary new protocol, the user needs multiple browsers. Needing several browsers is clumsy and defeats the synergistic cooperation that makes the WWW work.

With HotJava™ as a base, vendors can produce and sell exactly the piece that is their added value to what exists and integrate it smoothly with the products of other vendors. This seamless integration creates a final result that is very convenient for the end user.

Protocol handlers are installed in a sequence similar to how content handlers are installed. HotJava™ is given a reference to an object; a URL. If the handler for that protocol is already loaded, it is used. If not, HotJava™ searches, first the local system and then the system that is the target of the URL, for the protocol needed to interact with the object.

Freedom To Innovate

Innovation on the Internet follows a pattern: First a new technology is developed. Then the developers of the technology can freely experiment with the technology because no one else is using it and there are no compatibility issues. Gradually, people start using the new technology, and as they do an upgrade problem is created. Compatibility and interoperability concerns slow the pace of innovation. The Internet is now in a situation where even relatively simple changes that everyone agrees have significant merit, are difficult to make.

The dynamic nature of HotJava™ solves this problem. Within a community that uses HotJava, individuals can experiment with new facilities while at the same time preserving compatibility and interoperability. Data can be published in new formats and distributed using new protocols and implementations of these are automatically and safely installed. The upgrade problem does not exist.

Inventors of new technology are not the only ones who need these facilities. Almost all organizations and individuals need to adapt to changing requirements. HotJava's™ flexibility can greatly aid organizations to adapt to change.

As an organization requires new protocols and new data types become important, they can be transparently incorporated using HotJava™.

Sun spent considerable time developing their Java™ language to free the user from the rigors of making the Internet work for them. It is a marriage of C and C++ code. The details of which I will not attempt to describe. The "chef's secret" if you get my drift.

Internet Telephone Service

Look out, Ma Bell—a new competitor is in town. This competitor is not one of the global telecom super powers or even one of the traditional telephone companies. It is individuals and groups who will soon use the Internet as a backbone for world wide voice communications.

The network is in place. Individuals are regularly contacted at their Internet address. Add voice compression to this capability and you have the makings of real-time voice conversations over the Internet. The potential impact of this capability on domestic and international long distance service is enormous. You might ask how is this possible?

Through a technology known as Low-Bit Rate Speech Coding, speech data is compressed to a size that permits it to be transported over data lines very similar to standard data. A high quality form of Low- Bit Rate Speech Coding named CELP was developed by AT&T some years ago. CELP (Codebook Excited Linear Predictive) is an algorithmic process whereby human speech and other sounds are compressed while maintaining high sound quality.

To use CELP over the Internet both parties participating in the conversation need have the ability to compress their voice for outbound broadcast, and decompress the inbound voice prior to their hearing it. The compression used has to be the same at both ends of the conversation.

When using compressed speech, the sound quality varies according to the degree of compression used. The more speech is compressed the better it is for transmitting over a data circuit. However, there is a trade-off. The higher the speech compression, the more distorted it sounds. Sound quality is critical to the success of Internet telephone service.

Once compression is available on both ends of the conversation a very fast (14.4 or 28.8) full duplex modem or a digital interface to T-1, ISDN or other digital circuit must be used to provide the interactive quality of standard telephone service. I highly recommended the use ISDN. The larger the bandwidth used for transmittal, the greater the ability to conduct a coherent conversation by this method. Obviously, both parties need be on-line simultaneously to facilitate call completion. As is the case with standard telephone services messaging capabilities can be employed to store voice messages for later retrieval; thus enhance call completion.

Netra™ + SunSPARC™ CT Server = Internet Unified Messaging

The real "Killer" product for Internet use is Unified Messaging. In simple term it is a low cost replication of current messaging system technology operating over a different network topology.

The largest communications cost business and consumers incur today is the toll-charge paid for using traditional voice networks. This charge is based on a time and distance formula. The greater the distance, the larger the per minute charge. If the time and distance factors are removed and replaced with lower local charges or flat fees, communications costs are controlled and decreased.

Internet telephone service with unified messaging can be accomplished on the Sun Platform today. Linkon Corporation provides a complete line of DSP (digital signal processor) based technology for operation in a Sun Sbus environment. Using their CELP high compression algorithm will permit voice to be transported as data over the Internet. CELP is available in 9.6 Kbps, 7.2 Kbps and 4.8 Kbps compression levels with T-1, E-1 and ISDN connectivity. In addition, Linkon technology provides the ability to send and receive fax, send and receive E-mail plus use text-to-speech to access and retrieve fax, E-mail and data downloaded from another WWW site via more traditional telephone service.

Enhanced functionality is easily added to an Internet telephone by using a SunSPARC™ CTI server containing Linkon technology interfaced to a Netra™ server. This provides the capability of constructing a complete unified messaging system which will assist call completion and provide land-line and cellular telephone access for Internet and standard message retrieval.

I doubt that Internet Telephone Service will serve as a total replacement for long distance service. It will, however, provide a low-cost alternative to standard telephone service as we know it today.

Internet Telephone System Caution

The largest problem faced in conducting an Internet telephone conversation stems from a condition inherent to data networks; network latency. Network latency can provide conditions not particularly conducive to real time conversations. Today one can experience network latency of 5 to 30 seconds when using the Internet to retrieve data and communicate with others in chat groups and forums. As Internet traffic increases, so will the latency. The trade-off with latency is one's willingness to settle for this level of verbal interaction as a trade off in price versus standard telephone service. If, however, the Internet phone is used very much like voice mail, where verbal messages can be left in a mailbox along with E-mail and fax, the phone service will truly be a welcome low cost alternative to standard voice communication.

When you think about it, leaving a verbal Internet message rather than typing and sending an E-mail is much easier, less time consuming, and can be done with much more character. This service I have named VE-mail (vocalized E-mail).

|6|

Communications Networks

There is an assortment of technologies used in the public switched telephone network. Some technologies manage call handling while others route calls to their destination. CTI relies on its ability to interface, via common accepted standards, with the public switched network to the same degree that network standards are critical to the configuration of the equipment handling service demands.

Telephone networks are based on switched service technology. The switch located at the CO directs transmissions between the originating and terminating points of the transmittal. Telephone services rely on circuit switched and packet switched technologies to facilitate call completion and manage the network.

The telephone networks offered by the LECs have greatly expanded in capacity and complexity since the advent of Plain Old Telephone Service. We now use analog and digital network connectivity, available through our local CO, to send and receive voice, digital data and images. Most of this is done through the same twisted pair cabling we have been using for years. Compression and ingenious engineering made this all possible. Even highly touted Internet usage for the most part relies on the CO for its on-ramp and exits.

To better understanding this technological migration, let's review the assorted technologies that make it all possible. We begin with analog based telephone service and proceed through digital switched service to slow and fast packet switched networks.

Switched Services

Communication switched network services are composed of analog and digital services. Analog switched services are better known as analog or POTS circuits. This is the service most of us know as our standard telephone service. Digital service is more readily found in business applications. However, with the expanding SOHO (small office home office) market combined with the LECs' push towards ISDN, digital service is now making its way to the home.

Digital and analog circuits differ in the signaling protocol used. Digital service is based on bit streams of data. Analog relies on varying current to transport messages. As advanced as digital technology is not all forms of digital service can facilitate the real-time demands of voice transport.

Analog or Plain Old Telephone Service (POTS)

Analog or POTS lines comprise the vast majority of telephone connections we use today. This technology is as old as the telephone system itself and represents nearly 100 percent of all residential service. It is based on a two-wire or twisted pair connection. This service relies on low DC voltage (battery) supplied by the Central Office (CO) which varies in power to indicate line conditions. If a telephone is off-hook (handset is picked up) the circuit is completed and current flows through the wire. The current is known as loop current. The ends of the wires are known as tip (positive current side) and ring (negative current side). When the off-hook situation occurs, the CO switch reacts by returning dial tone that is a combination of 350 Hz and 440 Hz tones.

For the most part analog lines carry voice traffic. This has change with the tremendous growth of fax and now modem (E-mail). The cost of analog circuits is relatively low. Most of us pay a line fee that is less than $20 per line per month.

Digital Switched Service

There is a variety of digital service available. The following listing covers those services critical to the integration of computer and the telephone.

T-1 Digital Switched Service

T-1 is the most frequently deployed North American standard digital switched service. It is a signaling protocol that operates over two-wire twisted pair circuits divided into 24 time-slots. Each time-slot carries sound, digitized at 64 Kbps, and two signaling bits. The signaling bits are referred to as A and B bits. A single time-slot is referred to as a DS-0 signal; that is Digital Signal level 0. The total T-1 with 24 time-slots is referred to as DS-1 signaling.

The A and B signaling bits are not equal in all applications. However, the most common convention, E & M signaling, keeps the A and B bits equal and uses the A bit to indicate when the connection is active. Thus when the A bit value equals one the loop current is flowing. When the A bit value equals zero there is no loop current. Each of the 24 time-slots acts similar to an analog circuit as line status is determined via interaction between the CO switch, the caller and called phone.

T-carrier network service uses DS-0, DS-1 and DS-3 speeds. (See Table 6.1.)

TABLE 6.1 T Carrier Network Speeds

Carrier Type	Bit Rate	Circuits
DS-0	64 Kbps	1
DS-1 / T-1	1.544 Mbps	24
DS-1C / T-1C	3.152 Mbps	48
DS-2 / T-2	6.312 Mbps	96
DS-3 / T-3	44.736 Mbps	672
DS-3A / T-3A	89.472 Mbps	1,344
DS-4 / T-4	274.176 Mbps	4,032

Some adaptations of T-1 service require the use of a Channel Service Unit. This unit, also known as a Channel Bank, performs certain line-conditioning and equalization functions on the T-1 line. It also responds to loop-back commands sent from the CO and regenerates the digital signal. Many network interface cards for T-1 provide CSU functions as part of the interface.

E-1 Digital Switched Service

E-1 is the European version of the North American switched digital standard T-1. It does not require a CSU for connection. E-1, a CCITT standard, is based on network service segmented by time-slots into 32 channels of service. A single circuit, referred to as circuit level-0, operates at 64 Kbps. That is the same speed as T-1 service. Channels 0 and 16 carry the ABCD bits and synchronization (framing) bits. Only 30 channels are used for voice traffic. The ABCD bits are carried out-of-band in contrast to in-band robbed-bit used in T-1 service.

In North America the A and B bit signaling scheme has common agreement and used as a standard throughout that market. The E-1 ABCD bit conversion scheme, used to provide meaning for signaling protocol, varies widely from country to country and even within some countries. This has presented major problems for PBX and board level switch manufacturers.

Table 6.2 details the six levels of E-carrier circuit availability.

TABLE 6.2 E Carrier Circuit Availability

Carrier Type	Bit Rate	Circuits
0	64 Kbps	1
1 / E-1	2.048 Mbps	30
2 / E-2	8.448 Mbps	120
3 / E-3	34.368 Mbps	480
4 / E-4	139.264 Mbps	1,920
5 / E-5	565.148 Mbps	7,680

ISDN Service

There are three types of ISDN service. Each addresses a slightly different user and application requirement. In contrast to the existing telephone system, all ISDN services offer out-of-band signaling. Data and signaling information are carried on separate channels. This architecture provides greater flexibility for network control, operations, and for the possible services furnished by ISDN.

With ISDN, voice and data are carried on a bearer channel (B channel) occupying a bandwidth of 64 Kbps (kilobits per second). Higher capacity bearer channels, called H channels (consisting of a group of B channels), are also available. Each type of ISDN provides some number of bearer channels, together with a delta channel (D channel) that handles out of band signaling. The D channel is for bringing in information about incoming calls (i.e., Caller ID) and taking out information about out-going calls.

The three types of ISDN are:

⇒ **Basic Rate Interface (BRI).** The basic rate interface is designed to support simple applications such as telephony and basic computer communications. BRI consists of two B channels, 64 Kbps each and one D channel, at 16 Kbps. That is 2B + D. Two bearer channels can be joined, through a process called bonding, to a single 128 Kbps channel.

⇒ **Primary Rate Interface (PRI).** The primary rate interface is designed for volume requirements such as trunk lines for PBX systems and large scale computer connections. PRI consists of 23 B channels, 64 Kbps each and a single D channel, 64 Kbps also. That is 23B + D. In Europe, PRI is defined as 30 B channels instead of 23.

⇒ **Broadband ISDN (B-ISDN).** B-ISDN offers communications rates as high as 150 Mbps. That is significantly higher than BRI and PRI. It provides the capability to manage the higher bandwidth for applications such as video, high definition TV and graphics oriented services.

Broadband ISDN (B-ISDN) is still in development. Widespread use will not be seen for the next few years. To use B-ISDN a complete fiber optic network is needed. This network must run fiber from the CO to the residence or business using the service. This requires a substantial capital expenditure on the part of the LEC to offer the service.

Both BRI and PRI are limited to 64 Kbps per B channel. However, some telecommunications providers permit access to the aggregate ISDN connection thus permitting 128 Kbps for BRI and 1.544 Mbps for PRI. In Europe, aggregate connections for PRI are 2.048 Mbps.

The D channel is capable of 16 Kbps for the BRI interface and 64 Kbps for PRI.

ISDN's Historic Problems

Many segments of the telecommunications industry are engaged in an effort to make nationwide ISDN deployment a reality. Lacking a hardware and software standard plus the associated gaps in interoperability has slowed the nationwide deployment of this service. Under the guidance of Bellcore, the research arm of the RBOCs, a nationwide standard is being implemented through agreement by the LECs, IXCs and telephone switch manufacturers. This new standard is National ISDN-1 (NT-1).

Similar efforts are underway in Europe, through the leadership of the European Telecommunications Standard Institute (ETSI). Together with major equipment vendors and switch manufacturers, ETSI has established a plan for implementing EuroISDN. Their standard provides the same level of compatibility as National ISDN-1. The ETSI common standard was defined as part of the Memoranda of Understanding reached between all EC countries.

ISDN standardization is an evolutionary process that is accelerated by these new standards plus product and service availability. Over time, all countries, product manufacturers and service providers will meet ISDN standards.

Network Services

Network services, not traditionally used to transport telephone based messages, have been utilized as the need for speed continues. This need for increased bandwidth, specifically in distributed processing environments (i.e., client-server architectures), is the catalyst of this movement. In a sense, traditional corporate networks are evolving to function more like the traditional telephone network. We will now review some of these network technologies.

Frame Relay

Frame relay is at the slow end of fast packet switching network protocol. Its speeds range from 56 Kbps to 1.5 Mbps. It traditionally is used for data type services. Frame based services are composed of data bits, in a specific format, with a flag on each end to indicate the beginning and end of the frame. Frame relay uses smaller packets and requires less error checking than the faster more traditional forms of packet switching. Frame relay networks use band-

width only when there is traffic to send. Due to the variable length frame relay does not support voice.

Token Ring

Token Ring is a ring type of local area network (LAN) in which a supervisory frame, or token, must be received by an attached terminal or workstation before it can start transmitting. The workstation with the token then transmits and uses the entire bandwidth of whatever communications media the token ring network is using. A token ring is a baseband network. A token ring LAN can be wired as a circle or a star, with all workstations wired to a central wiring center, or multiple wiring centers. Regardless of the configuration, token ring always operates as if it is logically a circle with the token passing around the circle from one workstation to another. Through its Multi-station Access Unit technology it has the capability to automatically recreate the ring and continue operation when one on workstations crashes or is shut off for some reason.

Token ring operates at 4 Mbps to 16 Mbps and can accommodate up to 256 workstations.

Switched Multi-Megabit Data Service (SMDS)

SMDS is a fast packet switching Cell Relay protocol that uses fixed packet lengths in order to provide bandwidth as high as 45 Mbps. It has been called the connectionless network service because no set-up or tear-down overhead is required as with other fast packet switched technology. Cells are transmitted only with transmitting information. Error checking and handling information is provided by additional computers located at nodes in the network. SMDS is used to transmit groups of voice, data and image packets of all types.

SMDS nodes scan every packet of information transmitted through the network and filter only those with the correct address information for that respective site. SMDS and ATM (asynchronous transfer mode) are similar technologies based on cells of information encapsulated by high-level management information.

Cell Relay technology is similar to Frame Relay with the exception that the cell's length is not variable. It is short and constant and handles speeds in the Multi-gigabit range. The constant rate of cell relay technology permits deterministic bandwidth allocation. Thus, it is suited for the real-time transport of voice.

10 Mbps Ethernet

10 Mbps Ethernet, better known as conventional Ethernet, was designed when it was ambitious for an affordable departmental system to deliver 1 SPECmark of service. Provided as a standard feature or low-cost option with nearly every workstation, 10 Mbps Ethernet also known as Ethernet 10, is ubiquitous and simple to deploy. Its popularity has made possible economies of scale when connecting systems to share information. This capability has made Ethernet inexpensive. Much of Ethernet's early success is due to formalized specifications by the IEEE standards organization, but more importantly, it's a clear de facto standard.

IsoEthernet

IsoEthernet is a standard for putting voice, data and video on the same wire while retaining a high degree of compatibility with voice and data hardware and infrastructure. A single IsoEthernet link provides up to 96, 64 Kbps isochronous "voice channels" (ISDN B-channels) which is the equivalent of three T-spans over one piece of four-wire Cat 3 UTP. 8 kHz clocking on the voice channels is compatible with H.320 standard video codecs, as well as with T.120 data conferencing and G.711, 722 and 728 voice protocols. Full call control signaling, like ISDN, is supported. The same piece of wire also carries a full, standard, 10Base-T Ethernet link.

FDDI

Primarily used for data communications the Fiber Data Distributed Interface (FDDI) has become the standard for networks requiring 100 Mbps transmission speed. It provides media access control, data encoding, data decoding, drivers for fiber optic components and multi-layered network management schemes. Because of these capabilities FDDI often acts as a backbone for connecting servers in a large network with workstations using lower-speed network technologies. As discussed earlier, the most common lower-speed network used is Ethernet 10.

FDDI is more expensive to deploy than Ethernet, offers limited bandwidth and is currently not as widely embraced as other network technologies. It is good for users who require standardization, a high degree of interoperability, flexibility and robustness. FDDI is one of the most mature high-speed network technologies availability today. FDDI is a formal standard of the IEEE.

Fast Ethernet

Users needing a lower cost alternative to FDDI usually turn to Fast Ethernet. It is called Fast Ethernet because it expands the Ethernet 10 standard to 100 Mbps. It maintains the installed base of Ethernet 10 wiring yet has the ability to automatically sense 10 Mbps or 100 Mbps and operate on either speed, thus ensuring maximum compatibility with existing environments and applications.

SONET

Synchronous Optical NETwork (SONET) is a family of fiber optic transmission rates from 51.84 Mbps to 13.22 Gbps. It was created to provide the flexibility needed to transport many digital signals with different capabilities and to provide a design standard to which manufacturers may adhere. SONET was initiated by Bellcore on behalf of the RBOCs to help them be cost effective delivering existing end-to-end services, enable multi-vendor internetworking, create an infrastructure to support new broadband services, and service enhanced operations, administration, maintenance and provisioning.

ATM

Asynchronous Transfer Mode (ATM), as was the case with SMDS, is built on cell relay technology. It uses small fixed-length cells, 53-byte, with inexpensive high-bandwidth, low-latency switches. ATM operates with data rates from 51 Mbps to more than 2 Gbps. It provides the throughput needed by the most demanding interactive applications. Its use is adaptable to most wiring technologies and runs independent of the speed of the line on which it runs. ATM can run full duplex, permitting simultaneous sending and receiving of data without speed or throughput barriers. Its flexibility and low latency allow real-time applications to be supported along with phone, video and other services.

ATM allows the network to be tailored to the application. It is a scaleable, application transparent protocol that allows a mixture of data, voice and video to be used without compatibility issues. With its increased performance and ability to support advanced applications, ATM holds the promise of being the premier network solution going forward. The ATM Forum is an industry group establishing the standards for ATM.

Table 6.3 presents bandwidth alternatives for communications. Each has its merit for specific applications.

TABLE 6.3 Communications Bandwidth Alternatives

Technology Type	Speed
Switched 56/Switched 64	56 Kbps/64 Kbps
Frame Relay	56 Kbps to 1.5 Mbps
ISDN	64 Kbps to 1.5 Mbps
Switched Multi-Megabit Data Service (SMDS)	1.2 Mbps to 34 Mbps
T-1 / T-3	1.544 Mbps / 45 Mbps
Token Ring	4 Mbps or 16 Mbps
Ethernet 10	10 Mbps
FDDI	10 Mbps
SONET	51.84 Mbps to 13.22 Gbps
ATM	25 Mbps to 622 Mbps or more
Fast Ethernet	100 Mbps

SS7—Signaling System 7

SS7 is the out-of-band signaling protocol incorporated in the advanced intelligent network of carriers. It in itself is not a network service offering, but rather the underlying infrastructure with which many of the telecom carrier service offerings are based.

SS7 is the glue that makes it all work. This standardized signaling protocol was developed by Bellcore (the research arm of the RBOCs) and maintained in accordance to their specifications. It could easily be said that the Advanced

Intelligent Network is an architecture built around the SS7 standard. This out-of-band signaling system facilitates all basic network functions, including call set-up, connection and tear-down upon completion. For example, when a caller dials a number, the local central office switch immediately sends instructions to search for the best available voice path. While locating the appropriate open transmission path, SS7 automatically checks the integrity of the line, sets all switching mechanisms to reach the desired destination, and opens the appropriate account for billing purposes. All this is performed in less than 1/20th of a second. Pretty amazing. If a malfunction occurs an alternative path is instantaneously identified and the call is transferred to the operational line. The efforts of SS7 are assisted by connectivity to service-providing data bases.

Earlier information demonstrates that SS7 network connects SCPs with SSPs and Adjunct Processors with Intelligent Peripherals. Some networks use SS7 to connect Adjunct Processors to the network instead of using a high-speed LAN.

|7|

The Components

Building a Computer Telephony solution requires the selection, assembly and integration of the hardware and software components from many manufacturers. Selection of the appropriate combination is essential to providing the optimal solution for the task at hand. Larger integrators and original equipment manufacturers trend towards pre-packaged solutions that require very minor customization prior to installation. They have constructed systems after exploring and testing several combinations prior to settling on their standard design. In the long run this standardization simplifies support issues once many systems are in place. Smaller integrators tend to be more flexible in their offerings, constructing customized solutions from the ground up. They purchase computers, communications and network interface boards, and operating systems independently prior to writing the application software. Regardless of the route all systems are composed of basic hardware and software components.

Not all systems are created equal. There are many things to consider when designing a CT solution. The system need be robust, capable of continuous 24 hours per day 7 days per week operation. Developers quickly learn that system reliability is essential to one's sanity, ability to sleep at night and the preservation of life. Just like the telephone system we reliably use daily or your most critical network server; downtime is frowned upon by the users and not well accepted by management or the service provider. The degree of robustness is dictated by the targeted application and environment in which the system is deployed.

Systems for the Telco market and other "mission critical" applications obviously require a more robust system than a voice mail system for the SOHO market. However, in reality, no one wants to lose messages. In essence all CT systems are mission critical.

To enhance the overall understanding of the components comprising CT system technology we will work our way through the hardware components followed by a discussion of the software.

HARDWARE COMPONENTS

Computer Standards

Critical to the success of a CT application is the computer hardware used in the configuration. The computer must possess the capability of running continuously under load. As mentioned earlier, the application and the environment will dictate the selection of the computer box used. Most telco applications require a NEBS compliant computer and in some cases the system must be fault tolerant or high availability.

Hardware used for all network level services in the telco must adhere to stringent standards for ruggedness, reliability and redundancy. This equipment requires seven days per week, twenty-four hours per day capability complying for the most part to the "Five 9's" standard. The Five 9's is telco jargon for fault tolerant systems that must be operating error free 99.999% of the time. That translates to five minutes and fifteen seconds total down-time per year.

Bellcore's NEBS Standard

Bellcore periodically publishes Technical References under the umbrella of its Network Equipment Building System (NEBS) standard. NEBS standards detail the "minimum generic requirements" appropriate for all new equipment systems used in RBOC central offices and other telephone buildings. Many of these standards have been adopted and adhered to by interexchange, cellular and international carriers as well.

The NEBS standards address issues such as the electrical environment and electrical safety, space planning, test methods for temperature and humidity, fire resistance, earthquake protection, acoustic noise, clearing heights for equipment and cabling, floor loading limits for equipment, cabling and lighting specifications, requirements for totally floor-supported equipment, detail and use of standard floor plans plus compatibility issues for cable pathways.

The NEBS specifications for computer based technology consider serviceability; specifically while the network continues operation. Reflecting this philosophy is the "hot swappable" or "hot pluggable" nature of the components. This means that components of subsystems and subsystems themselves can be replaced, upgraded and serviced without disturbing system operation. These levels of operational requirements have been established due to the nature of RBOCs and interexchange carrier business. Even minor equipment failures could mean millions of dollars of lost revenue, damaged reputation and lawsuits from their customer base. To that end, equipment manufacturers designing equipment for use in telco environments have created Fault Tolerant and High Availability systems.

Fault Tolerant Systems

Fault Tolerant systems are designed for availability 99.999% of the time. To accomplish this, manufacturers have designed systems that contain specialized hardware to detect any hardware fault and instantaneously switch to a redundant hardware component. These redundant components are usually housed within the same chassis as the fails component. This instantaneous switching is virtually seamless. This architecture is generally built upon two or more central processors executing identical instruction streams in tight synchronization. This "trusted" processing system forms a self-checking, self-correcting core unit. In addition to fault checking and bad block mapping, the system verifies each read and write operation with a process called end-to-end checksumming. During any write operation a checksum is generated and written to both disks simultaneously.

The systems "trusted core" consists of CPU sets running in lock step combined with memory subsystems. Each CPU set normally drives two fault tolerant I/O buses with each bus having the capability of accommodating several I/O modules. Disk drives used in each system are configured as mirrored pairs. When a new drive is installed in the system, mirroring software automatically copies all data without interrupting normal system operations.

High Availability Systems

High Availability provides a high degree of system reliability through the use of networked system clusters at a lower cost than fault tolerant architectures. These system configurations have a cluster manager which tracks the status of all the servers which are cluster components in the network. When the cluster manager detects a server failure, it initiates a reconfiguration process which makes the data services of the failed server available as transparently as possible. Under normal conditions one server in a cluster configuration sits idle as a standby for one to three servers in operation. This idle server duty is rotated among the other servers in the cluster in accordance with a predetermined schedule. All active servers in the configuration concurrently work on the same data simultaneously.

The basic difference between the fault tolerant and high availability system approach is displayed in Figure 15.3 *(Chapter 15)*. Fault tolerant systems cost more but have 99.999% system operation levels. High availability systems provide minimal interruption.

Standard Systems

Less stringent system requirements permit a standard SPARC to be used lacking the capabilities to meet Fault Tolerant, NEBS compliant or High Availability standards. The majority of CT systems fall into this category. Standard systems must possess the capability of 24 hours-a-day seven-days-a-week operation, but are not required to provide disk mirroring, operate with dual hard drives or CPUs, and need not possess the capability to "hot swap" components without taking the system out of operation.

Board Level Technology

Network Interface Boards

In addition to standard Ethernet 10, Fast Ethernet and other technology linking high availability, client-server and other networked systems together there is interface standards specific to connecting with the PSTN (public switched telephone network). Analog, T-1, E-1 and ISDN Digital network connectivity are in this category. All are available for the Sun Platform.

Analog

Analog boards provide interface with POTS (plain old telephone service) to connect the computer containing the board with the PSTN. This connection is made similar to the way a telephone is connected to the PSTN when installed at home or into a non-digital office connection. The majority of existing telephone connections today are analog.

T-1

T-1 network interface cards join the CT processing capability to the network. T-1 is the most frequently deployed North American standard digital switched service. It is a signaling protocol that operates over two-wire twisted pair circuits divided into 24 time-slots. Each time-slot carries sound, digitized at 64 Kbps, and two signaling bits. The signaling bits are referred to as A and B bits. A single time-slot is referred to as a DS-0 signal; that is Digital Signal level 0. The total T-1 with 24 time-slots is referred to as DS-1 signaling.

The A and B signaling bits are not equal in all applications. However, the most common convention, E & M signaling, keeps the A and B bits equal and uses the A bit to indicate when the connection is active. Thus when the A bit value equals one the loop current is flowing. When the A bit value equals zero there is no loop current. Each of the 24 time-slots acts similar to an analog circuit as line status is determined via interaction between the CO switch, the caller and called phone.

T-1 line. It also responds to loop-back commands sent from the CO and regenerates the digital signal. Many network interface cards for T-1 provide CSU functions as part of the interface.

E-1

As was the case with the T-1 network interface cards, the E-1 network interface cards provide connectivity between the PSTN and the CT system. E-1 is the European version of the North American switched digital standard T-1. It does not require a CSU for connection. E-1, a CCITT standard, is based on network service segmented by time-slots into 32 channels of service. A single circuit, referred to as circuit level-0, operates at 64 Kbps. That is the same speed as T-1 service. Channels 0 and 16 carry the ABCD bits and synchronization (framing) bits. Only 30 channels are used for voice traffic. The ABCD bits are carried out-of-band in contrast to in-band robbed-bit used in T-1 service.

ISDN

ISDN network interface cards are recent arrivals in the CT integration strategy. There are three types of ISDN service. Each addresses a slightly different user and application requirement. In contrast to the existing telephone system, all ISDN services offer out-of-band signaling. Data and signaling information are carried on separate channels. This architecture provides greater flexibility for network control, operations, and the possible services furnished by ISDN.

With ISDN, voice and data are carried on a bearer channel (B channel) occupying a bandwidth of 64 Kbps (kilobits per second). Higher capacity bearer channels, called H channels (consisting of a group of B channels), are also available. Each type of ISDN provides some number of bearer channels, together with a delta channel (D channel) that handles out of band signaling. The D channel is for bringing in information about incoming calls (i.e., Caller ID) and taking out information about out-going calls.

Fixed Feature Boards

Fixed feature boards connect via analog or digital format with the PSTN and perform the function for which they were designed. Boards have been created to interpret DTMF (dual tone multi-frequency) signals while recording and playing-back speech, fax, modem, pulse code detection, speech recognition, video and text-to-speech. These boards utilize chip set technology or fixed point DSPs (digital signal processors) to perform their designated functions. Applications requiring a mixture of functionality require these boards to be connected via an internal bus (PEB, MVIP, SCSA see chapter 9) to facilitate the needs of the application. Developers desiring to add functionality to an existing system must add additional boards to construct the needed solution.

There has been a new set of fixed function boards entering the market; usually voice and fax combinations, employing chip sets and/or fixed point DSPs.

"Universal Port" Boards

Universal port boards utilize powerful floating point DSP based technology to deliver communications processing solutions via software algorithms. This highly sophisticated method of processing requires a DSP operating system to manage this multi-tasking environment and to provide very flexible communications processing. In essence, this is tantamount to running UNIX or Solaris at the DSP level.

Universal Port technology is very similar to standard computer processing. There is a CPU that serves as an engine. In this case the DSP is that engine. The engine requires an operating system to harness the processing power and manage functionality. The operating system on the Universal port board is the DSP-OS. The DSP-OS is a multi-tasking operating system very similar to Solaris. Processing functionality in standard computing is the result of applications software designed to perform a certain type(s) of activities (i.e., word processing, spread sheet, etc.). The applications software on the universal port board is also function specific (i.e., DTMF, voice record and playback, compression, fax, modem, speech recognition, text-to-speech, video, etc.).

The multi-tasking capability of the DSP based universal port board provides the ability to move from one technology to another, when needed, without requiring any switching, or any additional hardware, all within nanoseconds. For example, a caller can place a call from the telephone handset of a fax machine to a CT server containing a universal port board. Once the call is answered the caller can interact with the system using either DTMF or speech recognition to sort through menus of available information for topics of interest. This could be product information, technical support information, entertainment or anything else imaginable. Once in the appropriate category has been found, they could preview the information via text-to-speech and make their selections. When complete the caller then can press the start button on the fax machine they are calling from, and have their selections faxed backed to them immediately. This is all accomplished on the same telephone line without any switching involved at all. Now, that's amazing. One additional point of interest that brings a smile to the face of management: the caller paid for the call and for the paper on which the information was printed.

Universal Port boards are software upgradeable. They do not require the integrator to reconfigure their installed hardware after building the system to provide additional functionality. Functionality is loaded as software algorithms via floppy, CD, data tape or remotely via modem.

As was the case with Fixed Feature boards, Universal Port boards are connected to network interface hardware and other Universal Port boards via standard internal bus technology (PEB, MVIP and SCSA).

SOFTWARE COMPONENTS

There are several software layers in a CT solution. These start at the operating system, to the drivers interfacing with the assortment of hardware composing the system architecture, through APIs, application software,

Operating Systems

Operating systems allocate processor cycles, disk space, memory and privileged operations to applications and their users. However, there are characteristics and capabilities that make one operating system more suitable than another for a particular application.

Operating systems should ensure that the programs cannot cause a system failure or a failure of another program running on the same host. It must possess essential elements that act as security mechanism to prevent damage or data from unauthorized programs or users. Operating systems must have a complete set of tools to perform backup and restore, maintain high availability, and provide administrative and diagnostic capabilities. The operating system must provide time-tested reliability which only comes with extensive hours of operation in the field.

The power of the most sophisticated hardware is wasted without an operating system that can harness it. Server systems need true multi-tasking, large memory and file systems, turnability and tools to optimize system throughput.

The operating system must adhere to open standards. The CT industry is beginning to realize the value of a true open system architecture. Operating systems that embrace open standards philosophy will be well positioned to penetrate the CT market.

To date MS-DOS and Windows are the predominate hosts for lower end CT solutions. UNIX and Solaris dominate large system configurations. OS/2 has been deployed in some CT solutions with its strongest penetration in Europe. Windows NT has been used in some systems and should realize some additional market penetration as the product matures.

MS-DOS and Windows

Both Windows and MS-DOS have been extensively used in the CT market to date. For the most part this is due to the predominance of PC based solutions. Unfortunately both provide poor support for large memory applications, poor resistance to system failures and addressing violations and lack true multitasking. These limitations make both useful for low level CT tasks and supporting personal productivity application (spread sheets and word processing), but call into question their suitability for enterprise-level applications. Their limited network capabilities, specifically in WAN environments, make them subject to frequent crashes. They can only poorly support applications that require multitasking.

Windows 95

Though too early to tell, Windows 95 seems a better solution for CT applications than earlier generation Microsoft products. With the inclusion of TAPI (Windows Application Programming Interface, see Chapter 6), Windows 95 comes prepared for first party call control type features. Windows 95 will probably see successful use for desktop and workstation environments but will not be used for mission critical applications

Windows NT

With the introduction of Windows NT a few years ago it was evident that even Microsoft has recognized the failure of MS-DOS and Windows to address enterprise computing issues. Even though Windows NT shows promise in its ability to perform multi-tasking and demonstrate reliability, it lacks the proven maturity of a UNIX based and Solaris platform. Eventually, with the marketing dollars of Microsoft behind the product Windows NT will hold a prominent position in the CT platform market.

OS/2

Although OS/2 had an earlier start than Windows NT and Windows 95, use of this operating system in the CT market has been weak. IBM has achieved greater success in selling the product in the European market, but they have provided mixed signals as to their level of product support and continuation of their effort to market the product.

Solaris

Solaris is based on popular System V Release 4 (SVR4) version of UNIX. It has the necessary maturity, functionality, connectivity, market acceptance, and performance that make it a logical choice for hosting mission critical applications. In networked environments, such as call centers, Solaris is an excellent choice. Solaris has a variety of tools available to facilitate transparent local and wide area networking and can greatly ease administration duties; a major network cost. The ability to run on Sun's RISC architecture and Intel's *x*86 processors permits problem-free migration of product between both platforms.

The option of using Solaris for X86 is an important consideration for organizations that are committed to PC hardware for budgetary, technical or installed base reasons. Solaris for X86 offers the same functionality as Solaris for SPARC, giving Intel based systems the multi-tasking, memory management and integrated networking support to unlock their potential. As performance requirements grow, applications can be moved from Solaris for *x*86 to Solaris for SPARC.

Drivers

Drivers are software interfaces to hardware devices that provide instructions for reformatting or interpreting software commands for transfers of information to and from the hardware device and the CPU of the computer. All board level technology used within a system require drivers to interface with the operating system. In other words, a driver is a software module which "drives" the data interaction between the software layer and hardware layer of the CT system.

Driver Development Interface (DDI)

Driver development interface is a published set of specifications detailing the interface to the drivers for a particular piece of hardware. This published specification is usually part of a Driver Development Kit which also contains sample software to assist the driver porting effort and a means for testing the results of the programming effort.

Software Tool Kits

Software tool kits are operating system specific collections of C or C++ libraries that assist the developer in creating a CT application. They use a series of function calls which create software code for call processing activities in an application. For example there are function calls which create a mail box, out dial a number, receive a fax and recognize a far end disconnect. Sophisticated tool kits have a couple hundred function calls. Software tool kits are a lower level interface than an Applications Generator but higher than a DDI.

Applications Generator

An applications generator is a software tool that in response to the programmer's input generates computer code to perform a specific function. In simple terms, it is software that writes software. They tend to be either general purpose tools, providing support for specific applications or designed for a specific application. Applications generators are available in three different levels.

⇒ **Level One: Non-generator generators.** They do not actually create new software but they allow you to tweak existing blocks of code. There is no compiling and they are generally pretty easy to use.

⇒ **Level Two: Graphical User Interface (GUI).** Forms-based applications generators usually require the user to build a call-flow picture using either icons or easy to understand templates. When the call-flow picture is complete the user compiles it and it actually generates new software code.

⇒ **Level Three: Script Language Generators.** These use English language commands that are specific to the type of task they are doing. It is the most flexible of the three levels of applications generators as it permits a programmer to use their programming skills to develop code.

Applications

Applications are software programs that carry out some useful task. Voice mail, Fax Broadcast, Locators, Data Base Managers and Voice Activated Dialing are examples of application software. Industry standard applications are available off-the-shelf that require their installation on the target machine. Once live the user records the appropriate phrases to be spoken by the program and the system is ready for operation.

Applications requiring custom development are either created with an applications generator or manually coded by a programmer. Variations of these applications are limited only to the creative genius of the programmer writing the code. (Sample applications can be found in Chapters 1 and 14.)

Software Standards

There are several software and hardware standards that assist our efforts in assembling a Computer Telephony solution. They permit multi-vendor products to be assembled as one contiguous system providing communications processing functionality, and they permit servers to talk to switches and databases. The SunSPARC platform adheres to all internal box standards for connectivity and the SunXTL standards for communicating with external communications processing and databases. (Details on all standards discussed can be found in the appropriate sections of this book.)

|8|

Configurations

There are many ways in which computer telephony equipment can be configured. Some systems are stand-alone VRUs (voice response units) terminating several telephone lines, or used at the workstation level, connecting via a single analog POTS (plain old telephone service) line, to the telephone network. Others are components in a distributed processing configuration, interacting with database information as part of a call center solution or part of an Intelligent Peripheral (IP) which serves as an integral part of a telco's Advanced Intelligent Network (AIN). Whatever the need, the appropriate solution can be built from Sun Microsystems line of workstations and servers combined with the highly sophisticated communications processing technology available from Linkon.

To assist reader comprehension only the more common CT configurations are presented here. They are used to provide a template from which more complicated configurations can be built. Additional configurations, with accompanying business case and result, can be found in Chapter 14 of this book.

Stand-Alone Systems

Stand-alone systems are connected directly to the telco providing the service via analog or digital circuits. Analog connections use POTS lines. Digital connections use either full or partial T-1 circuits domestically, and full or partial E-1 circuits in most international markets. For the most part, stand-alone systems have traditionally performed dedicated computer telephony functions such as

IVR supplying information to callers. This information can be voice based, fax based information provided via a one or two call fax-on-demand system, or an E-mail hub where individuals are dialing-in via modem to download information.

As computer telephony continues its migration into the workplace, newer stand-alone systems are located on the very desktops employees perform their daily job functions. This results in more productive work and a cost savings strategy for the implementation of computer telephony processing. These new workstations run their more traditional job specific software plus a total communications package consisting of voice, data and image capabilities.

These workstations are configured with a single line, analog interface, Linkon FS-4000 communications board and the appropriate applications software to provide voice mail, fax mail and E-mail tightly integrated with the user's job specific software. These new capabilities provide each user with a degree of productivity and efficiency never before realized.

Figure 8.1 represents a stand-alone IVR connected to analog circuits, a stand-alone IVR connected with T-1 digital circuits and a stand-alone workstation connected via a single analog POTS line to the telephone network.

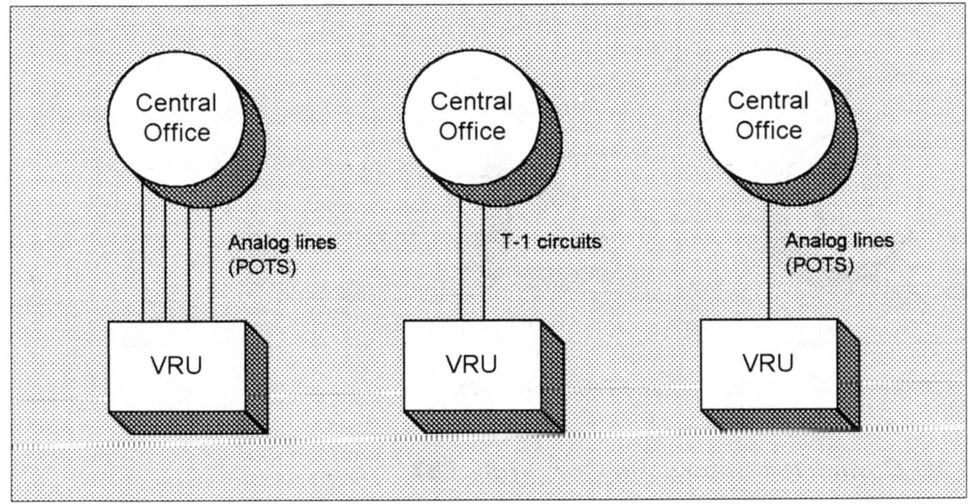

FIGURE **8.1** **Stand-alone system configurations**

Networked Systems

Networked CT systems serve as components of larger integrated solutions. Some configurations feature a single system, used as a server, providing the CT functionality for all client workstations networked in the configuration. Most networked systems interact with one or more databases, or are part of a LAN/WAN configuration. Some work as modules of a densely configured solution, requiring several systems to provide sufficient processing capacity to handle the large volume needs of a telco messaging system or a highly advertised electronic media campaign. Following are a few examples of networked CT systems.

VRU Connected to Corporate PBX

Most corporate voice mail (a/k/a VRU) systems are connected behind the PBX. In the typical configuration the VRU is connected with one or more extensions in a group attached to the company phone system. Using the flash hook transfer feature of the PBX, the VRU equipped with automated attendant functionality can transfer a call anywhere within the phone system. All incoming calls are routed to the hunt group extensions terminating at the VRU. If the intended extension is busy or unanswered, ensuing a set number of rings, the caller is allowed to transfer to another extension or leave a message.

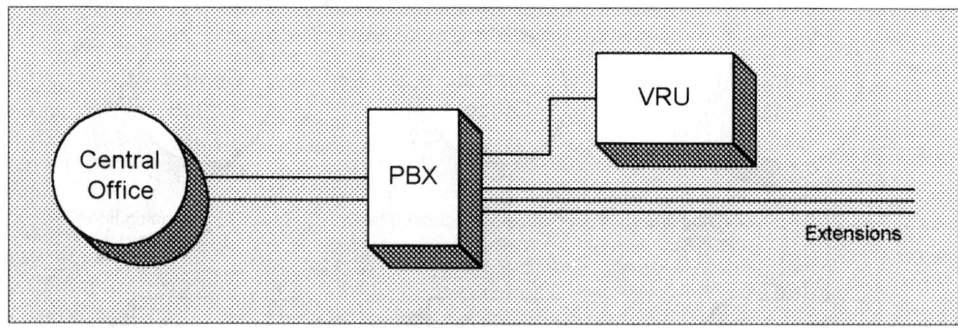

FIGURE 8.2 VRU connected to corporate PBX

VRU Connected Before the Corporate PBX

Situations where the VRU is connected before the corporate PBX are more rare than connecting it behind. It is usually done to facilitate a deficiency in the PBX or to capture ANI (automated number identification) or DNIS (dialed number identification service) information to be transmitted prior to the call. Capturing either lends the VRU the ability to handle inbound calls more intelligently. Using DNIS permits calls to specific numbers to be automatically transferred to specific PBX extensions. By capturing ANI customer information can be matched with the call prior to both being transferred to the appropriate agent handling the call. Figure 8.3 details this type of configuration.

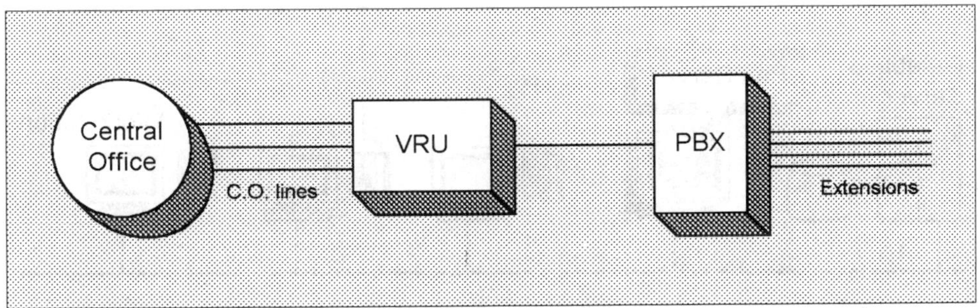

FIGURE 8.3 VRU Connected Before the Corporate PBX

Department Server

Computer telephony servers, used at the corporate department level, are usually function specific. An example of this is the traditional fax server connected to department LAN. All workstations on the network use this resource to send and receive faxes. Outbound faxes are queued for delivery in order sent, or scheduled for a specific delivery time. Large volume fax jobs can be scheduled for off peak delivery when calling rates are lower and the server is experiencing lower volume use. Inbound faxes, when using DNIS technology, route the fax to the appropriate network address or a central box for printing. Fax servers are usually more cost effective than configuring each workstation with individual fax capabilities, or using a stand-alone departmental fax machine with the typical accompanying human queue and the regular paper jams which further complicate the process.

When a departmental fax server is deployed, the voice mail system remains a separate system, centralized for use by the entire company. Figure 8.4 displays the typical department fax server.

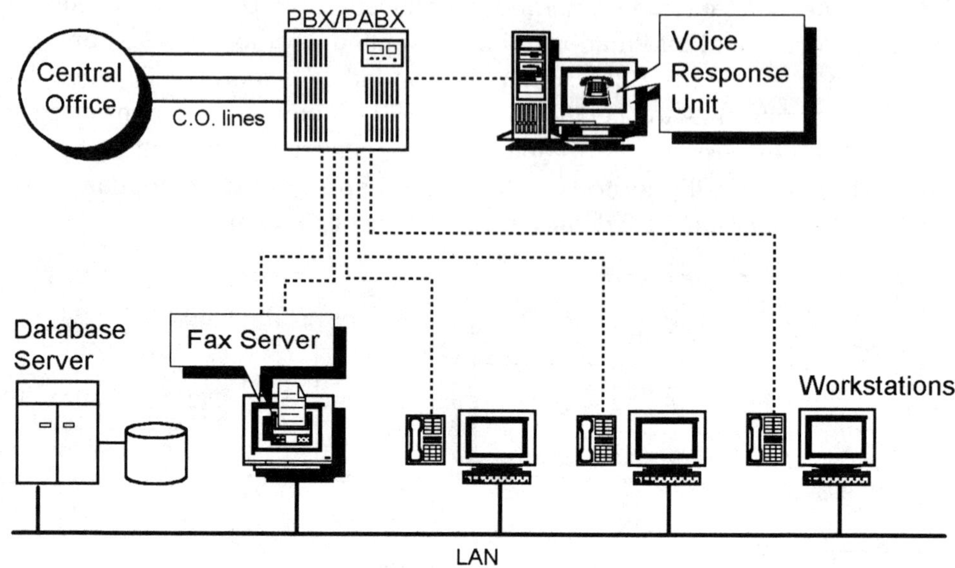

FIGURE 8.4 Departmental Fax Server

Call Center

Traditional PBX configurations are easily extended to accommodate the needs of an inbound/outbound telemarketing call center with ACD (automatic call distribution) and/or predictive dialing capabilities. The configuration normally used to handle this application requires one computer exercising overall control of the service. Connected via LAN, the configuration notifies the VRU and PBX of the call status. It monitors available operators and transfers calls based on the agents' capabilities and specialization.

DNIS and ANI technology can be used on inbound calls to interact with the appropriate advertising campaign data, to pull up customer information and the targeted telemarketing script which will be used by the agent fielding the call. Outbound calls are dialed by a predictive dialer, dialing numbers from a database, and only put through to an agent after a live contact has been made.

This configuration has the ability to capture call accounting information which provides up-to-the-minute status of the campaign by an assortment of statistical breaks. This information calibrates the effectiveness of the advertis-

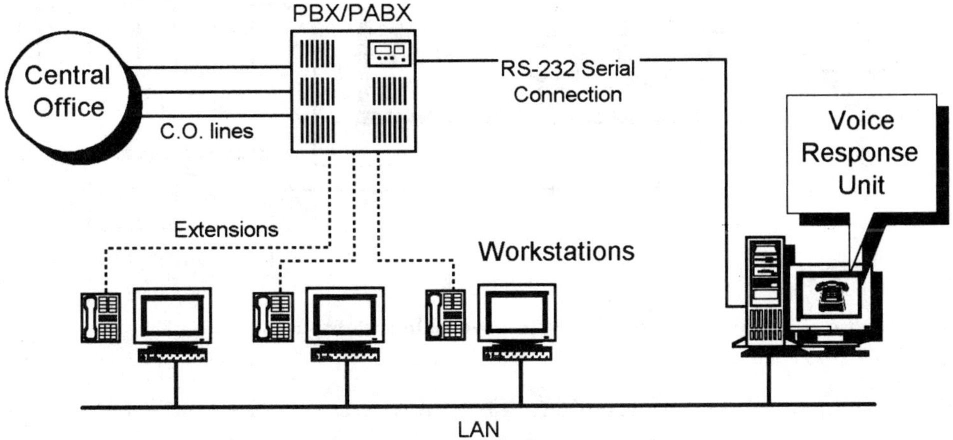

FIGURE 8.5 Call Center

ing campaign, details customer purchasing patterns, and provides call cater management with valuable data on agent productivity, work load and individual effectiveness. Call center integrated solutions have exponentially increased the productivity of the telemarketing business. They have quantified campaign effectiveness while providing call center management solid information.

Figure 8.5 details a common call center configuration.

VRUs Interacting with Databases

A VRU interacting with a database is perhaps the most common use of computer telephony technology. These databases can be located within the system containing the Linkon communications boards, be connected in a client-server configuration, be a host computer located within the facility but not interconnected as part of the configuration, or be located remotely and connected via dedicated circuits or dialed-up on an as needed basis. Whatever the configuration, the computer telephony technology accesses the information contained in the database to meet the real-time needs of the caller while interacting with the application.

For example, many databases used with applications contain customer information, inventory data, accounting and billing information, scheduling information and a sundry of other possibilities. These databases can be a series

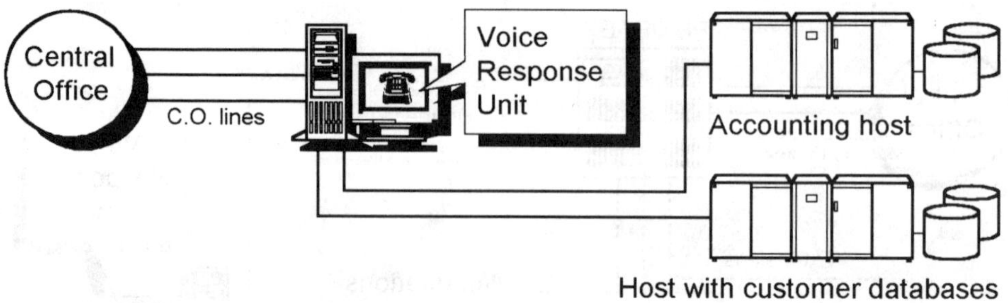

FIGURE 8.6 **Voice Response Unit with external host databases**

of computers networked together, large disk arrays, mini and main frame computers or any combination imaginable. Accessibility of the needed data is the only critical element to the application. Fig 8.6 graphically details a computer telephony solution interacting with a database.

The configurations detailed can be used with wireline and wireless telephone services. They represent a small collection of possible configurations but are common enough to provide the reader a base of knowledge to better understand the computer telephony industry. The intended use of computer telephony technology will dictate the design of the most appropriate configuration.

Computer telephony products are used as a mechanism for better access to information by the parties needing the information at that time of need. They traditionally are configured for use with existing databases which earlier were accessed solely via human interaction. Computer telephony technology is used to provide twenty-four hours per day, seven days per week access to information.

|9|

Computer Telephony Standards

Standards are essential to the growth of any industry. Each of us can easily cite numerous examples of organizations who did not move from their proprietary stance, who later met a brutal demise or who never quite realized their full potential in the market. Examples from the computer and telephone industries abound.

Most PBX (private branch exchange) manufacturers have realized, only too late, the penalty earned by their proprietary "go it alone and protect your market share" philosophy. While other organizations, at the same time, realized the need for a standard, or created a de facto standard by virtue of their market prominence and benefited.

History is filled with successful situations based on "common agreement" or a standard. Non-high tech industries have standardized on pipe sizes and fitting requirements, shoe sizes, screw sizes, fluid calibration. The highway system has standardized on signage and rules that somehow keeps us all moving in the right direction.

Bill Gates, with the assistance of IBM, standardized an operating environment (MS-DOS/Windows/Windows 95/Windows NT) which has reached the status of a household word, similar to the awareness Coke.

The telecom and computer industries are both built on standards specific to their respective industry. With the emergence of the Computer Telephony industry new standards need be established which consider those standards already in place yet expand the interoperability of the technology from both industries. This need has spawned much activity. The earliest activity arose from necessity. Newer standards efforts exhibit more of a collegial effort. Groups, committees and now organizations have been formed to facilitate this effort.

When developing standards, we have wasted no time on our cruise through an alphabet city of acronyms. We have PEB, MVIP, SCSA, TAPI, TSAPI, MAPI, VERSIT, ECTF and XTL just to name a few. The never ending list of de facto, du jour, delightful and delicious standards seems to have a life of its own.

The melding of mindshare created through discussions between organizations has witnessed an evolution from the earlier hardware emphasis to a combined hardware and software approach. As this industry evolves the predominate direction will be software. Standards will define common interfaces at the API level. This API will be operating environment independent with the ability to provide for the easy inclusion of new technology as it arrived in the market.

To be worthwhile standards need serve as a blueprint of the best way to match resources and services to an application in a distributed environment. They should simplify the development of computer telephony systems that can incorporate multiple technologies from multiple vendors through the use of a standard open interface.

It is foolish to assume one standard will be sufficient to meet the needs of all applications. Due to the youthfulness of the Computer Telephony industry and the speed at which it is growing it is unrealistic to expect a single, consistent set of standards in the near future.

They're All Similar

CT standards have two major areas of consideration; hardware interfaces with the telephone network and software interfaces between the standard computer operating environments which are comprised of drivers, applications and the service providers.

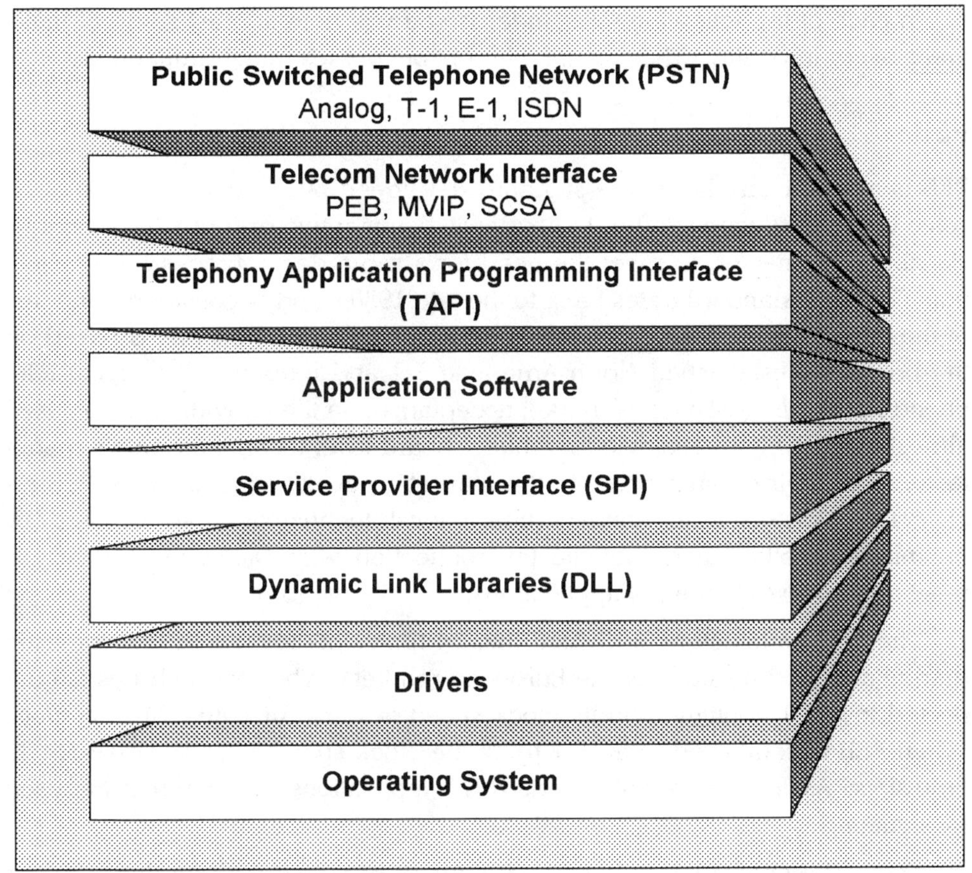

FIGURE 9.1 Macro view of CTI architecture

Figure 9.1 will serve as a road map through these common components. At the top of the diagram is the hardware connection to the public switched telephone network (PSTN), followed by the application programming interface (API) common to the specific internal bus (PEB, MVIP, SCSA) carrying the traffic. The service provider interface (SPI) translates the necessary communications processing functionality to the Linkon boards. Finally, the entire process is managed by the application software that interfaces with the operating environment and the hardware configuration via the drivers and the dynamic link library (DLL).

In each of the following CT standards you will note at least some of these areas of commonality. They are all different but posses many similarities.

PEB

PEB is a proprietary hardware standard developed by Dialogic Corporation, a voice board hardware vendor, to physically connect two or more voice boards, via ribbon connector, to meet the need for a more dense processing application. The PEB standard dates back to the late 1980's and is considered the first of the computer telephony standards. The design of the PEB bus is based on the previously established North American T-1 and European E-1 digital telephone standards, assisting its market acceptance. In the early days it was used almost exclusively by Dialogic to connect board level technology solely manufactured by their organization. As the market gradually expanded the PEB architecture was used to connect board level technology manufactured by organizations other than Dialogic for connection with Dialogic hardware to meet the processing need of applications.

The PEB architecture is divided into 24 time-slots for use in the domestic market and 32 time-slots for the European market. When the PEB bus is connected to a T-1 configuration its clock speed is 1.554 MHz and 24 time slots. When PEB is connected to an E-1 trunk the clock speed is 2.048 MHz with 32 time slots. As is the case with standard E-1 applications two channels are used for signaling.

Every PEB configuration begins with one or more network modules (T-1 or E-1) which connect the audio bus to the external telephone network. The network module is responsible for converting between the PEB format for audio and signaling information and the appropriate standard of the external connection. PEB is the precursor of the SCSA standard.

Signal Computing System Architecture (SCSA)

SCSA is a hardware and software standard that evolved from the use of the Dialogic PEB bus. It built on the experience that Dialogic and its developers had with PEB, and it provided a better solution to integrate the hardware and software from other vendors into a communications processing system. To understand this architecture we will first discuss the hardware specifications followed by the software.

SCSA Hardware

The SC bus. The Signal Computing System Architecture (SCSA) was designed to be the next generation PEB bus. SCSA is built on the SC bus; a higher capacity bus than its predecessor PEB. The SC bus can operate at data rates of 2.048 to 8.192 MHz (32.768 Mbps to 131.072 Mbps) with a default of 4.096 MHz (65.536 Mbps). It has 16 data lines for up to a maximum of 2,048 possible time-slots. Its default data rate supports 1,024 time-slots. Each time-slot is bi-directional and carries 8-bit, 8 Hz (64 Kbps) data, just like a T-1, E-1, ISDN, or MVIP time-slot.

SCSA signaling is handled on a separate data channel as opposed to the PEB standard which uses robbed bit signaling. Robbed bit signaling means bits are occasionally stolen from audio data to represent the signaling state of the line.

SCSA uses channel bonding techniques to accommodate the need for expanded bandwidth. Channel bonding refers the combination of two or more 64 Kbps B channels into a single logical channel which has a higher data rate.

The SC bus is a multi-master bus with automatic clock fall back. This means that any SCSA board in the system can be the master used to coordinate the activities of all other boards connected to the bus. Since the implementation of the multi-master feature is at the hardware level the switch to a new master is accomplished quick enough not to lose signal. This makes the bus configuration into a fault tolerant architecture.

Time-slot switching is also built into the bus hardware connectors. Through this it is possible to connect the receive half of one time-slot to the transmit half of a second time-slot to create a talk-path between two channels.

The SCx Bus. The SC expansion bus (SCx bus) is a ribbon cable that runs between different computer chassis connecting the systems together via the installed SCSA boards. This permits the creation of system nodes of up to 16 different computes. With this outside of computer box connectivity large multi-computer architectures are possible for use as unified messaging systems and other densely configured solutions. Redundant system configurations are possible where the disk storage in the configuration can duplicate the data stored very similar to disk-mirroring used in stand-alone systems.

SCSA Software

SC Firmware. This firmware directly controls the SC bus hardware and other board level components. It handles switching and the routing of time-slots. This standard permits board level technology from any vendor to easily be adapted for SC bus use.

Telephone Application Object (TAO) Framework. SCSA software is designed to accommodate the needs of both stand alone and distributed processing architectures. Therefore it can accommodate the needs of local and remote applications. With the input of over 250 vendors the Telephony Application Object (TAO) Framework was designed. It is composed of four components:

- ⇒ Standard **Application Programming Interfaces (APIs)** were established. They define the way the assorted technology is accessed for use with the messaging protocol and the system services. APIs for voice processing, speech recognition, text-to-speech, fax and other technologies. These APIs were designed by work groups composed of individuals from an assortment of companies providing these technologies. Efforts were made to standardize and accommodate the needs of all parties involved.
- ⇒ **Messaging Protocol** was developed which standardizes the way messages are passed among applications, system resources and system services.
- ⇒ Core **System Services** were established which include resource management, switching, event and error handling, configuration management, session management and security.
- ⇒ TAO set **Service Provider Interfaces** (SPIs) that define the way System Services and Resources access the Message Protocol.

SCSA TAO was designed to be scaleable and hardware independent. That is, it operates on PC bus hardware as well as on MVIP and other hardware architectures. It is an open software design with sophisticated application management capabilities. It is object-oriented and operating system independent. Its core System Services and technology-specific resource APIs are interoperable.

Multi-Vendor Integration Protocol (MVIP)

To create a standard undominated by a single competitor, a group of industry vendors, led by the voice board manufacturer Natural Microsystems and the PBX manufacturer Mitel Corporation, gathered and established the first MVIP standard in September of 1990.

MVIP is a standard architecture for interoperability among telephone-based resources, including trunk interfaces, voice, video, fax, text-to-speech, speech recognition and data for use within a computer chassis in an individual or networked configuration. The architecture is supported and driven by some 300 companies who are licensees of the MVIP technology. The movement is managed through GO-MVIP (Global Organization for MVIP).

MVIP is an architecture that provides a framework within which manufacturers of computers and communications systems can plan hardware and software products, complete systems and customer solutions to meet market needs. Its broad-based adoption provides an extensible platform upon which end-users can plan and implement their telephony enhanced solutions.

Layers Of The MVIP Protocol

Application Layer. The application layer is the user interface where the total solution is accessed. This layer can include fixed applications or configurable applications using parameter stacks or GUIs. Multiple applications from different vendors may be co-resident and are interoperable in a single system or integrated among many systems.

Application Programming Interface (API) Layer. The API layer includes a call control API for call set-up and tear-down, a connection control API for intra-chassis and multi-chassis switching, media services APIs for voice processing, speech recognition, video, fax, etc., and various support services APIs for managing events, prompt files, databases, and others in real-time environments. In addition, the API layer offers software bindings to specific environments such as TAPI, Germany's CAPI and France's IPAVE.

Client-Server Layer. The client-server layer employs industry standard structures with stand alone servers as well as across a network. This layer accommodates call control, connection control, media services and support services whether the architecture deployed is LAN, WAN or a combination using commercially available interfaces, servers, bridges, routers and gateways.

Due to its open-structure industry standard messaging capabilities are realized using standard, commercially available widely used techniques. The client-server layer uses the OSI transport layer interface on standard LANs for messaging (i.e., sockets or TLI under UNIX). TCP/IP, SPX/IPX, IPX modules or Net BIOS direct access is possible at a lower level. IBM's Communication Manager (on OS/2 systems) or Microsoft's LAN Manager can provide local or wide area support for MVIP's message architecture. This maintained industry computability.

Used in a stand alone node or chassis, interprocess messages use industry standard techniques such as shared memory and streams in UNIX or named pipes, semaphores and queues in OS/2 and Windows NT. By specifying the remote control procedures using established industry standards and formats, MVIP offers local and distributed resources that are operating system independent.

Device Control Layer. The device control layer is the intimate link between specific vendor products and the MVIP architecture. Each module is operating system specific, and integrates a specific device to the MVIP world. Device control modules use the techniques recommended for each operating system under MVIP standard such as device drivers and dynamic link libraries.

Switching Service Layer. The assortment of switch fabrics available are readily distinguished based on the method by which they are controlled. Single chassis MVIP is the lowest layer. At this lower layer all boards located in this chassis are controlled by a single software entity even though each board has its own resident switch capability. Higher switching layers provide for inter-chassis connections.

The second layer encompasses completely connected single layer, multi-chassis switch fabric and distributed control. Examples of this are MC1-MC4 discussed below. Control of the second layer switch fabric requires coordination of the independent processes running on separate computers. The third

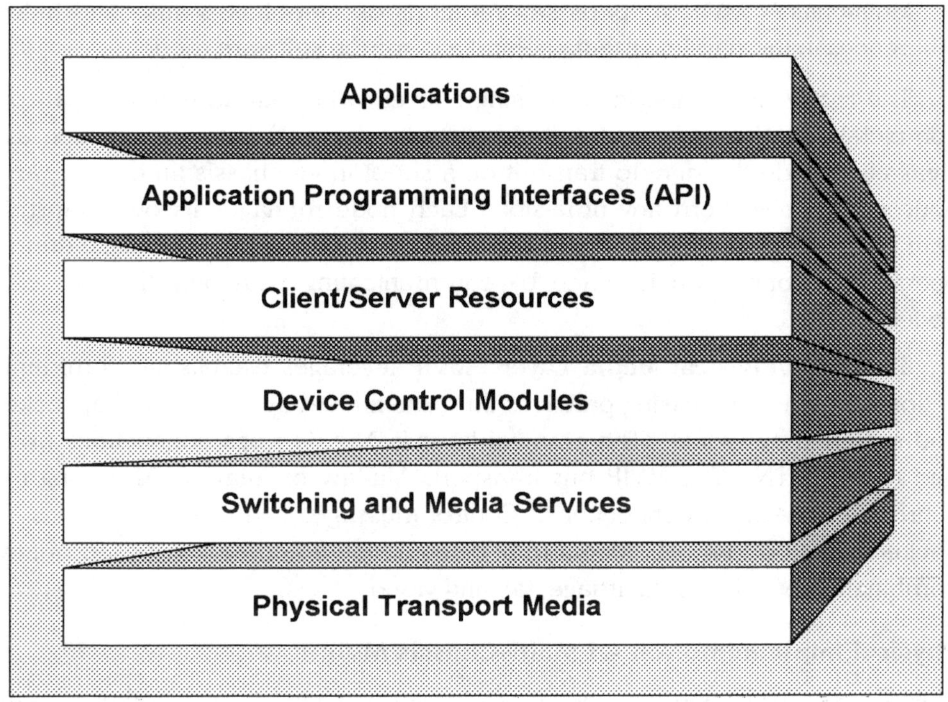

FIGURE 9.2 **MVIP architecture**

layer uses tandem routing using either star-connected central node or a sparse mesh of point-to-point links. Layer three software control of the switch fabrics requires a central database or significantly more complex distributed software. The MVIP Connection Control API provides a common interface for all three layers of the switch fabric.

There exists a fundamental difference between the control and switching of a single-chassis and multi-chassis MVIP configuration. The switch and central controlling software resides within one computer in a single chassis MVIP configuration. For multi-chassis applications a switch fabric interconnects multiple independent computers. Through this interconnection multi-chassis MVIP configurations can provide redundancy, and "hot swap" capability on a chassis-by-chassis basis, control and switching is distributed among all computers in the configuration.

Multi-chassis MVIP supports several kinds of inter-chassis links and hides differences in the physical media behind a common software model.

The major advantage to using single level switch fabric is that it can be completely controlled by distributed software. An administrative function assigns each node the right to transmit on a set of inner-chassis time-slots. Any node can receive from any time-slot. Each node manages its own transmit path on a call-by-call basis. There exists no need for a central database during operation. Connection is made by communicating with only the two end nodes.

Transport/Physical Media Layer. MVIP leverages established communications and data processing practices thus, maximizes access to media, packet and switched networks. The physical layer is N x 64 Kbps connections from and to the PSTN. The MVIP bus transports Mu-law or A-law PCM among resources and network connections. Control messages can use synchronous or packet switched formats. Media may be aggregated and be composed of any combination of voice, data, image, fax and video.

Physical Implementation Of Multi-Chassis MVIP

Multi-Chassis MVIP permits any chassis to be connected to every other chassis through its single-level switch fabric architecture. Following is a variety of the physical media that work.

⇒ **A mesh of T-1, E-1, IsoEthernet or other point-to-point links.**

⇒ **MC1: TDM on Twisted-Pair Copper.** MC1 is similar to the single-chassis MVIP bus in concept, but operates at 4,096 Mbps over twisted-pair cable using differently driven signals. The standard includes redundant clocks and is easily configured for fault-tolerant operation. MC1 is economical, however, its synchronous design limits cable lengths to no more than 15 meters. MC1 can support up to 20 attached nodes and provides up to 1408 time-slots per cable. Dual cable configurations provide 2816 time-slots.

⇒ **MC2: FDDI-II on Fiber or Copper.** MC2 uses the second generation of Fiber Distributed Data Interface (FDDI-II) to carry up to 1536 channels worth of 64 Kbps data between hundreds of chassis over a distances of up to 60 km. A dual FDDI-II ring can deliver up to

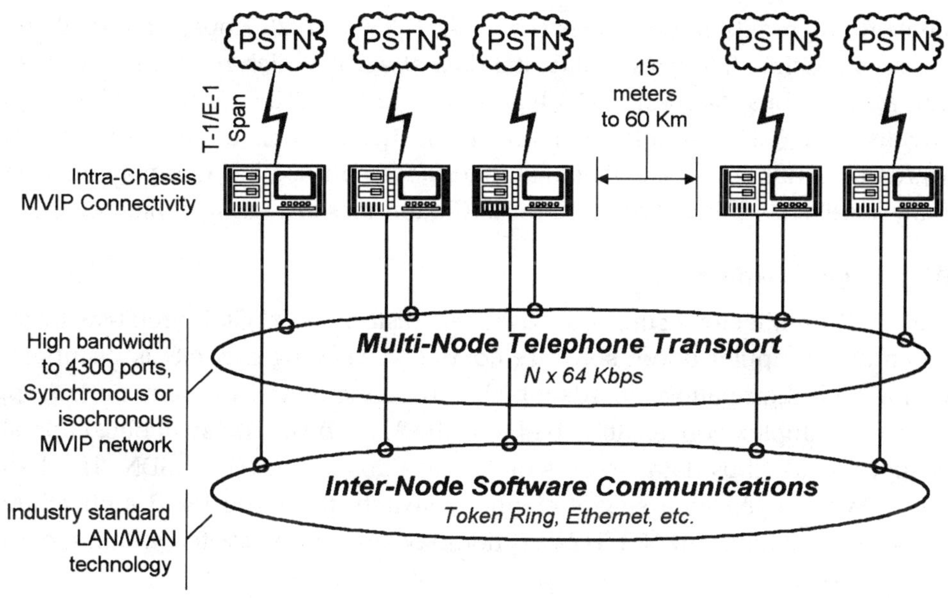

FIGURE 9.3 **MVIP Multi-Chassis Call Processing Architecture**

3072 x 64 Kbps of isochronous capacity, with fall back to 1536 x 64 Kbps in the event of a failure.

⇒ **MC3 : SDH/SONET.** SDH/SONET technology operates at 155 Mbps and provides either 2300 time-slots in a base configuration or 4600 in a dual ring configuration.

⇒ **MC4: ATM.** Traditional ATM is a cell relay technology transporting asynchronous packets rather than the constant bit transport traditionally used with voice. The MVIP standard requires the transport of 8 kHz framed isochronous data. This is not currently standard. Future deployment of this standard is planned.

MVIP Bus

The MVIP uses a standard digital telephony bus that allows connections of assorted technologies within a computer chassis. This multiplexed bus has 512 time-slots (512 x 64 Kbps capacity) which is equal to 256 full duplex communications paths. The bus also supports allocation of 64 Kbps bandwidth increments for high data rate services (i.e., 384, 1536, or 1920 Kbps). Internally the bus uses multiple 2 Mbps serial digital data streams which readily interoperate with Mitel ST-bus, Siemens PCM highways and the AT&T Concentration Highway Interface (CHI). In a PC with an AT bus these signals are carried between boards by a ribbon cable. Electrical interfaces can use LSI components from many suppliers. CMOS, PAL, GAL or ASIC interfaces are easily implemented.

MVIP Digital Switching

The MVIP architecture supports advanced distributed digital circuit-switching within the computer under software control. Switching capacity is distributed among MVIP compatible boards which can be expanded to many ports using the 256 full-duplex conversation paths on the MVIP bus. The switching capability supports 64 Kbps data streams or N x 64 Kbps groups (i.e., ISDN "H" channels). When used as a single telephony switch, the basic MVIP architecture provides the equivalent of a 512-port non-blocking circuit switch within a computer chassis.

MVIP Digital Clocking

The MVIP digital clocks design is sophisticated enough to support tandem switching between multiple digital trunks in a public network environment. All required timing signals are passed across the MVIP bus from a master clock selected under software control at power-up time. The master clock may be synchronized to one of any number of public or private network timing references. The MVIP clock architecture supports glitch-free reconfiguration in the event of network failure.

MVIP Software

MVIP software standards provide a common interface for MVIP elements that may be distributed on circuit boards from assorted vendors. MVIP software standards are in operating system independent form with specific implementations for UNIX, OS/2, MS-DOS, QNX and Windows NT.

MVIP FMIC Chip

The Flexible MVIP Interface Circuit (FMIC Chip) is an integrated circuit in distribution for MVIP Licensees. The chip provides a complete MVIP compliant interface between the MVIP bus and a wide variety of processors, telephone interfaces and other circuits. A built-in digital time-slot switch provides Enhanced Compliant-MVIP switching between the full bus and any combination of up to 128 full-duplex local channels of 64 Kbps each. An eight bit microprocessor port allows real-time control of switching and programmable device configurations. On-board clock circuitry, including analog and digital phase-lock loops, supports all MVIP clock modes. Local connections are supported using ST-bus, PCM Highway or CHI signal formats at programmable rates from 2 to 8 MHz. Local N x 64 Kbps channels are also available via parallel DMA through the microprocessor port.

Windows Telephony API (TAPI)

Windows Telephony Application Programming Interface (TAPI) was introduced by Microsoft and Intel, with the cooperation of many major telecommunications, PC and software companies in 1993. It is a defined package of hardware and software specifications designed to serve as guidelines for the development of PC based telephone applications. TAPI brings the utility and reach of the telephone network to the computer.

The idea for TAPI emerged from the experience Microsoft had interfacing their operating environments with the sundry of printers that continually appeared on the market. They realized early out that if it was difficult to enable an external peripheral to work with their operating environment it would slow the entrance of these new computers containing their software into the market. With much work and hand-holding, these incompatibility problems were gradually solved through the interaction of Microsoft engineers with the engineering groups of the printer manufacturers. This was a manageable solution because Microsoft was dealing with a finite audience and the peripheral device was rather simple in comparison to our telephone networks. With the user group expanded to potentially anyone owning a computer and the technology to be interfaced with became exponentially more complex, it became apparent that a standard need be created. Windows TAPI represents the fruit of this effort.

TAPI was designed to simplify the enormous complexity involved when integrating telephone based technology into the computer. When used, TAPI permits CT applications to be written without knowledge or concern for

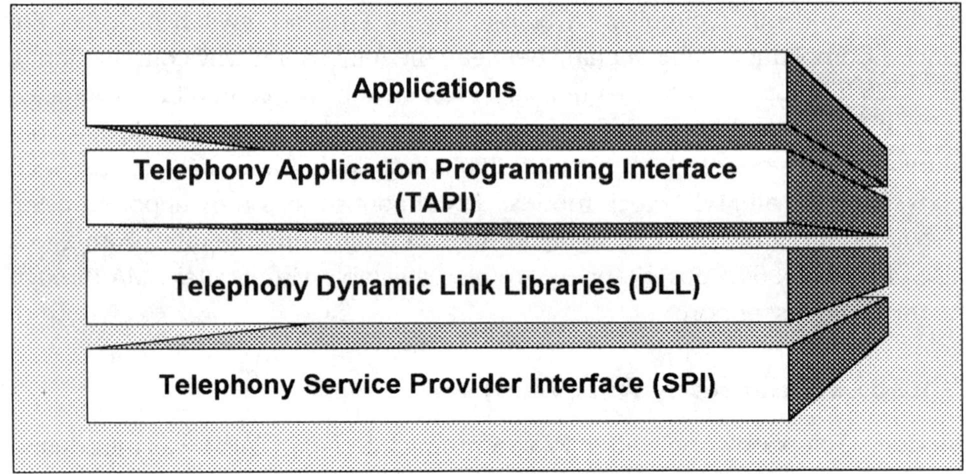

FIGURE 9.4 **Three TAPI layers interacting with applications software layer**

the enormous number of incompatible issues when interfacing with telephone based equipment.

To date, TAPI has been embraced by hundreds of organizations and is currently included in the recent Windows 95 and Windows NT release.

Although our topic of discussion here is TAPI, it is better understood when discussed in light of the other APIs from Microsoft. Windows TAPI is one of several APIs that comprise Microsoft Windows Open Services Architecture (WOSA). WOSA provides a single-set of open-ended interfaces to enterprise computing services which developers can easily access to create applications. WOSA APIs include services for data access, software licensing, connectivity, financial services and messaging.

TAPI is considered first party call control in that it works at the computer desktop or single computer level to manage CT activity. It allows programs to set up calls and access telephone networks through a standard interface. To do this TAPI was designed with three major elements:

⇒ **API.** This Application Programming Interface contains all the calling protocol and specifications to which a calling program must adhere.

⇒ **Windows DLL.** The Windows dynamically linked library (DLL) is a set of routines that can be loaded when required at runtime. The Windows Telephony DLL acts as an intermediary, translating calls from an application's functional need into Service Provider interface calls.

⇒ **Service Provider Interface (SPI).** The SPI is an interface standard that hardware and network providers write their drivers to for communicating with the Windows Telephony DLL.

Figure 9.4 represents the applications layer and the three TAPI layers.

Messaging Application Programming Interface (MAPI)

MAPI is a set of API functions and OLE interface that lets messaging clients such as Microsoft Exchange interact with various message service providers such as Microsoft Mail, Microsoft Exchange Server, Microsoft Fax and several CT servers running under Windows NT. Overall, MAPI helps Exchange manage stored messages and defines the purpose and content of messages.

The MAPI service provider architecture lets you mix different types of recipients in the same message. It is possible to send a message simultaneously to destination addresses America Online, the Internet, Microsoft Fax and standard fax as long as the profiles of these destinations have been defined within Microsoft Exchange Personal Address Book.

MAPI allows messages to be processed based on the transport protocol used to send them. The transport protocol chooses the correct modem connection, uses TAPI to create a dial string and sends the message into the appropriate form and print.

Telephony Server API (TSAPI)

The TSAPI standard was developed by AT&T and Novell to provide a means of creating server based communications processing. With LANs (local area networks) and WANs (wide area networks) coexisting with the corporate telephone system as two discrete information and communications sources, AT&T and Novel agreed that the two networks should be talking. Client-server applications, specifically those operating in call centers, need a means to tightly

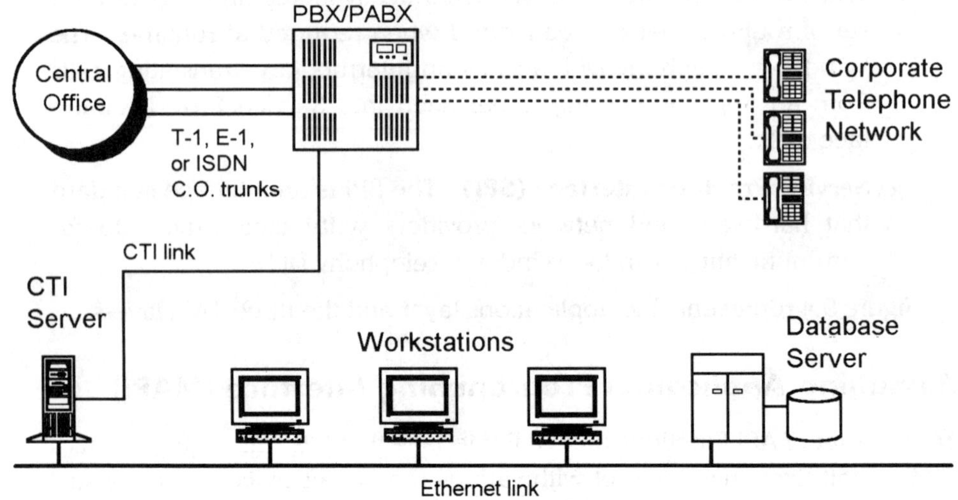

FIGURE 9.5 **A typical TSAPI configuration**

couple the PBX with the host database(s). TSAPI is critical to the successful implementation of this process.

TSAPI is primarily used as a link on device between the server and the PBX. It does not possess the capability of managing media functions. The media functions control interactive voice response features such as voice record and playback, reception and interpretation of DTMF tones, handling of talk-off and situations requiring echo cancellation. For these features TSAPI relies on the Computer Supported Telecommunications Application (CSTA) standard. This European standard was adopted to interface, through function calls, with TSAPI to perform message passing between the server and the PBX. This communication is accomplished usually over a standard Ethernet link or a proprietary hardware/software technology specific to the PBX manufacturer.

Figure 9.5 depicts a typical TSAPI configuration.

TSAPI Components

Following are the software and hardware independent components that comprise TSAPI:

CTI Link. CTI Link is the hardware specific connection between the network server and the PBX. This connection is typically made via standard Ethernet cables an RS-232 serial link.

CTI Link Hardware is located inside the server that is required to make the connection. This hardware could be a standard Ethernet card, a com port or a proprietary card made by the PBX manufacturer.

Switch Driver. The switch driver is a set of NetWare Loadable Module (NLMs), running under NetWare control, which communicate between the hardware independent components of TAPI, provided by Novell, and the hardware specific components provided by the phone system vendors. The switch has the ability to translate proprietary CTI protocol specific to a PBX manufacturer to TSAPI.

Switch Driver Interface. The switch driver interface is the software layer running under Novel NetWare, passing messages between the application and the switch driver.

Telephone NetWare Loadable Module. The telephone NLM resides at the driver level and provides the conversion between LAN message packets and messages to and from the switch driver. Usually the telephone NLM participates in the security restrictions defined in the telephony services users database.

Telephony Server. The telephony server is the computer housing and processing all of the hardware and software components discussed here.

Telephony Server API. TSAPI itself is a set of function calls and messages which can be used on a Windows and NetWare client PC or workstation on a network with a telephony server.

TSAPI Client Library. The TSAPI client library is usually written in C and C++. It is used for linking with applications running on Windows in order to use TSAPI. Understanding how to use this library is critical to developers using TSAPI.

TSAPI Library. Because the TSAPI Library contains function calls available to an NLM, application components may be run on the telephony server or another NetWare server located on the same LAN. This would all be transparent to the application.

SunSoft's XTL

Details on XTL are discussed at length in Chapter 10 of this book.

Standards Summary

After reviewing all the standards available to guide our efforts for building computer telephony solutions, which require the productive use of assorted software and hardware technologies from an assortment of vendors, it becomes apparent that COMMON AGREEMENT between the standards is critical. Without such an agreement we have essentially created standards that are proprietary to specific vendors or groups of vendors. Closed standards are exactly that. We all need move to an area of further common agreement. A set of rules that will essentially create superior standards. This is actually underway.

Enterprise Computer Telephony Forum (ECTF)

The ECTF was created by a group of computer and telephony suppliers in an effort at working towards interoperability between the existing computer telephony standards. The desire of the group is to promote a worldwide extremely open, competitive market for computer telephony integration.

ECTF is a non-profit organization composed of approximately thirty companies who realize the need for common agreement to provide this industry a clear path for growth. The members agree that due to the infancy of this industry it is unrealistic to think that there could be one strict standard to which all organizations would adhere. There currently exists several standards all of which have their APIs, SPIs and hardware specific to their connectivity. For example, Microsoft / Intel's Telephony API (TAPI) and AT&T / Novell's Telephony API (TSAPI) have achieved market momentum, but there is no clear way these piecemeal standards can actually be implemented in a variety of environments. In addition, products developed to these standards need to coexist with many types of PBXs and switching systems, many of which do not currently offer standard open interfaces, or which have disparities on how the open interfaces are implemented.

The goal of ECTF is to focus on standards-based multi-vendor computer telephony services that can be handled a bunch of ways to meet all of the needs of the enterprise. The work of the ECTF is critical to our survival.

VERSIT

VERSIT is an informal association formed between Apple, AT&T, IBM and Siemens Rolm with the goal of enabling diverse communication and computing devices, applications and services from competing vendors to interoperate in all environments. They want the universal ability to communicate and collaborate with anyone, anytime, anywhere. July 1995, VERSIT merged with ECTF.

|10|

The SunXTL™ *Teleservices Platform*

The SunXTL™ Platform architecture realizes a solid foundation for both desktop telephone and enterprise-level, client-server telephony applications. Through object-oriented designs, modular interfaces, and extensibility, the SunXTL™ TELESERVICES platform provides standardized interfaces and services for software and hardware developers.

Goals of SunXTL™ Teleservices

The design and architecture of SunXTL™ are driven by five principal goals:

⇒ **Provide a development platform for call center management applications.** Applications such as customer service center, call router, telemarketing center, etc. need to integrate the ability to control PBX call control functions into a robust, distributed computing environment.

⇒ **Provide a development platform for desktop CTI applications.** Applications such as "Feature Phone" GUI, remote access via DTMF, personal voice mail, etc. can share access to a workstation's telephony hardware.

⇒ **Provide transparent porting between PBX, analog, ISDN, ATM and other communication media technologies.** Applications should not be needed to be recompiled as network technology changes or is upgraded.

⇒ **Provide simple access to common voice services.** The building blocks of voice applications, like DTMF and silence detection must be available easily and efficiently, using "onboard" resources if available.

⇒ **Enable specialized or non-voice services.** Telephone interface hardware also provide fax, modem, video, speech recognition, speaker verification, compression and text-to-speech capabilities.

SunXTL™ Platform Architecture

The SunXTL™ Platform is a multilayered software architecture based on a client-server computing model. SunXTL™ consists of four key components:

⇒ **SunXTL™ Libraries.** Provide a high-level, object oriented programming interfaces for creation of applications—the SunXTL™ Application Programming Interface (API), and providers—the SunXTL™ Media Platform Interface (MPI).

⇒ **SunXTL™ Server.** Providing multi-client and multi-device support, the server is the central point of contact for all teleservices, resource management and security are provided by the server.

⇒ **One or More Providers.** Providers are the device-specific software entities that enable the telecommunications device to communicate with and/or be controlled by the SunXTL™ platform and its applications. Providers ensure a consistent mechanism for applications to be set up, control and terminate phone calls in a consistent manner.

⇒ **A Data Stream Multiplexor.** The universal multiplexor (Umux) provides a uniform means for applications to access and share data channels associated with a telephone call. Umux is a streams pseudo-device driver used to connect data channels to applications.

A SunXTL™ application is linked with the SunXTL™ API library (known as XTLS), through which it communicates with the server. The XTLS library uses an asynchronous symmetric messaging protocol which is implemented using the Solaris loopback transport mechanism.

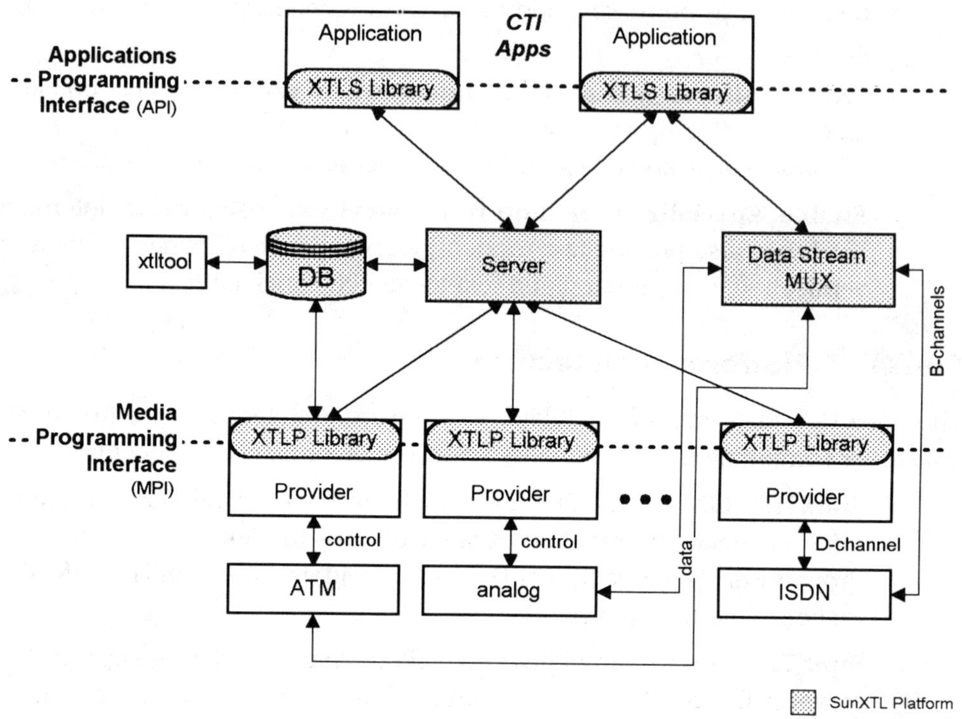

FIGURE 10.1 **SunXTL Platform Architecture**

On behalf of the application, the SunXTL™ server communicates with the SunXTL™ provider object to place and receive telephone calls. An application may access the data associated with a call by acquiring a data stream from the API. The API teleservices server and SunXTL™ provider work together to deliver the stream to the application.

The API first connects to the data stream multiplexor, Umux. Through the XTLS library (and its protocol), the API then instructs the SunXTL™ server to connect the data channel associated with the call to the API's Umux connection. The server accomplishes this by asking the SunXTL™ provider to link the data channel under the Umux, and then instructs Umux to splice the data channel to the API's Umux connection. The application can now use the Umux to send and receive data from the data channel associated with the call by using the usual STREAMS interface. This can be seen in Figure 10.1.

SunXTL™ Programming Interface Layers

The lowest layer of the API is a message passing interface between an application and the teleservices server. The server application interface includes the call control functionality of the provider interface as well as support for data channel access, support for multiple client applications accessing the server, security management, resource contention and sharing.

Message passing within the teleservices environment is implemented using remote procedure calls similar to the ONC RPC/XDR standard. Messages are 'one-way' and assume reliable connection between sender and receiver.

The access method of the lowest layer of the interface is like a Remote Procedure Call (RPC). To send a message, the user calls objects with names corresponding to the message contents. When a message is received, a callback object is invoked. The callback object name corresponds to the name of the message, and the arguments of the method correspond to its contents.

To more easily use the RPC interface, a library containing default callback functions is supplied. An application developer can override these defaults by writing application specific functions and linking them ahead of the default library. In fact, the next higher layer of the teleservices interfaces, the object oriented interfaces, is constructed in that manner. The default library also provides functions for handling security interactions with the server, functions for acquiring and accessing the data channel of a call, and other low-level actions.

The object-oriented interface manages connection to servers via Server objects, handles the creation/ answering of phone calls via Provider objects, and the control or status reporting of individual calls via Call objects. Instead of supplying or overriding the callback functions using the linker, callbacks are provided as methods of Provider or Call objects. Thus, an application can handle individual classes of calls with methods specific to those calls.

The object-oriented layer hides the interaction with the lower-level protocol messages, although advanced users and developers may still access that layer using Message objects. The object-oriented interface makes it easy for an application to manage multiple calls or multiple kinds of calls and even calls using multiple servers.

The SunXTL™ Applications Programming Interfaces

The SunXTL™ Application Program Interface (API) defines the model and control mechanisms required to develop CTI applications. The interface provides the high level programmatic model for accessing the functionality of telephone devices, together with managing security and concurrency.

The SunXTL™ API includes functionality to:

⇒ *Place a call*

⇒ *Register with the SunXTL™ service to answer a call*

⇒ *Enable call control features*

⇒ *Send and detect DTMF tones*

⇒ *Enable security and the sharing of calls between processes*

The SunXTL™ API is used by applications to manage the communication between the application program and the teleservices server process. As such, the API presents objects which the application programmer may utilize to send and receive messages.

The Teleservices API and MPI Libraries

The Teleservices Application Library, XTLS, provides an interface defined as an extension to the Teleservices Provider API and MPI and their internal protocol, designed to allow multiple applications to share access to multiple providers through the teleservices server.

The XTLS library defines messages for managing application requests passed to the server, including:

⇒ *Call control and creation*

⇒ *Data stream control, allocation and interconnection*

⇒ *Call ownership and security—resource sharing*

⇒ *Management of call monitoring and status logging*

The Teleservices API includes a C++ client library for defining messages, which is completely independent of the protocol specification. The Teleservices API provides the complete mapping—developers do not interact directly with the protocol or protocol layer. Figure 10.2 illustrates the SunXTL™ API and its relationship to the XTLS library.

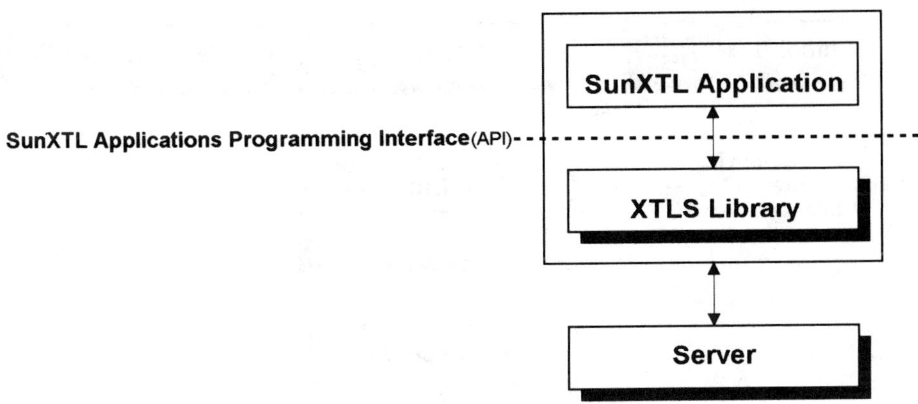

FIGURE 10.2 SunXTL Application, API, XTLS Library, and Server

The Teleservices Provider Library

The Teleservices Provider Library, XTLP, manages the set up, control and tear down of telephone call connections in a device independent manner. The hardware and internal protocol specifies on the underlying telecommunication technology are hidden beneath the XTLP's Media Programming Interface (MPI) which provides a standard interface, independent of the communications transport mechanism.

A different XTLP process supports each service, device, or subsystem installed on a system. The providers illustrated in Figure 10.4, provide support for ATM, Analog, ISDN (Sun's DBRI implementation), and Ethernet. XTLP library implementations can be developed to support:

⇒ *Local switch-dependent protocols*

⇒ *Vertical applications or software extensions*

⇒ *Alternative telephony technologies such as analog or ATM*

When the XTLP library is used with ISDN, the workstation functions as Terminal Equipment (TE), allowing the system to act like a telephone, provided appropriate telephone hardware is present.

The XTLP's Media Programming Interface provides a system programming system interface, requiring substantial knowledge of system internals, and is

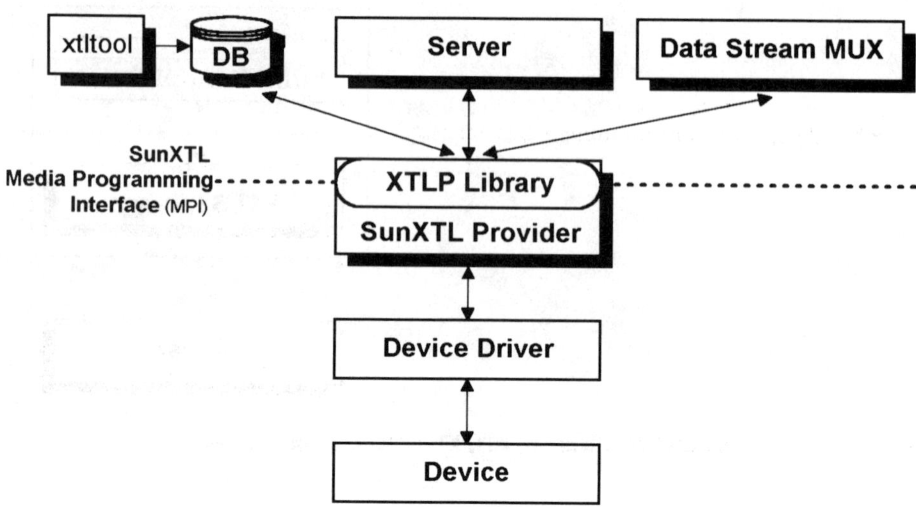

FIGURE 10.3 **SunXTL Provider and Device Driver**

principally of interest to developers of devices or software communications protocols. Figure 10.3, illustrates the SunXTL™ provider and media platform API, and their relationship to the SunXTL™ server, Umux, physical device and configuration tool and database.

Dependent upon the underlying communications technology, the programming abstractions supported by the XTLP library in developing providers can include support for:

⇒ *Call hold*

⇒ *Call transfer*

⇒ *Dropping call*

⇒ *Forwarding*

⇒ *Call conferencing*

If additional features beyond these are available, the XTLP library can access them through the extension mechanism.

A SunXTL™ provider is designed to support a single client application, the SunXTL™ server. Therefore multiplexing and de-multiplexing of multiple clients using the XTLP library must be performed by software above the MPI interface. Figure 10.3 illustrates the SunXTL™ provider, and its interface to a user (client code), configuration tool (and database), and their relationship to the communication device.

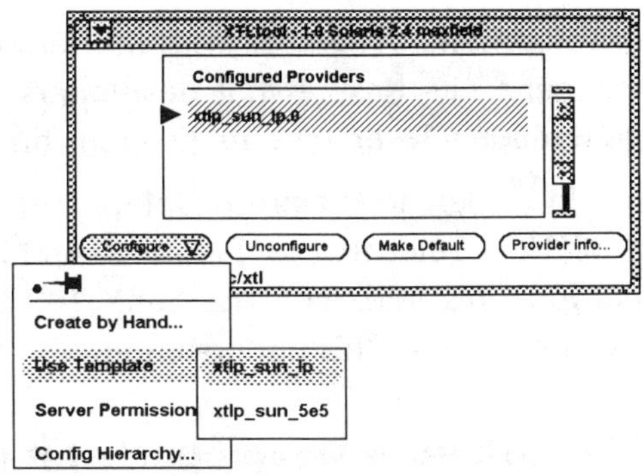

FIGURE 10.4 XTL tool provides the user interface for configuration of providers

The Configuration Database

The SunXTL™ database contains information on providers available in the local SunXTL environment. SunXTL™ Provider objects maintain configuration files in the database as lists of keys and values that define specific provider characteristics necessary for the successful operation of each provider.

SunXTL™ Provider objects establish connections over telephone lines using lower-level protocols that control the telephone hardware and manage these connections. The entities that implement the lower-level protocols are also called providers. The provider object must be able to connect to a valid provider in order to perform a given telephone function.

SunXTL™ applications can use the SunXTL™ database query functions to ensure that the provider object can successfully establish these connections. Applications can use query functions to:

⇒ *Check that the database is appropriately set up*

⇒ *Access a list of available providers and their names*

⇒ *Get free key-value lists*

⇒ *Access provider key and value information*

The SunXTL™ environment includes an Open Windows-based tool, called **xtltool**, which provides an easy to use interface for managing the database.

Developing CTI Applications and Hardware

Developers of CTI applications and hardware under Solaris™ should utilize the Software Developer Kit (SDK) and Driver Developer Kit (DDK) to assist in the development. Together with the teleservices runtime environment, these developer's kits ensure that developers have the tools necessary to build next generation teleservices applications and hardware.

In order to begin development, developers need to understand the SunXTL™ Teleservices Platform Architecture, as well as how its individual components interact. The SunXTL™ architecture section of this chapter provides a basis for this understanding.

The Software Developer's Kit (SDK)

The Software Developer's Kit provides the basic tools to develop applications under Solaris™; versions 2.5 and earlier. The SDK is available on CD-ROM and includes documentation and sample code for development of Solaris applications using virtually any Solaris™ technology from window systems and graphics to multithreading and interapplication integration. For the development of teleservices applications, the SDK provides:

⇒ *SunXTL™ Teleservices Environment.* (Server, sample providers, etc.)

⇒ *SunXTL™ Teleservices Developer Tools.* (XTLS and XTLP libraries and header files, manual documentation, etc.)

⇒ *Examples* of application, provider, and device driver code.

The Driver Developer's Kit (DDK)

The SunXTL™ DDK provides the necessary documentation and sample code for the development of device drivers under the Solaris operating system.

The DDK includes specific code to facilitate creation and testing of device drivers for SunXTL™ Teleservices and provides source and binaries of a sample SunXTL™ provider, designed for application testing purposes (provider "stub").

SunXTL™ providers can be developed to integrate SunXTL teleservices systems into ISDN and non-ISDN Private Branch Exchange (PBX) systems and, to provide for analog telephone support for third-party controllers as well as advanced network integration. Network technologies such as Fast Ethernet and Broadband ISDN (B-ISDN) offer high bandwidth alternatives to the networks of today. Broadband ISDN employs Asynchronous Transfer Mode (ATM)

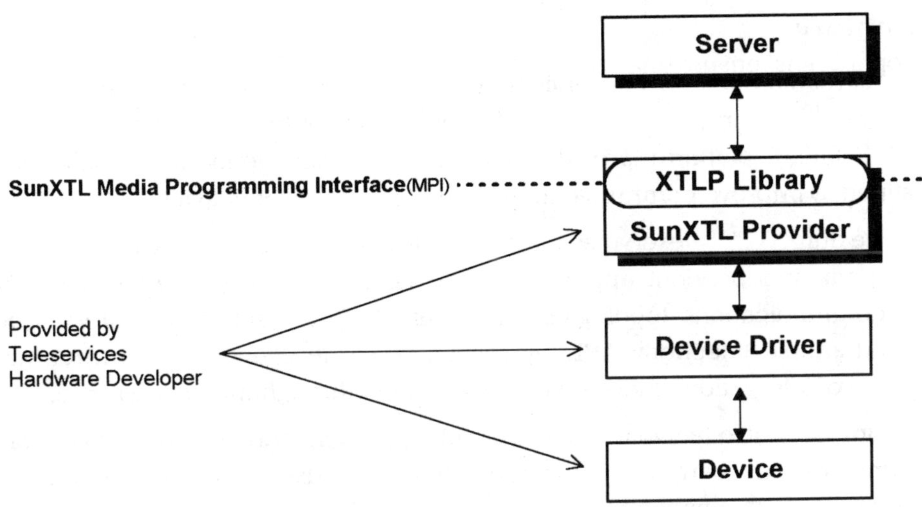

FIGURE 10.5 SunXTL Provider and Device Driver

or the Synchronous Optical Network (SONET) as a means of packaging and transporting data.

Developers of teleservices hardware devices to be integrated into Solaris™-based systems require two components—a device driver and provider. The device driver handles low-level details of the underlying network transfer scheme or telephony devices (e.g., ATM, Fast Ethernet, analog telephony, etc.), and is written to conform to UNIX System V Release 4 standards.

The provider is the code necessary to interface between the SunXTL™ server and the device-specific driver. As such, all providers must present a common appearance to correctly map the providers protocol, via the XTLP library, used to communicate with the server. (See Figure 10.5.)

Where devices have additional features not defined by the teleservices API, the provider extension mechanism can be employed to map the devices features to client applications. Once a provider is written and tested, it may be installed in a user's environment with the xtltool—which provides access to the provider configuration database.

Summary

Sun Microsystems sees teleservices as a key component in the future of distributed, client-server computing. The integration of teleservices into the standard Solaris™ platform provides the basis for next-generation applications destined to improve communications and corporate productivity.

The SunXTL™ Teleservices Platform delivers a stable robust base for device and media-independent applications and flexible support for new hardware and communications topologies. The SunXTL™ Teleservices Platform goes beyond simple telephony APIs by providing a true client-server environment capable of integration of teleservices application throughout an enterprise.

SunXTL™ also achieves a balanced approach between data and voice communications, permitting their integration at the user or application interface layer and thereby enabling new applications which effectively integrate telephony with mainstream data-oriented applications and processes.

Finally, SunXTL™ incorporates a flexible and extensible architecture capable of unifying data and public network technology, regardless of whether they are analog voice, ISDN, ATM, LAN, WAN, T-1 or other emerging technologies. SunXTL's™ well defined application and provider interfaces enable developers to build teleservices applications today, confident with the knowledge that they will be supported as new communications technology emerges.

|11|

Assistive Technologies
for CT Applications

The first wave of computer telephony integration left its mark by establishing acceptable standards whereby the product expectation level of the user matched, at least reasonably well, the existing product performance level. However, once in service, inadequacies discovered through product use drove the need for increased flexibility and desired open access to these new automated technologies. Innovators, in response to human factors considerations, technological advancements and frustration with available products (upon their escape from voice mail jail) birthed CT Wave Two; Assistive Technologies.

Assistive technologies are more natural interfaces which provide a new level of intelligence plus solve many of the problems existing in earlier CT Wave One products. They move computer telephony's usefulness into the 1990's. These second wave of CT technologies include:

⇒ *automated speech recognition*

⇒ *speaker identification*

⇒ *speaker verification*

⇒ *text-to-speech*

At one time or another we have all cursed the wonderful voice mail or interactive voice response system which just couldn't provide the appropriate option to accomplish what we so calmly set out to do. We couldn't seem to remember whether we needed to press the star (*) key on the DTMF tone telephone keypad, or was it zero to escape the endless loop we somehow got trapped in. Perhaps we weren't smart enough to operate the equipment and didn't really need to place that $1 million order for the product that the company so successfully kept us from ordering. Unfortunately, these examples go on forever.

We needed systems which seemed more thoughtful. Provided the option to ask for help. Let us effortlessly search for any human contact, still breathing in the organization, who could converse or somehow assure us that what we set out to accomplish had a chance of taking place in the near foreseeable future.

Don't despair this new wave of CT technology permits developers a method to move standard familiar products in this more user friendly direction. Although we cannot have a totally meaningful interactive conversation with a computer today we are steadily moving in that direction through increasingly innovative use of Assistive technology.

Those of us who will admit to it, will fondly recollect Hal, the voice driven computer from the mid-1960's film "2001: A Space Odyssey", who astounded us with his ability to understand human voice commands, perform functions and provide vocal feedback. The technology seemed almost human and required no training on the part of the users to accomplish the required tasks.

The people working on the "Hal dream" have been doing it from around the same time period as the movie, and have made some major inroads towards highly usable technology that actually works. Of course, this has been an iterative process where improvements have been gradual and continue to come.

We all have our own impression of speech recognition and how it should work, and on more than one occasion offered our professional critique of the "drunken Swede" representing text-to-speech.

Well, the drunken Swede has survived the twelve step program and is now sober enough to interact with assorted forms of speech recognition capabilities sufficient to provide some of the most advanced, user friendly CT technology ever before available. This is quite an exciting time on the CT technology front.

Speech recognition, speaker identification and verification and text-to-speech are quite complex technologies which could easily take an entire book to explain. Following is a fairly in-depth, yet concise, discussion of each accompanied examples of each of these Assistive Technologies in use.

SPEECH RECOGNITION

Automated speech recognition has experienced a very bumpy road to market and is still not widely used. It's accuracy and reliability are critical to its usefulness in any CT application and salability as a product. Speech recognition technology is based on software algorithms which have taken years to develop and fine-tune. Once complete and deployed equally important is the design of the application software harnessing the power and making it easy to use by callers to the system.

Algorithms for Speech Recognition

All recognizers use software algorithms to analyze speech input and return a result to the application. The majority of algorithms are based on Hidden Marcov Method also referred to as Hidden Marcov Modeling(HMM), Neural Network (NN) or a hybrid technology combining HMM and NN.

HMM technology uses probabilistic techniques for recognizing discrete and continuous speech. Neural network technology is designed as a processing web where the interaction of the web creates a pattern of data and generalizes from it. Each use of the Neural Network further trains the processing web therefor increases the accuracy of the algorithm.

Speech recognition algorithms need compensate for assorted telephone network conditions, a wide range of input quality from the hundreds of handset types available and the assortment of speaker types using the CT product. The algorithm must provide a degree of filtering to minimize the effect of ambient noise and assorted background sounds while performing consistently under a heavy workload.

Due to the complexity of this underlying technology no further discussion of algorithms will be provided in this text. Books and other research data on the algorithms are readily available through speech labs and research libraries.

Speech recognition is available in two basic varieties: speaker dependent and speaker independent. Speaker dependent speech recognition requires the user to train the system to recognize his or her voice speaking a specific word(s) or phrases. This training builds a user specific database as it is used. Speaker independent speech recognition is more universal in nature because it is based on a scientifically designed database which is representative of the population using the CT system. No training is required with speaker independent speech recognition.

Speaker Dependent Speech Recognition

Speaker dependent recognition is the oldest of the automated speech recognition (ASR) technologies. It requires the user to make a series of recordings repeating words, numbers, phrases or whatever the target application needs. This recording process is called training. Caution must be exercised when training the system that the environment in which the speaker dependent technology is trained is the same as the one in which it will be used. For example, a training session conducted in a quite office environment would probably not be sufficient for that same individual to use for a voice dialing application from the noisier environment of his or her car.

Speaker dependent technology creates a voice template for each word or phrase recorded at the time of training. These templates are stored in the computer memory as a database for that specific individual. The stored templates are referred to as reference templates.

When a speaker dependent application is uses the caller initiated the process by activating the system. This can be done via ANI information or the input of some identifying information. At that time the appropriate user database is activated. In response to a voice command a template is immediately created. That template is then compared to the reference templates in the users' database. If a match exists the command is accepted as valid input and the desired activity begins. Simultaneously this new template is used to update the reference template in the individual user database.

Speaker dependent technology is continually learning and adapting via training. It however restricts use to only the individual(s) who have trained the system.

Template construction is based on a refined feature extraction model which need be robust and allow for the use of different handsets, handset positions and ambient noises.

Speaker dependent technology can support large vocabularies but it requires training for each word or phrase to be recognized in the vocabulary which will be used. This can be quite an onerous task.

Speaker Independent Speech Recognition

Rarely will you find speaker dependent speech recognition deployed by itself in an application. It is usually accompanied by speaker independent speech recognition which is used for command words. This technology has achieved a degree of notoriety in voice dialing applications. Due to the user specific nature of this technology it is highly accurate for use by the individual who trained the system in quiet as well as noisy environments. Even though the system is easy to use it must be used only in the environment for which it was trained.

Voice dialing applications for the home as well as mobile use are capable of supporting numerous individuals. Reference templates are created and stored in user directories within a database. Due to this individualized approach you and your wife / husband / significant other can use the same command word for entirely different calling numbers and never experience a problem. For example, *"call mom"* would be a different number for each of you.

Voice dialing applications use speaker independent speech recognition for the command words which accompany the user specific vocabularies.

Speaker independent recognition requires no training and can accommodate a limitless number of users. The only caveat being the user's speaking dialect and accent is represented in the database supporting the application. Speaker independent recognition is a category which covers an assortment of vocabulary types including digits, letters of the alphabet and specific words in any combination or order.

Unlike speaker dependent recognition, speaker independent vocabularies do not require any adaptation after system installation. The database used for speaker independent recognition is built on speech samples which are geographically balanced from the target area of users segmented by age and sex.

Databases used for speaker independent recognition can only be used for applications which the sample was designed and collected. A land line telephone database should not be used for a noisy environment such as a cellular telephone application. If a database is used in an environment for which it was not created its accuracy will be compromised.

Recognition Accuracy

Recognition accuracy is the most critical element of ASR use. A system which produces an unacceptable accuracy rate cannot be used for commercialized applications. Recognition accuracy refers to the percentage of time the recognizer correctly classifies an utterance.

Recognizers generally make three basic error types: substitution errors, rejection errors and spurious response errors.

Substitution errors are the most critical because the recognizer substitutes an incorrect word for a spoken word. If you have used a recognizer at all this has probably happened to you. For example, a substitution error has occurred when you are speaking your telephone to number to a CT system requesting this input, and the number 5 is recognized for your spoken number 9. Both are valid input to the recognizer because they are digits between zero and nine. However, five was not the number you gave as input. The industry norm for acceptable performance dictate that substitution error rates must be 2 percent or less. To minimize the effect of these errors most applications repeat back digit and other short utterance input for confirmation prior to final acceptance.

Rejection errors occur when the recognizer does not classify the target word when spoken. When a rejection error accrues the application usually requests the speaker to repeat the word input until the recognizer identifies it. These errors are obvious to the used and somewhat annoying while interacting with the application. Ideally the rejection error rate is 3 percent or less.

A spurious response error occurs when the recognizer classifies a sound or invalid word as a valid word. For example, the recognizer identifies a cough or sneeze as a valid vocabulary word. Or the speaker says a word that is not in the recognizer's vocabulary, but the recognizer acknowledges it as a valid word.

Both substitution and rejection error rates are used to determine a recognizer's accuracy. The average recognition rate is defined as the sum of the substitution error rate and one-half of the rejection error rate. The basic definition of recognition accuracy does not incorporate spurious type errors.

The Database

All speaker independent speech recognizers rely on a database to look-up the spoken word for identification. The database however is not merely a dictionary type listing of words. It is a composite of sounds which comprise an entire language. Non-tonal languages (Asian languages are tonal) are composed of phonemes. These phonemes are categorized into diphones and triphones. Phonemes are phonetic representations of spoken words. For example, phonemes for particular words can be found in a standard dictionary. They are the phonetic representation, located in parentheses directly following the alphabetically listed word. There is a finite number of phonemes in any language. English for example has 48 phonemes which are used in a variety of combinations to speak any English language word.

Database building begins with the design of a sample. For American English the sample would need regional representation to accommodate the assorted accents, age segmentation to reflect the distribution of the population within the region and sex segmentation to accommodate the differences in pitch between the male and female voice.

To build the speech sample each respondent is required to read a paragraph of phrases which is scientifically designed to contain all of the known phoneme combinations of that spoken language. During this session the respondent is the given a local newspaper to read. This is to provide a free-form environment where regional pronunciations and speech nuances can be recorded. Finally the respondent is asked questions which require no reading to obtain a more relaxed delivery of every day speech which include slang , phrasing peculiarities plus sounds and word formations germane to that area of the country.

It has been statistically proven that larger databases tend to be more accurate and accommodating when used with a well designed, robust speaker independent recognizer. Some very high quality American English databases are available based on recording sessions of 5,000 speakers. On average databases of 1,200 speakers are known to be quite accurate.

Once the recording sessions are complete linguists cut the recorded words into the phonemes comprising the target language and label them. Database building is a very difficult process. Its accuracy will dictate the accuracy of using the algorithm.

For specific applications some databases are further fortified with whole words which are common to that particular use. For example, a database used for voice mail or voice messaging would be fortified by the inclusion of command words, numbers, yes, no and words specific to the organization using that particular CT application.

Personal Vocabulary Editor

Some speech recognizers have a Personal Vocabulary Editor (PVE) capability to create custom words when needed. These PVEs work particularly well with large speaker size databases. PVEs provide the capability to build a custom vocabulary without collecting speech samples. They access the existing database by typing in the phonetic transcription of the words required. The underlying phonemes are assembled and the target vocabulary is created within minutes.

Types of Speaker Independent Speech Recognition

Discrete and Continuous Word Recognizers

Word recognizers share the reputation of providing some of the most reliable and accurate speaker independent speech recognition available today. They fall into three basic categories according to their requirements for input. They are:

⇒ **Discrete word** recognition requires more than 250 milliseconds of silence separating each word in the series spoken. Discrete word recognition is usually characterized by a beep between each utterance to provide the proper timing of the input.

⇒ **Connected word** recognition of a series of spoken words where at least 50 but not more than 250 milliseconds of silence separate each word.

⇒ **Continuous word** recognition where less than 50 milliseconds of silence separate each word in a series. This technology permits digit strings and phrases to be input.

Response Time

The response time of some recognizers is a problem which must be cleverly masked with the application. This is done by playing a speech file of information while the analysis portion of the recognition is taking place.

The recognition response time refers to the time it takes the recognizer to recognize a word after the end of the word is spoken. In actuality, the response time could be longer than anticipated because there is a certain amount of silence which must occur before the recognizer declares the end of the word and begins the analysis.

Key Word Spotting

Key word spotting technology is based on highly advanced algorithms which identify specific words spoken during natural unconstrained speech. This technology possesses the ability to ignore extraneous sounds and words during recognition sessions. All irrelevant surrounding words and non-speech sounds, such as coughs, clicks and ambient room noise, are automatically discarded. (See Figure 11.1.) Users do not need to utter key words in isolation, but can speak fluently, which greatly enhances the naturalness of the man-machine interface. Key word spotting offers highly robust and flexible speaker independent speech recognition for a wide range of applications.

In real life applications, key word spotting techniques have a higher accuracy rate when compared to traditional isolated word recognizers. Many times the cause of poor recognition lies not with the actual recognizer but with the discipline required by the user to utter command words in isolation. Often people are unprepared to talk to a recognizer. Rather than saying the key word only, they embed it in a sentence or unrelated sounds. For example, a caller to

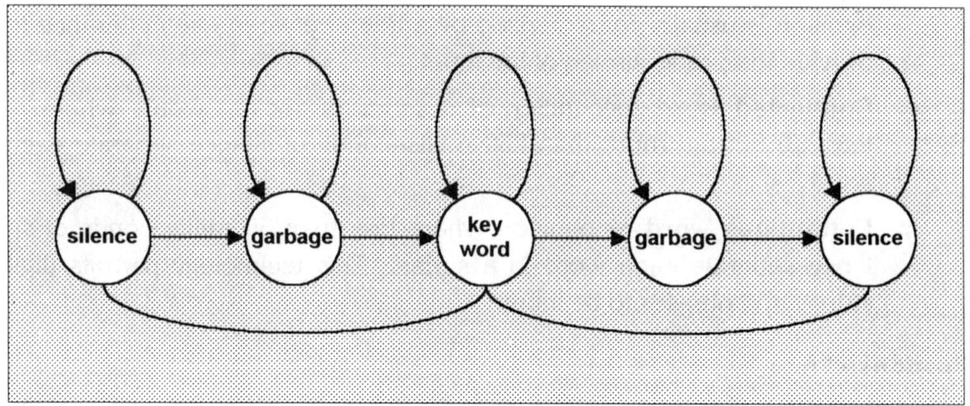

FIGURE 11.1 Surrounding sounds cause recognition errors with traditional recognizers, but a *key word spotting* algorithm distinguishes key words from other words and sounds.

an automated attendant, when presented with call routing options, tend to say *"Give me the operator"* or *"Uh, operator, please"* rather than simply *"operator"*. This however is the way we actually speak. Uttering isolated words is not natural. Unfortunately surrounding sounds cause recognition errors with traditional recognizers.

It has been found that a recognizers' accuracy increases with use. Most of these increases however turn out to be the recognizer training the speaker rather than the speaker training the recognizer.

Recognition of key words is based on speaker independent phoneme models. The basic phoneme models are trained via discrete utterances of complete words, sampled from a large group of native speakers. The key word recognition models are established by concatenating the appropriate phoneme models according to phonological rules.

Alphanumeric Recognition

Alphanumeric recognition technology recognizes vocabularies composed of alphabet letters and the digits "zero" through "nine". Due to the small utterances length of alphabet letters and digits accurate recognition of these characters is quite difficult. As is the case with standard speaker dependent and speaker independent recognizers alphanumeric recognizers are based on both

HMM and NN technology. Due to the finite nature of the alphabet letters and the digits to be recognized the perceived difficulty in recognizing these strings is reduced. Character strings and sounds with other content are automatically eliminated.

Upon analysis it has been determined that longer character strings are easier to recognize than shorter ones because more information is available to determine which string was spoken. The number of valid string candidates is critical since the best results are attained with fixed length strings. With this technology as the number of candidates increases, the accuracy decreases while the required computation time increases.

One method used for recognizing a spoken alphanumeric strings is called score-based recognition. It involves assigning recognition distances between each spoken input and the corresponding letter or digit in the same position within each string represented in the database. Each recognition distance is a measure of the acoustic dissimilarity between a spoken input and a hypothetical character. For example, if an "A" is spoken, then the recognition distance for "A" is expected to be quite low. It is also likely that the distances for characters that sound similar to "A", such as "8", "H", "J" and "K", will be higher but also fairly low, and the distance for highly dissimilar characters such as "9", "Q", and "W" will be high.

The alphanumeric strings contained in the database are named reference strings. After the caller says the first alphanumeric character, each reference string is assigned a distance value equal to the recognition distance between the spoken character and the first reference character of that string. After the caller says the second alphanumeric character, each reference string distance value is incremented by an amount equal to the distance between the second spoken character and the second reference character of that string. This process continues accumulating distances for each reference string, until the caller says the last character. At this time the reference string with the lowest cumulative distance is the recognized string.

Talk-Over or Cut-Through

Talk-over or cut-through are names for a technology available for use with both speaker dependent and speaker independent speech recognition. It provides the caller with the ability to override a prompt, while it is playing, and input speech for recognition. The caller virtually interrupts the prompt and his or her words are recognized at the same time.

This is quite a useful capability as repeat callers familiar with the application need not wait for the prompt to speak. It makes the system more human like and much easier to use.

It is interesting how it works. When this technology is active in the CT system the application plays the outgoing message, when the caller speaks the voice prompt is reflected back to the recognizer input where it is effectively canceled, through echo cancellation technology. The input to the recognizer is the sum of two signals, the word spoken by the caller, and the residual signal from the reflected prompt after cancellation. To achieve reliable recognition accuracy, the residual signal needs to be small relative to the signal strength of the caller's spoken word.

SPEAKER IDENTIFICATION & SPEAKER VERIFICATION

Speaker identification and speaker verification are similar technologies as they both analyze characteristics of the human voice. They are however quite different in application.

Speaker Identification

Speaker identification is used to determine the identity of a known speaker. It is accomplished by taking spoken input and searching a database of all known system users for a match. Due to its speaker dependent recognition characteristics you must first be enrolled as a user prior to using the system.

To enroll as a user an individual is required to speak one or more password phrases which are recorded. These phrases create a reference templates which are stored in the system user database for later use during identification sessions.

When in operation the individual using the system is prompted for a specific password or password phrase. When speaking the prompted password as input it creates a new template. For analysis this template is then compared to all reference templates in the system for that particular password. The reference template with the closest match is selected. The uniqueness of each user's voice and the finite number of users of the system makes the identification accuracy quite high.

With speaker identification the speaker does not claim to be a particular individual. He or she is identified from a group of common users. For the most part this technology is used for hands free operation of a system where messages and other information specific to that identified individual is pulled-up for use at that time.

Speaker Verification

Speaker verification, on the other hand is an automatic process which uses characteristics of the human voice to decide whether the speaker is the person he or she claim to be. The speaker is asked to record a sample of his voice, and manner of speaking by saying a password sentence, which they are prompted for, after stating their identity. The characteristics of his or her voice are compared to voice samples they have previously given. If the match is close enough, the speaker verification system will decide that the speaker is the correct individual and that speaker will be accepted. If the sample and data-

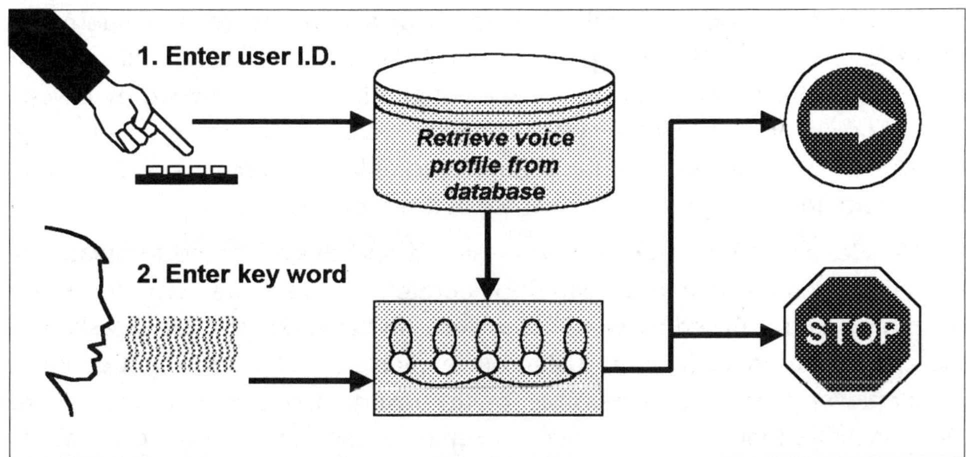

FIGURE 11.2 **In *speaker verification,* the speaker first identifies himself, perhaps using a PIN code, and subsequently utters a key word or phrase. The algorithm determines how much the utterance matches the speaker's voice profile previously stored in a database.**

base information are different, obviously, the speaker is not the individual they claim to be and they will be rejected.

The automatic acceptance / rejection decision is based on biological features unique to each human voice. Each human voice has intrinsic characteristics which cannot be duplicated or stolen.

Most speaker verification systems have registration, training and verification modes. Each mode performs a specific function which makes this technology work. A description of each mode follows.

Registration. To register with a verification system a speaker is first provided a means to establish their identity. This can be an employee identification number or perhaps a bank account number provided by the insertion of a smart card into an automated teller machine (ATM). Once an identity has been established the speaker is required to record a password sentence. The data recorded from the session is processed and stored in the speaker verification database. This recording and processing session is repeated a few times to capture the variations of the speaker's voice and possible differences of the recording conditions caused by the telephone network and environment. Depending on the application requirements for which the system will be used, the registration recording can be complete with only one take.

Training. During training the system builds a summary reference template of the user's voice. This template contains information detailing individual vocal characteristics, their manner of speaking, differences between their voice and all others enrolled by the system plus the quality of the telephone handset used. This template is called the reference template. Training is performed by the system; the user does not participate in this process.

Verification. After the user is registered and the reference template has been created the system is prepared to operate in the verification mode. This is the main and most used mode of the system. In this mode the user is asked for the appropriate identification. Using our earlier example, this input could be an employee identification number or the insertion of the ATM smart card. Once identified the user is provided a sentence they are prompted for as the password utterance. This utterance creates a template which is compared to the reference template created during training. An acceptance / rejection decision is made at this time.

Due to periodic changes in the human voice the reference template is further trained each time a system is used. This updating process insures continued accuracy over time and is transparent to the user.

Speaker verification and speaker identification systems are text and language independent. They provide a easy means of enrollment. They readily adjust to the environment in which they are used and possess the ability to have adjustable security levels.

Accuracy of Speaker Verification

As with other forms of speech recognition technology, speaker verification accuracy is critical to its usefulness. This accuracy is better described as a balance between two acceptance / rejection errors. They are:

⇒ Error Type 1: *Rejection of an authorized user.*

⇒ Error Type 2: *Acceptance of an unauthorized user.*

Speaker verification systems permit the adjustment of the acceptance / rejection threshold so that the number of errors from wrongly rejecting true speakers versus wrongly accepting impostors is selectable. This makes it possible for the targeted use of the system to determine how strong the security conditions need be, and to which type of error the system should be made more resistant.

Speaker verification provides a feasible solution to long distance calling card and telephone based credit card fraud. In addition it provides a more secure access for any type of telephone based voice application.

Carrier loss due to long distance credit card fraud is in excess of $3 billion annually. Any savings would be found money to these carriers.

If calling card company used speaker independent speech recognition to input the account information and speaker verification to verify card use by the appropriate individual substantial dollars would be immediately saved.

TEXT-TO-SPEECH

Some ten years ago the first versions of speech synthesis, better known as Text-to-Speech, were introduced to the market. Since that time these products have gradually evolved, but shown limited commercial success. This less than spectacular product commercialization is due in part to its inferior intelligibility and unnatural sound quality when compared to natural speech. For the most part the market perception of TTS has been that it sounds like a "drunken Swede".

There exists a magic performance threshold that must be reached before TTS is accepted by the general public. This quality level is now attainable with some of today's technology. With the increase of CPU power and the use of DSPs (digital signal processors) the use of TTS for CT applications is poised for a market explosion.

Today, more and more companies are interested integrating text-to-speech technology into their product. With the explosion of E-mail, the use of unified messaging coupled with the increased use of the Internet and other on-line services the day of TTS as an assistive technology has arrived.

Traditional interactive voice response systems are limited in the amount of data they can "give a voice" using pre-recorded speech phrases which are concatenated together to speak needed digits and phrase combinations. Using TTS, a voice can be given to a limitless amount of data.

Regardless of which company produces the algorithm, the general outline of most text-to-speech systems is very conventional. There are three main phases: linguistic, phonetic, and acoustic.

The TTS process begins with the input of ordinary text into the system. A linguistic module converts this text into a phonetic representation. From the phonetic representation, the phonetic processing part of the TTS system calculates the speech parameters. Finally, an acoustic module uses these parameters or generate synthetic speech signals. (See Figure 11.2.)

Linguistic Processing Module

The linguistic module of most text-to-speech systems consists of four parts. There are: text normalization, orthographic (spelling) to phonetics, lexical analysis and syntactic analysis. Systems utilizing current technology only perform a limited amount of semantic processing.

Text Normalization

A text-to-speech system must have the ability to read aloud any written text, even if it contains abbreviations, dates, money and time indications, addresses, telephone numbers, bank and security account information as well as esoteric things such as symbols. For example, to solve the abbreviations problem, an abbreviations dictionary need be used. Abbreviations that do not occur in the dictionary are then pronounced as single words or are spelled out, depending on the phonotactic structure of the abbreviation.

A TTS system should process digits according to the syntactic and semantic context in which they appear. In English, digits such as 1995 are pronounced differently according to the context; number or year. In Spanish, the conversion of digit strings also needs lexical information because the pronunciation of the digit string sometimes changes depending on the gender of the noun or the abbreviation that follows the digits. Text normalization should also correctly handle quotation marks, parenthesis, apostrophes and other punctuation marks. To handle text normalization, text-to-speech systems use a lot of orthographic (spelling) knowledge frequently phrased by linguistic context dependent rules.

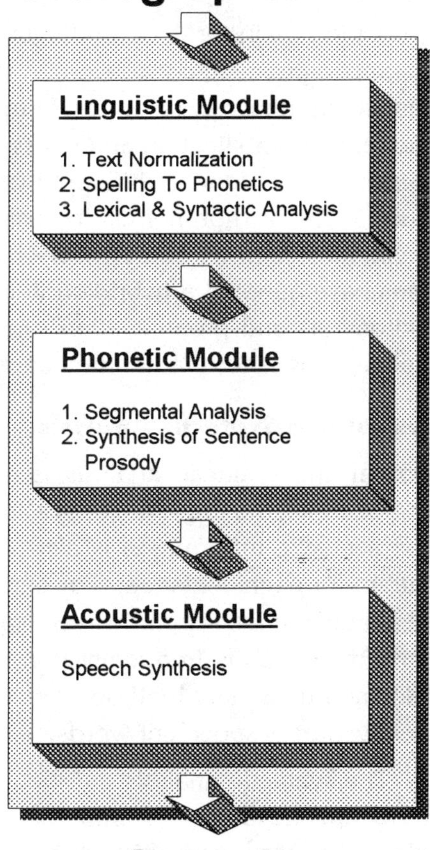

FIGURE 11.2 The *text-to-speech* algorithm is used to convert text into natural sounding synthetic speech. The algorithm specifies three major phases of processing.

Orthographic-to-Phonetics

One of the main tasks of the linguistic processing part is to perform the orthographic-to-phonetic conversion. A text-to-speech system needs substantial pronunciation knowledge to perform this task, which includes grapheme-to-phoneme, syllabification and stress assignment. Orthographic-to-phonetic conversion can be done in different ways: by consulting dictionaries containing full word forms or morphemes, by using a set of pronunciation rules or by using techniques such as neural nets or decision trees. Most multilingual TTS systems have adopted a hybrid strategy combining word dictionaries, morpheme dictionaries and pronunciation rules. Although there are many similarities in the way languages are handled each language has its peculiarities.

Lexical and Syntactic Analysis

Lexical and syntactic analysis is needed to solve pronunciation ambiguities. For example, the English word *re'cord* can also be pronounced *'record.* Lexical and syntactic information is also very important to create a correct prosodic pattern for each sentence. For example, important syntactic boundaries entail intonational changes and vowel lengthening. Tagging isolated words with their parts of speech is frequently done by using a combination of morphological rules and dictionary look up. For example, particular word endings help predict the part of speech of words.

The syntactic analysis can be performed with different parsing techniques frequently developed within the field of natural language processing and adapted to the needs of text-to-speech synthesis. Parsing techniques should meet the real time requirement of the application. Most of the commercially available TTS systems do not perform the full syntactic analysis. They do not always construct a full syntax tree but rather perform a local syntactic analysis. For example, context-dependent rules are sometimes used to solve part-of-speech ambiguities and to divide a sentence into word groups and prosodic phrases.

Phonetic Processing Module

The phonetic module performs two main tasks: the segmental synthesis and the creation of good prosodic patterns.

Segmental Synthesis

Segmental synthesis is responsible for synthesis of the special characteristics of the synthetic speech. In most cases the amplitude is also handled by the segmental system module. There are two different approaches to segmental synthesis: phoneme synthesis (rules based) and synthesis based on segment concatenation.

The concept of phoneme synthesis can be briefly described as follows. For each phoneme some target speech parameters needed to drive a speech synthesizer are stored. During synthesis, the system starts from these target values and then use rules to create correct spectral transitions. The resulting speech parameter vectors are then used to drive a speech synthesizer which is frequently a format synthesizer.

Other text-to-speech systems use segment concatenation. They use small speech segments taken from human speech to create synthetic speech. With a finite set of well-chosen speech segments, it is possible to synthesize any text by concatenating segments. The selection of the elementary building blocks is an important factor and determines to a great extent both the complexity and the quality of the system. TTS systems using the segment concatenation technique frequently use diphones as elementary speech units.

A diphone is a small segment of speech starting somewhere in the middle of one phoneme and ending in the middle of the next phoneme. This means that the transition as well as most of the coarticulation effects between the phonemes, are preserved inside the unit. The use of diphones as elementary building blocks for speech synthesis is based on the assumption that coarticulation effects are local effects. However, this is not always the case. This is the main reason why in some systems, segments of different lengths are used.

Prosodics

To synthesize intelligible and natural sounding speech is essential to create good prosody characteristics. The synthesis of prosody involves two steps: the production of good intonation contour and the assignment of a correct duration to each phoneme. As mentioned earlier, the creation of a correct amplitude contour is frequently handled as part of the segmental synthesis module.

Intonation

Each sentence contains one or more dominant words. In most languages the dominate word is verbally marked via intonation accent realized by a pitch change on the lexical accented syllable. Intonation is also used to mark a phrase type and to mark important syntactic boundaries. In most Asian tonal languages, the word meanings and grammatical contrasts are conveyed by variations in pitch. In pitch accent languages such as Swedish and Japanese, a particular syllable in a word is pronounced with a certain tone. This is in contrast to languages such as English where each word has a fixed lexical stress position, though there is less restriction on the use of pitch.

A text-to-speech system should include a language-specific module that models the perceptually relevant intonation effects of the target language. Different approaches are possible. One approach is to define pitch contours in terms of rises and falls. Rules specify how these elementary pitch movements can be combined to create intonation contours for whole messages. These rules take into account the number and the location of the dominate word(s) and the major syntactic boundaries.

Phoneme Duration

It is near impossible to assign a correct duration to a phoneme. Measurements of speech data as well as perceptual experiments prove the relevance and the importance of good duration models. Phoneme durations are influenced by many factors. A duration model should take into account the phonetic context, the stress level, the position within the word, the syntactic structure of the sentence and the opposition between content and function words.

Phoneme models are developed and implemented in assorted ways. Two common implementations are rule based and neural net based. Some of the models are phoneme-oriented while other models predict the duration of the syllables before assigning duration to phonemes.

Acoustic Processing Module

The final portion of the text-to-speech system performs the acoustic processing. The speech data created in the previous stage are now converted into speech signals.

A text-to-speech system based on phoneme synthesis usually uses a format synthesizer to create speech output. However, a system using segment concatenation does not necessarily need a synthesizer.

Concatenation of speech segments and synthesis of prosody can be done in the same domain using speech synchronous methods such as PSOLA (pitch synchronous overlap add) technique. Another possibility is to use pitch synchronous synthesis techniques in combination with a residual excited LPC model. In this case prosody manipulations can be done in the residual domain.

TTS Applications

TTS technology is particularly well suited for integration into a wide range of large volume voice applications:

- ⇒ *Telephone system access to databases, bulletin boards, Internet, and electronic mail*
- ⇒ *Multimedia input/output applications, such as computer aided training*
- ⇒ *Automation of telephone network services such as directory inquiries*
- ⇒ *Various applications in hands-busy or eyes-busy environments*
- ⇒ *Speaking data base look-up requests. For example, TTS can be used in interactive voice response systems to speak the data retrieval portion of the information.(i.e. airline flight information, train schedules, bank and security account balances and telephone directories).*

In summary, the use of assistive technologies moves the resourcefulness of earlier CT technology into the 1990's. These technologies will soon impact all our lives with highly useable technology literally at our beckoned call.

|12|

Sun's Products

Sun Microsystems, Inc., provides solutions that enable customers to build and maintain open network computing environments. Widely recognized as a proponent of open standards, the company is involved in the design, manufacture and sale of products, technologies and services for commercial and technical computing. Sun's SPARC workstations, multiprocessing servers, SPARC microprocessors, Solaris operating software and ISO-certified service organization each rank number 1 in the UNIX market.

Building superior system level technology since 1982, Sun holds the lead market share in workstation technology and servers and has set standards by which others are judged.

Sun has led the UNIX industry since its early years. They were the first organization to fill the need for reasonably priced workstation level computing, the need for true open standard computing, the need for an industry standardized file sharing (NFS) capability and the need for highly advanced network management; Solstice.

They have ceased the number one position in Internet servers. They introduce Netra™, a pre-packaged Internet solution with their HotJava™ browser and now the hardware independent Java™ language which is such a sensation that it was recently the cover story of "Business Week". It will continue. UltraSPARC, zero-administration client....

Sun's product line is quite comprehensive and continues to evolve at a fairly rapid rate. It is a living demonstration of the philosophy espoused by Sun's legendary hardware guru Andreas Bechtolsheim, "You have to obsolete your own product, anytime you try to protect your past, you're in trouble because the future is cheaper".

Sun Hardware

Sun Microsystems Computer Corporation manufacturers a full family of SPARCservers and SPARCworkstations. These range from single processor systems to multi-processor power houses. They have been configured as stand-alone workstations or part of a client-server architecture.

Due to the rapid development and introduction of new Sun products I have not provided system level information. Those desiring information on the current product offering should contact their nearest Sun Microsystems sales office or telephone: 800-USA-4-SUN.

SunSoft Software

SunSoft does not directly manufacture or support any application software specifically designed for computer telephony. This software is available through third party vendors which is detailed in Chapter 17.

SunSoft does provide their SunXTL™ Teleservices software, detailed in Chapter 10, Sun's platform for desktop and enterprise level client-server telephony applications. Linkon Corporation as well as an assortment of major telecom switch manufacturers have written providers which operate with SunXTL™.

Solaris 2.X, is the multi-threaded operating system is used for the design and operation of many applications. It's flexibility and power has been harnessed for use in many computer telephony solutions. The Solaris operating system is covered in Chapter 7.

Java™, Sun's newest addition to their software offering, has achieved immediate notoriety. It is an object-oriented language which is operating system and hardware independent. It has currently been licensed for use by such organizations as Microsoft, IBM, Toshiba and Netscape. A detailed discussion of this new language can be found in Chapter 18.

Access Worldwide Channel Integration, Inc.

Access Worldwide Channel Integration (800-730-6462), Sun's largest master reseller, recently introduced a line of Computer Telephony servers. Each server comes already configured with Linkon hardware and application software. With very little work a system integrator can have the system operational in just a few hours.

There currently are three basic server designs available:

TeleFusion IVR. The TeleFusion IVR servers include everything you need to create a telephone-enabled interactive voice response (IVR) without programming. Everything you need is in one server. No PBX upgrades or additional hardware is needed to begin using the system. Everything is configured and tested in advance. You assured a complete working system out of the box. The system features user defined menus, prompts and applications which can be customized to meet the specific need of the user. It is available in 8, 24 and 30 port configurations for use with analog or digital telephone network connections. ANI and DNIS capabilities are included. The system is capable of network connectivity via TCP, NFS, FTP and Telnet. TeleFusion IVR has database access for Oracle and Informix. Included in the system capabilities are voice record and playback for audiotex applications, fax mail, automated attendant and several system reporting for administration.

TeleFusion Fax. TeleFusion Fax servers provide a low-cost powerful method to link telephones, fax machines and corporate databases. As is the case with TeleFusion IVR, this system comes ready to use right out of the box with very similar features. Available in 8, 24 and 30 port configurations and connectivity to analog or digital telephone circuits. The system comes loaded with application software which performs fax on demand functions with one and two call faxing capabilities. Fax messages in fax mail are handled the same as voice messages. Fax messages remain confidential and can be forwarded to local or off-premises fax machines and network printers. TeleFusion Fax servers can be interfaced with standard networks via TCP, NFS, FTP and Telnet. All operational data is available by using the system reporting and administrative functions.

TeleFusion Messaging. TeleFusion Messaging provides seamless integration of voice and fax with E-mail onto a single platform. This permits the user company to handle more calls giving customers direct access to the critical information they need. As is the case with the two servers detailed above, TeleFusion Messaging comes ready to use right out of the box. The system is available in 8, 24 and 30 port version with connection capability for analog or digital telephone networks. Both ANI and DNIS is available for use by the application. Data network connectivity is possible via TCP, NFS, FTP and Telnet. TeleFusion Messaging voice features include configurable pager activation and scheduling, message delivery to external phones, multiple greetings for external callers and group messaging. With the fax and E-mail messages are handled the same as voice messages in linked mailboxes. The system has automated attendant capabilities and can easily be customized to meet the needs of the user. Advanced administrative and reporting features are available with the server.

In addition to the Sun hardware and software listed above, Access is a distributor of Linkon hardware and software, detailed in the next chapter.

As part of their Team CTI program, Access regularly trains VARs for Computer Telephony.

|13|

Linkon's Computer Telephony Products

Linkon Corporation empowers Sun's Customer Management Solutions™ with a truly innovative line of modular design computer telephony hardware and software. The products' unique design provides developers with a cost effective platform for delivering a variety of highly flexible solutions.

Others talk about it and even allude to it in their advertising, but Linkon's approach to computer telephony technology features a true open architecture, designed as easy to use "building blocks". Any developer can deploy as much technology as they currently need, and even after the system is installed and operating, upgrade it to include enhanced functionality, such as speech recognition and text-to-speech, merely by adding software via a CD-ROM or remotely via modem.

Linkon, known for innovation, has manufactured computer telephony hardware and software since 1985. Sun Microsystems discovered Linkon's technology in 1992, and requested an Sbus product be designed and developed for their SPARCservers and SPARCworkstations.

In 1993, Linkon delivered on the request with the FS-3000, a single channel "universal port" analog DSP (digital signal processor) based solution, integrated with powerful C tool kits, TeraVox and ProVox (see chapter 17), operating in a Solaris environment. That was just the beginning. Mid-1995, Linkon introduced the FS-4000 Maestro™ series of powerful multi-channel universal port DSP based computer telephony processing engines for Sbus. These engines power the broadest suite of communications processing algorithms, delivering the most feature-rich product offering, in the industry.

The universal port capabilities permit the processing of several communications technologies simultaneously during the same call session. It is a true multi-tasking offering at the board level.

You, as most people that are presented this concept, are probably wondering how this all works. History has proven that the best way to explain this universal port technology is through an example.

When using a system configured with universal port technology an individual can call the order entry system of a retailer from the telephone handset on a fax machine. They can respond to the menu selections played for them with either the DTMF tone buttons of their phone or through speech recognition. They can then review information on products of interest via text-to-speech, and have additional product information plus confirmation of their order faxed back to them all during the same phone call by simply pressing the start button on the fax machine they are calling from. This is all accomplished on the same telephone line without any switching required. This type of solution not only saves cost when implementing a computer telephony solution of this type but the system owner also saves because the entire session is paid for by the caller. Now, *that's* cost effective marketing.

Building Block Modular Architecture

Linkon's building block type architecture is reflected in the design of the hardware as well as the software. The theory being developers need total flexibility, in an open architecture, when bringing "living" applications to market. These applications must meet the current processing demands and provide a clear path for expansion of capacity and functionality without major hardware

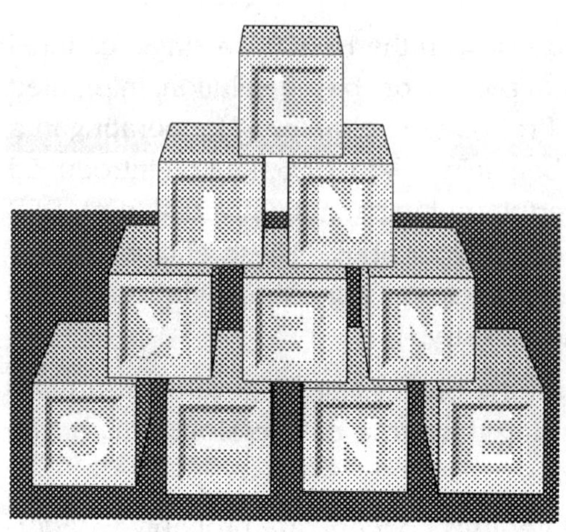

FIGURE 13.1 Building Block Architecture.

changes. A system in the field need stay in the field processing the application for which it was designed.

When reading this chapter you will soon realize that with Linkon's flexible open architecture, a developer can expand system functionality and capacity via software without any major hassles. In fact, most changes can be made remotely with a simple data download.

As is the case with Linkon Maestro™ series computer telephony solutions for other computer buses, the Sbus offering consists of a foundation and switching card, and a series of modules which attach to the foundation and switching card, like building blocks, to provide the desired processing capacity. Any combination of modules can be used on the foundation and switching card.

Building upon the hardware structure is a board level operating environment, software licensing technology for security, and software feature modules for call management and communications processing functionality.

Foundation and Switching Card

The foundation and switching card is available in two-slot-Sbus width and three-slot-Sbus width versions. Both versions have software controlled switching capabilities which permit any channel on the card to be switched to and/or connected with any other channel. The two-slot-wide foundation and switching card has the capacity of handling 32 independent channels. The three-slot-wide version can handle 48. The foundation and switching card has connectivity to MVIP and SCSA internal computer telephony bus architectures. (Figure 13.2.)

The switching capabilities of the foundation and switching card provide the ultimate in flexible architecture design and cost savings. Conferencing and switching are easily accomplished without the call leaving the foundation card on which it was originally terminated. Several foundation cards can be connected together via internal computer telephony bus (MVIP and SCSA) to facilitate the need for larger configurations. When connected together, any channel on any of the foundation and switching cards can be interconnected or switched to. In addition, a group of callers can be connected together to form a conference call. With all this capability Linkon's technology certainly minimizes, and in some cases eliminates, the need for a PBX.

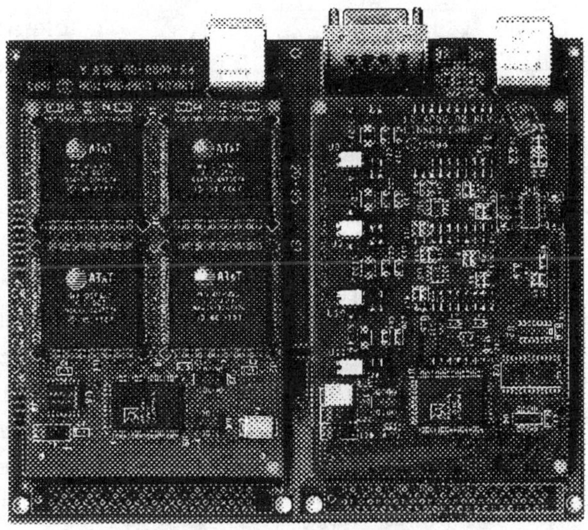

FIGURE 13.2 **Double-slot width version of Sbus foundation and switching card (bottom) with a 4-DSP module (left) and an analog module (right) installed.**

DSP "LinkEngine" Modules

Linkon's DSP modules are a complete processing system at the board level. They are comprised of hardware, Link-OS, a multi-threaded operating system, Link-OS Software Feature Modules, and LinkLogic, Linkon's software license control mechanism. (See Figure 13.2.)

The DSP hardware engines are available in two basic configurations. Version One features one megabyte of RAM per DSP; Version Two has four. The maximum configuration of both versions is four DSPs per module. Applications requiring less processing power are accommodated by using modules populated with three, two or one DSP depending on need. The DSP modules are easily attach to the foundation and switching card, like building blocks, when assembling the solution.

Each DSP module has its own software based power switch. Any module in the configuration can be powered-up or powered-down, as needed, without taking the system out of service. This permits integrators deploying fault-

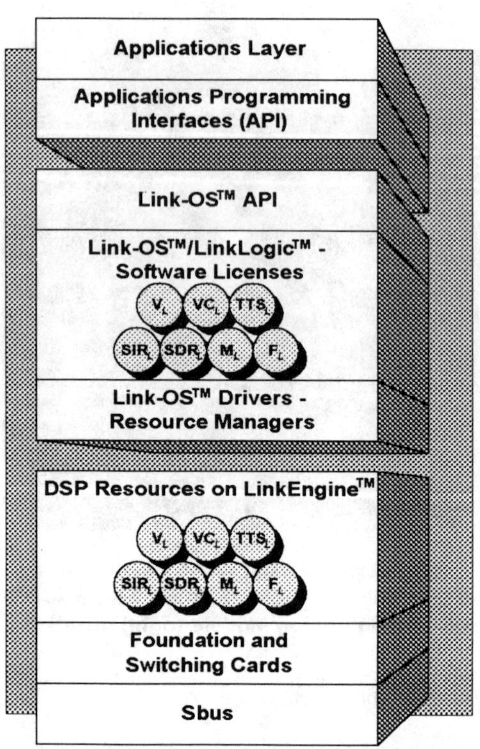

FIGURE 13.3 DSP Engine with Link-OS, LinkLogic, Link-OS Feature Modules, and Application and Sbus Interface Layers.

tolerant and high-availability systems to provide their customers spare processing power which can be activated via simple software command. This is truly a unique feature which can save an organization deploying a mission-critical system substantial money and aggravation. Some creative integrators have written code into their program management software which upon discovery of a processing abnormality, seek out the processing engine with the potential problem and automatically shut the module down and power-up a new DSP module. The system then resumes operating the application at full capacity.

The DSP module has its own multi-threaded operating environment, Link-OS, which manages the 30+ possible software algorithms operating on the engine at any point in time. Link-OS allows the full power of the DSP managed universal port architecture to work as a standard computing environment, operating with the host software and DSP processing power, to dynamically implement assorted software processes. This universal port capability provides developers the freedom of concentrating on the application layer without having to worry about resource allocation, call control and management.

Link-OS is an operating system and API designed to control the DSP module operations and interface with the Link-OS Feature Modules (voice, fax, speech recognition, speaker verification, modem, text-to-speech and CELP). The operating system is the transport layer to the command and control features of both software and hardware. It enables the application layer to recognize all Linkon system components while directing their functions. An application calls a Link-OS function, and the operating system completes the task and returns the appropriate status to the application. (See Figure 13.3.)

Working with Link-OS is LinkLogic, Linkon's software license control mechanism. It is an encrypted software object which monitors all system operations. One of LinkLogic's unique monitoring functions regulates the number of software feature processes available for use at one time. This number is determined by the port allocation made with the purchase of Link-OS feature modules. This mechanism allows Linkon's customers to custom configure software features to cost effectively meet their current need. These software features are easily modified. LinkLogic tracks each DSP module serial number and maps the software licenses to function with only the DSP module for which it was licensed.

Operationally LinkLogic is similar to a client-server; the telephone lines or channels being clients and Link-OS and LinkLogic being the server. The cost structure is also implemented in the same manner as client-server licenses. For example, given the server has 24 clients and only 10 data base licenses, only 10 simultaneous accesses to the data base can be accepted at one time. However, all 24 have general access to system wide functions. If a given system has 24 lines, but only 10 fax licenses all 24 lines have access to general software like voice but only 10 have simultaneous access to fax license. In both cases the number of software licenses can be increased with the purchase of additional software licenses.

Link-OS can be upgraded with additional software features as needed. The maximum number of licenses per feature equals the number of telephone lines or ports on the system. As indicated earlier, upgrades can be accomplished by distribution of floppy disks, CD-ROM and electronically downloaded to the system via the Internet and other data communications paths. The upgrade would include the LinkLogic security module, software feature modules, software licenses, and serial number(s) of the DSP modules to be upgraded.

Digital Network Interface Module

The network interface module provides connectivity to T-1 digital circuits. As is the case with the DSP module, the network interface module attaches to the foundation and switching card. The module provides network connectivity via

RJ-48C and supports twenty-four time slots terminated in up to eight or twenty-four channels. It provides D4 and Extended Superframe (ESF) framing.

Analog Network Interface Module

The analog network interface module provides connectivity to four standard POTS (plain old telephone service) lines. This module attaches to the foundation and switching card. (See Figure 13.2.)

Station Set Module

The station set module is used in configurations requiring interactive voice response integrated with live agents. For example, this module would be used by call centers and telemarketing firms to provide the call to the agent at the same time as the account information. This module provides ring and telephone connection to agent handsets and is easily attached to the foundation and switching card.

Link-OS Software Feature Modules

(Communications Processing Power)

Link-OS feature modules provide the dynamically deployed communications processing functionality. There are currently over 30 algorithms operating on Linkon's DSP module. These algorithms provide call control functions as well as communications processing functionality. Most algorithms are running and interacting with many others at an given time.

Call Control Features

Many call control functions are taken for granted when constructing an application. However, most are critical to the interaction between analog or digital connections and the telephone network. Call control algorithm activities are not as pronounced as fax, modem, speech recognition and text-to-speech but without their involvement none of the earlier mentioned functions would be possible. These functions include:

- ⇒ Automatic gain control
- ⇒ Full call progress monitoring
- ⇒ Echo cancellation

- ⇒ Standard DTMF detection and generation
- ⇒ Programmable tone detection
- ⇒ Caller ID
- ⇒ Onboard bridging and conferencing
- ⇒ Full channel to channel switching
- ⇒ Multi-channel DSP sharing

Voice Compression Features

To permit voice to be stored by a computer, transported across computer buses and sent down data circuits compression is required. Compression is an arithmetic sampling process where samples of the voice are taken, as representations of the whole voice, for recording, transmitting and storage purposes. The higher the compression rate the smaller the voice sample taken, and the less it sounds like the whole voice heard in person by the human ear. Due to recent advancements in this technology many of the high compression techniques used today sound quite good.

Voice compression is not a new technology. Any voice which is stored and transported by a record and playback device uses compression. The most common example being the standard telephone circuits we use daily to connect two or more calling parties use compressed speech.

Linkon's Link-OS feature module for voice compression has the broadest offering in the industry. The compression suite includes the following:

- ⇒ 128 Kbits/second 16 bit linear
- ⇒ 32 Kbits/sec. ADPCM CCITT G.721
- ⇒ 32 Kbits/sec. OKI/ADPCM
- ⇒ 24 Kbits/sec. OKI/ADPCM
- ⇒ 24 Kbits/sec. Sub Band Coder
- ⇒ 16 Kbits/sec. Sub Band Coder
- ⇒ 9.6 Kbits/sec. CELP
- ⇒ 7.2 Kbits/sec. CELP
- ⇒ 4.8 Kbits/sec. CELP

Fax Features

Fax has become as commonplace in communications as the business telephone. It is now realizing substantial growth at the household level with the expanding SOHO market and the ever increasing multi-media households market segment. Linkon's functionality for it's Group III fax offering include:

⇒ V.17 compression

⇒ Error correcting mode

⇒ Binary File Transfer

⇒ Text-To-Fax on Card with loadable fonts

⇒ Bitmap Input/Output

⇒ PCX file input/output

⇒ Any combination of Group III bitmaps and PCX can be concatenated to the same page

Modem Features

To date, Linkon's modem offering has been used primarily for system upgrades and integrators interested in the lower speed applications such as in-store credit card processing terminals, and other in-field devices for sending and receiving short data transmittals. The Link OS Modem feature module operates at the following modulation speeds:

⇒ Bell 103

⇒ Bell 202

⇒ Bell 212

⇒ v.21

⇒ v.22

⇒ v.22 bis

As ISDN connections become more commonplace, BRI, available cheaply through most RBOCs, will eliminate the need for high speed analog modems.

Speaker Verification Features

Linkon's speaker verification boasts some of the highest accuracy scores in the industry. This technology reduces the risk of unauthorized entry to a system, application and even a door. It can be utilized as a screening device for a building floor, screening at an automated teller machine, used by an on-line telephone based securities trading system for verifying the authenticity of the user, and a telephone calling card fraud prevention system.

Similar to speaker dependent speech recognition technology, speaker verification requires all users to train the system by recording speech samples. Further accuracy is obtained through continued use of the system as it continually trains itself, refreshing the reference template each time the system is used. A detailed discussion on speaker verification technology can be found in Chapter 11 of this book.

Speech Recognition Features

Linkon offers the widest selection of speech recognition technology available in the market. Both speaker dependent (individual specific; requires training) and speaker independent (anyone can use without training) technologies are available from an assortment of speech recognition vendors.

All speech recognition technologies have had a slow-start in the market. Continual advancements made by many of these specialty houses has made some of the product offerings commercially viable. Linkon engineers have identified some of the best technology and made these available as Link-OS speech recognition Features. They continually conduct worldwide searches of speech recognition technology to identify the most viable and commercially sound algorithms for integration with the product.

No discussion of speech recognition technology is complete without covering the accuracy rate of the algorithm. Most speech recognition technologies boast accuracy rates from the mid-90 percentile and higher. Optimal accuracy levels are achieved through the proper use of the technology in an application. Even the most accurate speech recognition algorithm will not produce optimal results in a poorly constructed application.

To fairly assess the value of speech recognition in an application one must look at alternative technologies which could be used by the application. Where available, DTMF tone is the largest alternative technology. Rotary dial, even with pulse conversion technology, does not work well in applications requiring the user to input information or make numerous selections.

Several studies of DTMF tone use indicate that correct connection is only accomplished 85 percent of the time. The 15 percent deficiency is due, for the most part, to dialing inaccuracy. When comparing speech recognition to DTMF tone, mid-90% accuracy rate doesn't seem so bad after all.

Linkon's speech recognition technology features recognizers which run totally on the DSP and others which are host based and managed by the DSP. Each speech recognition algorithm has its strength and weakness. None is appropriate for all applications. Some are good for alphanumeric recognition, others are better at continuous digits. Whatever the need, Linkon provides a speech recognition technology which will meet the need of the developer.

Some developers are achieving greater levels of success by using an assortment of speech recognition technologies to better meet the needs of the application. Speech recognition is considered an assistive technology.

Following is a listing of the speech recognition technologies currently available through Linkon. Due to the continual expansion of this offering, please contact Linkon (203) 319-3175 for the most current listing of available technologies. A detailed discussion of speech recognition technologies can be found in chapter 11 of this book.

⇒ **Speaker Dependent Speech Recognition.** Speaker dependent speech recognition technology attained its highest degree of visibility through its use in voice dialing applications. Candice Bergman demonstrated it when making phone calls for SPRINT by merely speaking the name of the intended called party. Use of this technology requires training by individuals prior to its use. Each user creates a reference template for each word they train. These templates are stored in an individual specific database which is accessed at the time of use. The system changes as the user changes because the reference templates are continually updated through use. Linkon has speaker dependent technologies available from three different vendors.

⇒ **Speaker Independent Speech Recognition.** This recognition technology is designed for universal use. It is based on scientifically compiled databases which are statistically representative of the intended user group. Speech samples are recorded which reflect the age distribution, geographic peculiarity and sex of the potential user group. No training of this system is required prior to use. Linkon has speaker independent recognition offering from five vendors. Several languages are currently available; others are scheduled for introduction soon.

⇒ **Alphanumeric Recognition.** Based on speaker independent technology, this algorithm is quite useful for catalog ordering, credit card and warehouse order-entry applications. As indicated by its name alphanumeric recognition recognizes the letters of the alphabet and digits spoken in natural language format. Due to the small size of these utterances this recognition is very difficult to perform. This technology is available through only one vendor.

⇒ **Large Vocabulary Recognition.** This speaker independent based technology has an available vocabulary of 2,500+ words. The algorithm is highly accurate and currently deployed by an RBOC in several applications. It is currently available from one vendor.

Text-To-Speech Features

This technology provides the capability of putting a voice to data. This data can be in the form of ASCII text, a fax or E-mail message. It is great for use in a unified messaging platform or an application with continually changing information (airline, bus and train schedules). At present Linkon offers high quality real-time text-to-speech capabilities for American English and an assortment of major European languages. The offering is available from two vendors. Both algorithms are easily adapted to disclose any type of written information contained in various environments.

Direct conversion of written text into spoken messages assures high flexibility for all kinds of voice response applications. Updates or changes to written input text can be made accessible in spoken form almost instantaneously.

The use of text-to-speech technology opens possibilities of integration in a wide range of large volume applications. These include:

⇒ over-the-phone consultation of assorted data bases, plus information on the Internet and bulletin boards

⇒ over-the-phone review of fax and E-mail received in mailbox

⇒ multi-media input/output applications such as computer aided language training

⇒ automation of telephone network services such as directory assistance

⇒ various applications in hands-busy and eyes-busy environments

A detailed discussion of text-to-speech technology can be found in chapter 11 of this book.

Additional Hardware

In addition to the FS-4000 Maestro™ product detailed above Linkon also resells a T-1/E-1/ISDN interface board from Newbridge Microsystems, and provides a Linkon Sbus Expansion Center, for building larger capacity solutions.

The Newbridge NM212 Sprite Card

Newbridge Microsystems manufactures a single slot Sbus card, NM212, which provides T-1 and E-1 connectivity to digital telephone networks. The card, originally designed for use with data circuits, was enhanced for use in computer telephony solutions. The Sprite card connects to the Linkon Maestro series Sbus cards via MVIP internal computer bus. It is available exclusively through Linkon.

Sprite E-1 Features Include:

⇒ Network Connector—DB9 to BNC adapter for use with 75Ω Cox and RJ-48C or DB9 for use with 120Ω twisted pair

⇒ Framing—Channel Associated Signaling (CAS), Common Channel Signaling (CCS) and 31 Channel Mode

⇒ Line Interface Module (LIM)—75Ω or 120Ω

⇒ Timing—Line or Internal Oscillator

⇒ Statistics—Link, Channel and G.821

⇒ Diagnostics—Loopback Tests for Equipment, Line, Payload and Per Timeslot

⇒ Network Approvals—CCITT G.703, G.704, G.821 Statistics and OTR001

⇒ EMI Conformance—EN50081-81 and EN50082-1

⇒ Safety—EN60950 and EN41003

Sprite T-1 Features Include:

⇒ Network Connector—One RJ-48C connector

⇒ Channel Supported—Twenty-four slots terminated in up to eight or twenty-four channels

⇒ Framing—D4 and Extended Superframe (ESF)

⇒ Line Code—B8ZS, JB7 or clear channel

⇒ Alarms—Yellow alarm, red alarm, exceeded error rate, alarm indication signal, etc.

⇒ Timing—Line or Internal Oscillator

⇒ Diagnostics—per channel loopback, local line loopback, local and remote framer loopback and CSU remote loopback

⇒ Specification Compliance—AT&T Pub 62411, 54016, 43801, FCC Part 15 and 68, ANSI T1.107, CCITT G.703 and G.704, UL 1459, CSA 225 and ISO 9001

Linkon Sbus Expansion Center

The Linkon Sbus Expansion Center offers a wide range of server solutions in a convenient stand-alone or rack mount configuration. In one small package, the Expansion Center provides industry leading price/performance breakthroughs, high availability, easy system management and tremendous expandability.

As is the case with the entire Linkon product line, the Expansion Center is modular and assembled as building blocks. It consists of a SPARCstation 10 system tray, Sbus expansion trays, a load sharing power supply, disk storage modules, and a cooling fan assembly. All of these devices slide into an integrated system back plane which supports a single system tray, up to three Sbus expansion trays, two RAID drive modules, and four 5.25" devices, two redundant power supplies, and a fan module. Because of its unique modular design, the Expansion Center can easily accommodate emerging technologies as the next generation of systems and processors become available.

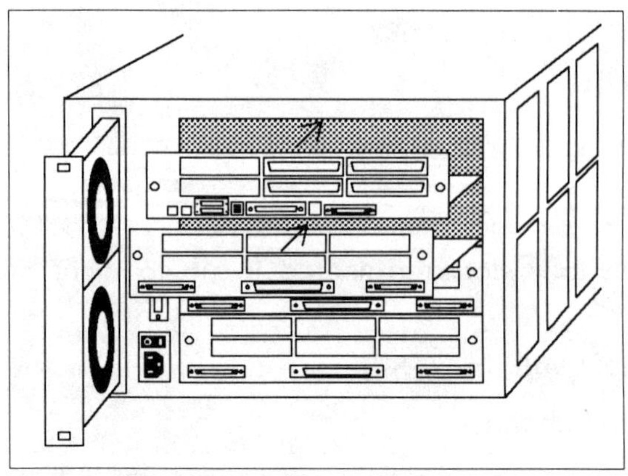

FIGURE 13.4 Linkon Sbus Expansion Center.

Reliability is the first and foremost concern for the Linkon Sbus Expansion Center. An early warning system consists of an emergency signal alarm, a remote serial link, and panel status indicators which constantly monitor the status of the Expansion Center. Through this any hardware failure can be identified instantly, on site or remotely. In the unlikely event a failure should occur, all vital components of the system are "hot swappable" for easy replacement. It is an effortless task to replace a cooling fan or the 450W load-sharing power supply system without interruption. The Expansion Center protects a business by ensuring steady operation for mission critical applications.

Up to 8 GB of data is kept safe and secure with up to two 5.25" RAID drive sub-systems. In the event of a drive failure, the RAID system can automatically rebuild the damaged drive without interruption or data loss. Four additional 5.25" drive bays provide space for CD-ROM drives, tape backup systems, optical drives and removable hard disks. (See Figure 13.4.)

Linkon Sbus Expansion Center: Options and Specifications

The Processor

Compatibility .... 100% binary compatible with
Sun SPARCstation 10 and 20

Architecture .... Superscalar SPARC Version 8

Memory Management .. SPARC reference MMU with 65,536 contexts

Internal Cache .... 20 K instruction and 16 K data on chip

External Cache .... 1 MB optional

CPU Interface .... Two 64-bit 40 or 50 MHz Mbus slots

Standard Interfaces

Ethernet	10 MB/sec twisted pair standard (10Base-T)
	AUI available with optional adapter cable
SCSI	10 MB/sec SCSI-2 (synchronous)
Serial	Two RS-232C/RS-423 synch. serial ports
Parallel	Centronics parallel port (cable required)
Sbus	Up to 19 32-bit Sbus master slots

Additional Options

Mounting	19" rack mounting assembly

Environment

AC Input Voltage	100 to 240 VAC
AC Input Frequency	47 to 63 Hz
DC Input Voltage	48 VDC (in optional rack mount system only)
Operating	0°C to 40°C (32°F to 104°F)
	5% to 95% relative humidity, non-condensing
Nonoperating	-40°C to 75°C (-40°F to 167°F)
	5% to 95% relative humidity, non-condensing

Dimensions and Weights

Height	12.5"
Width	19.0"
Depth	21.5"
Weight	97 lb.

Application Software, C Tool Kits and Driver Development Interface (DDI)

In addition to the DSP based communications processing technology Linkon provides an application suite which works with TeraVox™ and ProVox™, the company's tool kits for the C programming language. Developers desiring a lower level approach will find Linkon's well documented DDI quite useful.

The application suite provides developers a base structure from which they can create customized applications which meet their unique specifications. The suite is composed of LinkFax™, Linkon's fax-on demand application and LinkMail™ a full featured integrated messaging system for voice, fax and E-mail.

Application Software

LinkMail™

LinkMail™ is Linkon's messaging system which integrates voice, fax and E-mail messages in a single virtual mailbox. LinkMail™, written in C, is available in three forms. Level One works with voice messaging only. Level Two supports voice and fax. Level Three supports voice, fax and E-mail (modem) interfaces. All messages are treated equally. It doesn't matter whether they are voice, fax or E-mail; truly Unified Messaging. Most companies take the vanilla LinkMail™ and add their own flavoring, integrating the result with their product or as their product. Source code is what Linkon sells; there are no royalties, licenses or other fees. An in-depth discussion of LinkMail™ can be found in Chapter 17.

LinkFax™

LinkFax™ is Linkon's software for fax-on-demand and the complete integration of fax services. It is written in C and available in source code form. Most companies use the fax-on demand capabilities of LinkFax™ to provide information to clients and potential customers. As is the case with LinkMail™, many organizations start with the vanilla LinkFax™, add their own flavoring, integrate the result with their products and sell the result. Additional details on LinkFax™ are covered in Chapter 17 of this book.

C Programming Language Tool Kits

Linkon Corporation has developed two C tool kits for communications processing. These tool kits have been used by many developers to create their applications. Traditionally developers have reduced their applications development time by 50 to 75% through their use.

TeraVox™

TeraVox is a fully featured application development environment capable of supporting any type of application requiring interactive communications. The API is a set of powerful, high-level C programming language libraries of complex computer telephony functions. TeraVox™ provides an optimized file management system for voice data storage. Some of TeraVox™'s unique features grew directly out the experience of operating a large service bureau supporting 1,500 telephone lines, and thousands of hours of message storage. The API is completely scaleable from the small 4-port systems to the very large multi-node systems required by large corporations.

With TeraVox™, the normal UNIX file system for all voice and database input/output is bypassed. A completely new application-specific file manager is provided which is custom tailored to the desired audio response environment. This file manager differs from standard file systems in a number of ways. All control structures are memory resident for high performance. Control structures are replicated and checksummed on file storage. Raw disk I/O is used so that the UNIX disk caches are not disturbed, and "elevator" seeking is performed to reduce disk arm movement. The TeraVox™ file manager allows a developer to create and maintain lists of objects like mailboxes, or messages in a mailbox. These kinds of features and many others give TeraVox™ robust, high performance voice storage transfer rates. Typical read/write rates are 700 KB to 1 MB per second.

TeraVox™ has over 25 command line utilities for system administration and reporting. Standard formation captured includes call counts by application by hour, system call counts by hour, system call statistics and event logging.

TeraVox™ contains a C function library that contains over 90 entries, through which user applications communicate with the TeraVox™ system tasks. In this way, all application development is performed using standard UNIX tools. C programmers have virtually no learning curve and can begin developing voice applications almost immediately. The TeraVox™ library is simple enough to provide the productivity increases usually associated with applications generators, but powerful enough to create truly innovative voice applications.

An Intertask Communications (ITC) facility tightly couples TeraVox™ drivers and system tasks. The TeraVox™ ITC is streams based at the task level, providing for deep queuing of commands.

Special extensions to the UNIX kernel are supplied in the form of device drivers, eliminating the need for custom kernel objects. These extensions supply high resolution timers and some address conversion services required by TeraVox™ system tasks.

ProVox™

ProVox™ is an open client-server implementation of C programming language libraries for complex computer telephony functions. "Open" is defined by the fact that it can use any UNIX database for file management, and processes run both on the telecommunications gateway as well as on the server. Built on TeraVox™, ProVox™ is designed for applications that need to work with very large databases and complex file management. ProVox™ was developed under the direction of a major telco, and was designed for WAN telecommunications.

Direct Driver Interface (DDI)

Linkon provides a well documented DDI for developers desiring integration of Linkon technology into an existing application, or developing a custom application from a lower level interface.

Additional Software

There is a significant amount of application software which utilizes the dynamic capabilities of the Linkon technology. Some developers desiring a more custom solution use one of the available application generators to accomplish this task.

As was the case with LinkMail™, LinkFax™, TeraVox™ and ProVox™, a detailed discussion of available application software and application generators, for use with Linkon's technology, can be found in Chapter 17.

Linkon's Sun XTL™ Provider

Linkon, with the assistance of Sun has created an XTL Provider which incorporates all the functionality available from Linkon. Rather than presenting detail on the provider here, please see Chapter 10, of this book for a detailed presentation of Sun XTL™.

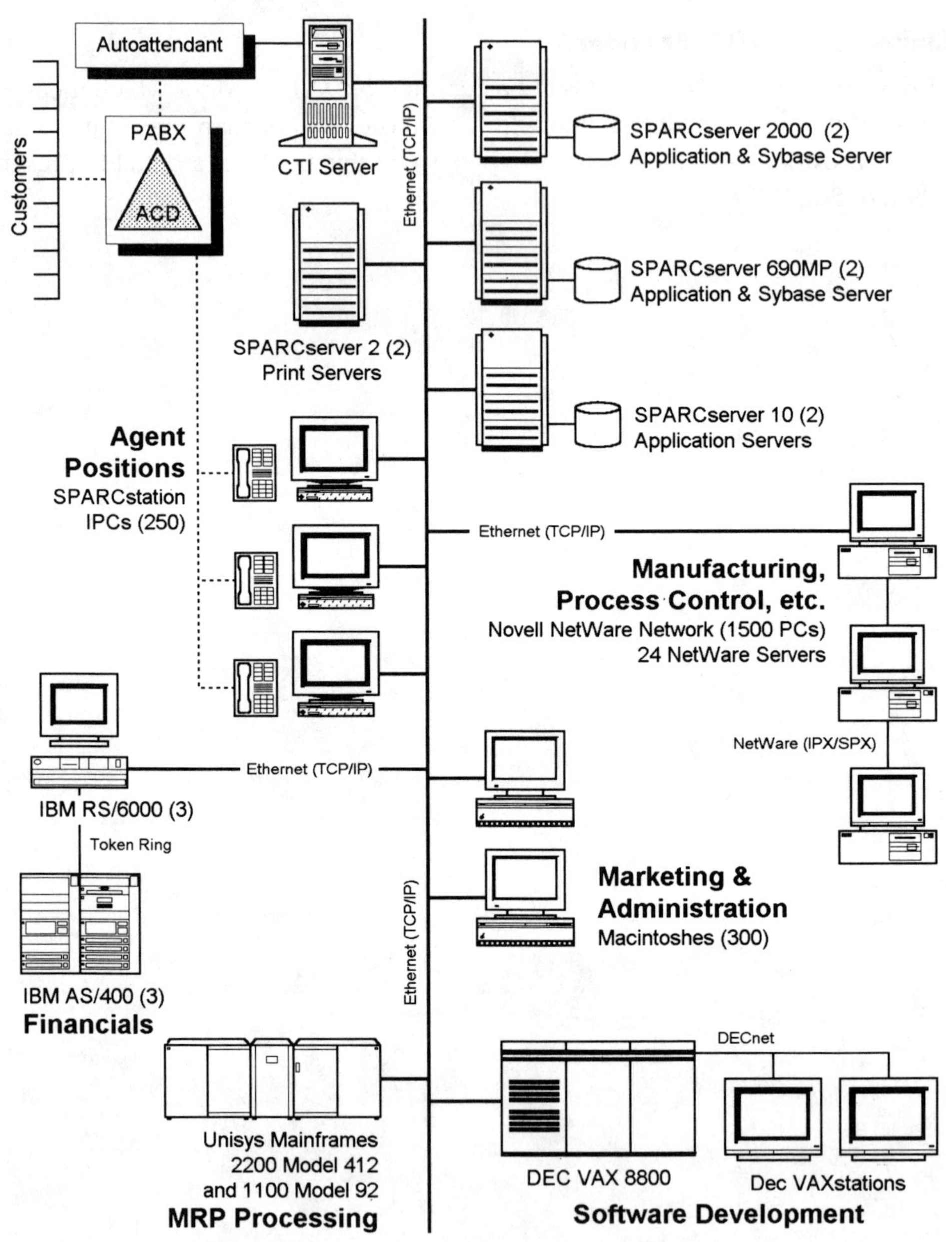

FIGURE 14.1 The Pharmaceutical Manufacturing and Distribution Solution

|14|

Sun Solutions for Computer Telephony

With the current tool kit of Linkon Corporation "Universal Port" open architecture hardware and software combined with an assortment of application suits, applications generators, custom software designs and Sun's processing power, development of a near limitless variety of Computer Telephony solutions are available to the able hands of system integrators, original equipment manufacturers and value added resellers. Following are ten solutions made possible through these combinations. These examples are provided to shed light on some of the computer telephony solutions possible and stimulate your ingenuity to develop some of your own.

SOLUTION #1 PHARMACEUTICAL PRODUCTS MANUFACTURING AND DISTRIBUTION

A large pharmaceutical company in the Midwest manufacturers and distributes hundreds of products worldwide. It operates in dozens of countries providing products to scores of large retail pharmacy companies and smaller outlets.

The company began a downsizing effort back in 1988, long before rightsizing and reengineering became leading industry trends. There was early recognition among the company's management that, to remain competitive, they needed greater and more up-to-date access to manufacturing and distribution information. Their efforts began with the initial migration of business applications from their mainframe systems to IBM minicomputers. This effort was followed by the addition of numerous Sun servers and desktop systems to provide an infrastructure supporting robust client-server computing, which, in turn, furnished a more flexible basis for their expensive operations.

Business Issues

⇒ *Applications required months to develop and were costly and difficult to maintain.*

⇒ *Systems were too slow for on-line transaction processing applications.*

⇒ *No graphical user interface support.*

⇒ *It lacked sophisticated systems integration; data resided solely on the mainframes.*

⇒ *Information on orders and their fulfillment was not always easily accessible.*

⇒ *Robust connectivity to the company's mainframes and minicomputers was required.*

Distributors of the company's pharmaceutical products used the call center to order products and check on shipment status. Due to the centralized data processing approach, the complexity of computing requirements, manufacturing, orders processing, and decision support plus the volume of products and orders, getting up-to-date information was problematic. Assuring quality customer service to its distributors became critical to remaining competitive.

Solution Profile

⇒ *Redesign of a customer call center to a distributed environment.*

⇒ *Provide instantaneous information (order status, reconciliation, product information).*

⇒ *Reduce downtime and dependency on mainframes.*

The solution profile was driven principally by the need to preserve an existing data processing investment while developing a distributed environment for information processing. The company's heavy investment in mainframe systems, personal computers and internally developed applications led to a solution that permitted the graceful migration from mainframe customer service operations to a distributed client-server environment supporting heterogeneous systems.

Existing systems include several Unisys mainframes supporting hundreds of terminals and dozens of PCs. Two dozen Novell NetWare servers, serving hundreds of PCs, were also employed. They however provided no connectivity to the mainframe, its applications or data. Additionally, several hundred standalone Macintosh systems were used for a variety of purposes, including marketing, publishing and related tasks.

The company's centralized computing environment needed to evolve, but had to maintain support for existing applications and systems while new ones were brought on line. Supporting hundreds of PC and Macintosh systems, integrating multiple network protocols including TCP/IP, DECnet and Novell's IPX/SPX appeared to be a daunting task. With so many of these systems in place, only an open systems approach appeared to offer promise of relieving the information processing bottleneck while providing the necessary flexibility for growth.

The company's call center operations employed more than 200 customer service representatives, using terminals connected to the mainframes. No computer telephony integration was in place or planned because the number of customers, most of which were distributors, was limited to a few hundred, and services were generally order fulfillment, status determination and reconciliation in nature.

The rightsizing solution implemented by the customer included two IBM AS/400 minicomputers, each with IBM RS/6000 systems serving as front ends to the corporate network and supplying database gateway software for the Unisys-based in-house databases. A variety of Sun systems were added providing support for manufacturing applications, migrating from the mainframes, order processing applications, decision support software, print management utilities, file services and other facilities. An existing DEC VAX minicomputer and terminals were integrated into the environment to support software development.

Existing PC and Macintosh systems were upgraded to include Ethernet interfaces supporting TCP/IP as well as ongoing support for Novell NetWare. All ASCII terminals connected to the mainframe systems were replaced with Sun workstations or PCs.

Although the graphical user interface was considered a valuable feature of the move toward workstations and PCs, initially, the text-based applications running on the mainframe were simply converted to run within the Windows environment. Over time, the customer plans to migrate its applications to take full advantage of windowing capabilities inherent in a GUI.

Implementation Issues

⇒ *Orders must be processed in real time, immediately acknowledging the caller's orders and their anticipated ship dates.*

⇒ *Manufacturing analysis needs to be more accurate, with greater access to demand estimates.*

⇒ *Integrated decision support activities must be based on manufacturing data, order run rates and other variables.*

⇒ *Higher system availability, with less dependency on mainframes, was required.*

⇒ *A migration tool and platform availability based on open systems were essential.*

⇒ *Call center agents made more productive with calling party information on screen at the time of call and queuing for next available agent for all inbound calls.*

The overall rightsizing plan reflected a need to migrate mainframe applications to a distributed system over time. In the interim, connectivity to the mainframes and minicomputers was a requirement. At the same time, the company's many PC and Macintosh systems, after long languishing as isolated systems, needed to be integrated into an overall MIS environment where users had unencumbered access to the mainframe's data and applications and where PC-based tools could provide front-end processing of financial, manufacturing and order processing data.

A key consideration in developing an infrastructure for the company entailed implementing a corporate network. TCP/IP and NetWare IPX/SPX were selected for the backbone. Token ring and DECnet were used in some areas (AS/400 to RS/6000, DEC VAX to VAXstation) for several reasons, including the existing heavy investment in DEC equipment and unacceptable performance of IBM's TCP/IP product for the AS/400. PCs were incorporated into the network using ICC's TCP/IP product while Macintosh systems were connected through TinyTerm TCP/IP. The Unisys mainframes employed a Unisys product known as a Host LAN Controller, which provided an Ethernet interface supporting TCP/IP.

Benefit Analysis

The company was capable of handling more call efficiently with fewer agents. The next stage of implementation will include IVR's (interactive voice response) systems to provide non-agent access to order status and ship dates. There was a gradual reduction in maintenance costs as mainframe systems were retired. An increased quality of overall information management, resulting in better decision making, more accurate determination of product demand, less inventory, greater manufacturing efficiency and increased call center agent productivity resulted. Improved systems management was realized through the use of open systems, including standardization of networking protocols, application access, client-server application development and window systems.

SOLUTION #2 TELECOMMUNICATIONS COMPANY REPAIR SERVICE CALL CENTER

Telecommunications is among the fastest growing industries today. With service becoming one of the few differentiators, telecommunications providers are increasingly focusing on delivering the highest possible availability and, where problems occur, correcting them as quickly as possible. The trade-off for most telecommunications providers is in meeting customer service requirements while maintaining efficiency in the repair centers.

This solution describes a telecommunications company's repair service call center. It is based upon the implementation of an actual customer management solution for a large telecommunications provider in a large western state.

This telecommunications company call center includes or directly interacts with several operations, including front-line repair, dispatch, customer service operators desk and remote test operations. The requirements for improving its call center operations were extensive.

Business Issues

⇒ *Poor orders and order resolution tracking.*

⇒ *Unacceptable call abandonment and overflow.*

⇒ *Poor customer history recording and retrieval.*

⇒ *Lack of personalized service.*

⇒ *Ineffective problem screening.*

⇒ *System too difficult to navigate (too many screens and options).*

⇒ *Inefficient technician dispatch (wrong skill set).*

⇒ *Ticket status difficult or impossible to determine.*

⇒ *Sales staff unable to navigate or query the system efficiently.*

⇒ *Inadequate sales history.*

⇒ *No proactive outbound telemarketing in place.*

⇒ *Technicians lack remote access to system when servicing problems.*

⇒ *No geotracking capability (inefficient dispatch of technicians).*

Traditional repair call centers are staffed by minimally trained agents who answer customers' calls and prepare trouble reports based on what the customer indicates. The agents generally will not know if that customer has reported the same problem earlier or if this the fifth problem in as many months. Customers dislike having to repeat their story and they want a personalized acknowledg-

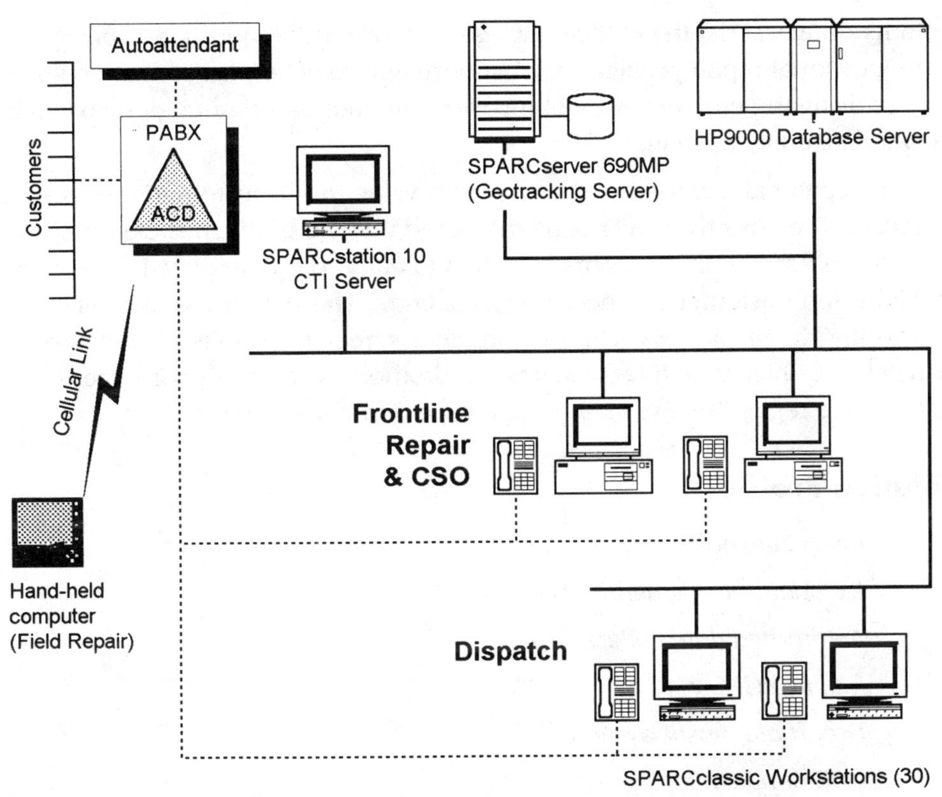

FIGURE 14.2 The Repair Services Call Center Solution uses geotracking and hand-held computers for field personnel.

ment of their difficulty. Unfortunately, the systems used by the agent only contribute to the depersonalized service. The underlying computer based or manual systems were not designed to enhance customer relations; they were designed for efficiency or to replace the paper and pencil process.

Because the agents are generally only minimally trained on the system, they cannot interpret the information to make the appropriate diagnoses. They merely give the customer a standard "repair by" date and time and pass the trouble report to the maintenance center where the reports are screened, circuit tests are performed and the unresolved complaints are passed to an outside technician, who is dispatched to fix the problem. Because the existing systems do not assist the call center agent in asking the questions appropriate to an individual repair problem, vital information may be missed that could assist in solving the problem without involving another work group or dispatching a highly skilled technician.

The repair call center generally adds no value to the customer's service experience. The customer will not be reassured that the problem is clear and will be fixed on time. If the customer has had to call more than once for the same problem, the customer's concern is amplified. The call center actually adds cost to the repair process when incomplete screening results in unnecessary dispatches of highly paid technicians. Customers who have poor experiences with service repair are less likely to purchase additional services.

Solution Profile

⇒ *Integrated voice response unit.*

⇒ *Exceptional call identification.*

⇒ *Automated data collection.*

⇒ *Intuitive GUI.*

⇒ *Expert diagnostic system.*

⇒ *Outbound calling.*

The solution allows customers with routine problems to report them through an intelligent voice response unit. They enjoy the same or better service that they would from a traditional call center, without being placed on hold. The voice response system accesses the customers records and acknowledges the customer by name using text-to-speech, assuring the customer that the system is working. The VRU and the CTI puts the customer in control by allowing them to create and forward trouble report.

Based on the customers' response to the VRU's questions and database inquiries, customers with unusual or repeated problems are transferred to an agent along with their records, data collected by the VRU, repair history, test results and other pertinent data. This information is collected by screen scraper programs, which extract specific fields from existing, transaction-oriented screens.

A graphical user interface displays the information for the agent, with an appropriate greeting, so the agent can give a personalized response acknowledging what the customer has already told the VRU. An expert system helps perform diagnostics by prompting the agent to ask additional questions and providing a script that enables the agent to act as a technician before sending the trouble report to the maintenance center for review. The expert system performs appropriate line tests and makes a diagnosis based on all available information. A trouble ticket tracking and dispatching is automatically generated and sent to the appropriate systems and work groups.

An outbound calling procedure contacts the customer after service is restored to ensure it is working satisfactorily. If not, the call is transferred to an agent for handling. This allows immediate action for problems not properly resolved, eliminating an additional call from the customer.

Existing T-1 facilities, routers and bridges were sufficient for the call center. An existing HP9000 was considered sufficient for the database server platform. New 486 PCs running Windows were ordered for front-line repair and CSO.

Implementation Issues

⇒ Sizing agent workstations and server configurations.

⇒ *Access to existing mainframe database*

⇒ *Development of expert system database*

⇒ *Selection of development tools for dispatch, Front-line repair and CSO front-ends*

Sizing the Sun systems in the solution was very straightforward. The desktop system (dispatch) needed to support the Voicetek CT software, several applications employing graphical user interfaces and the X Window System. The geocoding system, based on Sun technology, had to support the large geographic database for a major portion of the state. Because no additional data was typically saved on this system, there were no significant growth requirements.

As the data for customer records resided in an ORACLE database on an IBM mainframe, ORACLE was selected to run on the HP9000. This allowed an easy transition from the IBM mainframe and facilitated the momentum of the existing database to the HP system. The front end applications development tools have interfaces to ORACLE, UNIX and MS-DOS systems. The GainMomentum software used interfaces well with existing systems and has a screen building and high-end graphics capability. The Remedy trouble tracking package runs on UNIX and Windows systems and has an open interface which allows easy integration with telephony equipment.

Benefit Analysis

⇒ *Implementation of a new system in 14 months with total pay back in 6 - 10 months.*

⇒ *Better telephony integration.*

⇒ *Integrated customer history recording and retrieval.*

⇒ *Reduction in unnecessary technical dispatch.*

⇒ *Installation of comprehensive telemarketing.*

⇒ *New geotracking capabilities for more efficient dispatch.*

The new system, implemented in phases, added geotracking, service history, integrated telephony, more effective call forwarding, service representative performance analysis and a host of other capabilities to the company's call center.

Pay back was possible within ten months, assuming no increase in outbound telemarketing, and in only six months assuming a modest gain in telemarketing. Many of the benefits were less obvious. Employee morale saw significant gains, despite some staff reduction facilitated by the elimination of redundant responsibilities. Overall call center efficiency increased by 30 percent in the first four months.

The addition of geotracking and the ability to maintain service history raised customer satisfaction levels -- first by providing a better skill match with the problem and second, by lowering the customers' frustration with problem resolution.

SOLUTION
#3
BROKERAGE FIRM

Business Issues

⇒ *Increase the firm's capacity to offer longer hours and more service*

⇒ *Promote and support higher transaction volumes*

⇒ *Identify and retain priority customers*

⇒ *Control cost of personalized service*

⇒ *Provide integrated information and transactional services*

⇒ *Supply timely information regarding existing portfolio and potential purchases*

⇒ *Cross-sell and promote development of a brokerage relationship*

Brokerage firms compete either on the basis of price or service. Full-service brokerage firms have made significant investment in training their people to understand a client's financial goals and provide investment advice to support these goals. Discount brokerages assume that clients are after the best deal in commissions and an acceptable minimal level of service. Both groups benefit from providing high-speed, efficient execution of stock trades and other securities transactions.

To retain their best customers, the firms are offering an expanded list of products and services. They now are composites of bank, insurance company, and pension fund manager, as well as traditional stockbroker. Financial services customer service call centers (CSCC) help clients access their expanding set of services 24 hours per day.

Marketing managers determine the proper level of service for various categories of customer. They may use the CSCC to initiate a call to tell a customer when an individual stock has moved sufficiently to warrant a buy or sell order. During idle times, agents may be prompted to perform outbound telemarketing duties.

Solution Profile

- ⇒ *Integrated voice response*
- ⇒ *CTI-based coordinated call transfer/call diversion*
- ⇒ *ANI screening*
- ⇒ *DNIS screening*
- ⇒ *Distributed RDBMS, access to customer files, transaction files and product information*
- ⇒ *Expert systems to script interaction with customers*
- ⇒ *Facsimile response*
- ⇒ *Distributed image processing*

A voice response unit may answer the incoming call and prompt callers to identify themselves, enter a personalized ID number (PIN) or use speaker verification technology for validation. The VRU then prompts callers to enter an extension number or menu choice either by using the touch-tone keypad or through automatic speech recognition. Calls are transferred according to the caller's instructions. Computer telephony integration software and communications links ensure that information about the caller is delivered along with the voice circuit. An alternative is to use automatic number identification (ANI) to identify callers and perform initial routing of calls according to the brokerage's status of a caller. For example, a high net worth individual might be transferred directly to a broker while people who have exceeded their margin limit might go straight to a collections officer.

Dialed number identification service (DNIS) can also provide initial routing. DNIS allows individual numbers to correspond with specific extensions, departments or work groups.

Broker/agent workstations use a GUI-based based display to furnish a number of sources to promote agent productivity. In conjunction with database and network servers, information from a number of sources may be displayed simultaneously in support of a query response or transaction. Expert system software can be implemented to lead agents through scripted interactions with customers and provide suggestions for products or services.

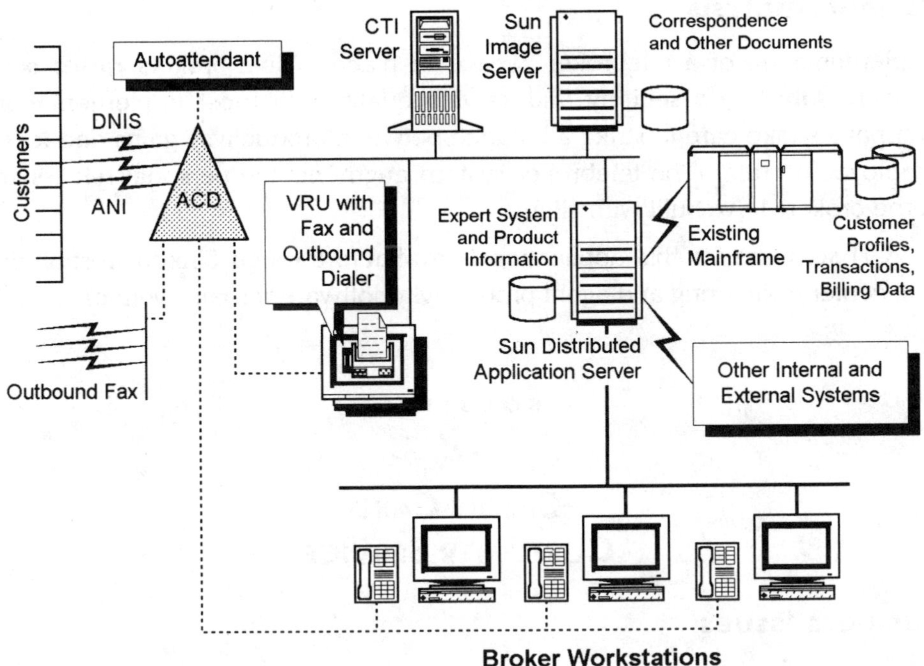

FIGURE 14.3 **Brokerage Solution**

Transaction or order confirmations may be transmitted by facsimile responses if required, although facsimile jeopardizes user confidentiality. Image processing can display previous statements in support of reconciliation or resolution of disputes. Workload is dynamically allocated among agents based on such factors as time of day.

Implementation Issues

⇒ *Agent morale*

⇒ *Managing heterogeneous switching systems*

⇒ *CTI interfaces*

⇒ *Fraud prevention*

⇒ *Selection of image processing vendors*

Benefit Analysis

Implementation of a telebroker service in place of live brokers raises some concerns about job security and service quality. In most implementations, companies take care to strike a balance between productivity gains and loss of employee morale. The telebroker system augments the personal relationship some brokers have built with clients.

CTI software for this application is available through Linkon Corporation. Call center monitoring and agent productivity software remains vendor specific.

SOLUTION #4 CREDIT CARD CUSTOMER SERVICE

Business Issues

⇒ *Eliminate the need to call several numbers for service*

⇒ *Reduce paperwork*

⇒ *Provide 24 hour service*

⇒ *Supply service around the clock in the users native language*

As the number of firms providing credit cards increase, the companies are being driven to consolidate and differentiate their cards from all others. Many companies are choosing to do so by serving their customers better which helps with customer retention.

Customer's don't want to call several numbers to solve a problem or get information. They want one point of contact with the company for all of their needs. Once they have reached the company, customers don't want to be forced to fill out endless paperwork to get something accomplished. Too many forms and letters also bog down service agents who are trying to help.

Limiting customer contact to the hours of 8 AM to 5 PM is too restrictive for many people's lifestyles. Customers also want support from their credit card

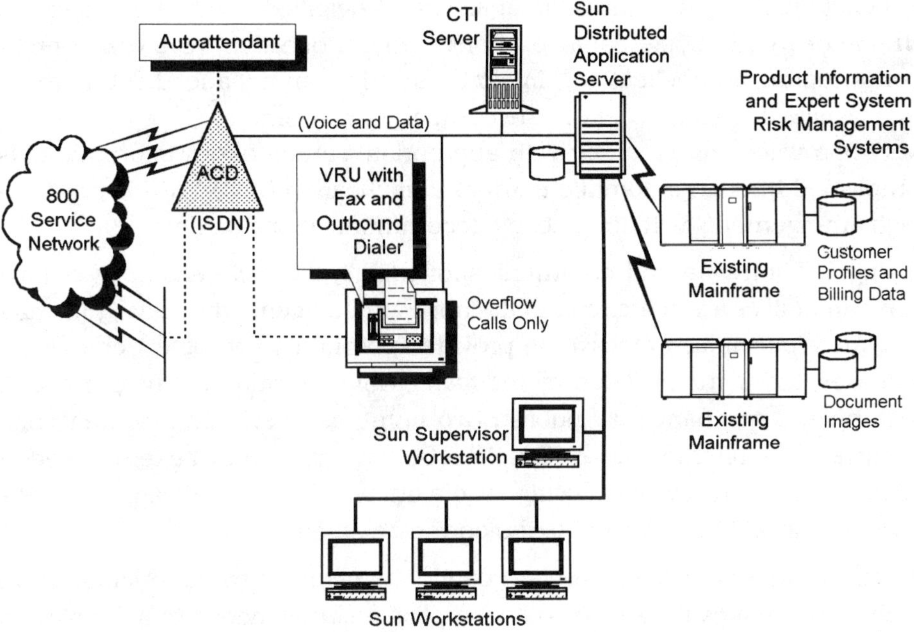

FIGURE 14.4 **Credit Card Customer Services Solution**

company, wherever they may be traveling. Twenty-four hour worldwide assistance is critical.

Solution Profile

⇒ *Enhanced 800 number access to multiple sites*

⇒ *Automatic number identification (ANI)*

⇒ *Dialed number identification service (DNIS)*

⇒ *Transfer of voice and screens using CT integration with ISDN*

⇒ *Distributed relational database management system (RDBMS)*

⇒ *Expert systems to prompt representatives through qualification & application processes*

⇒ *Facsimile response*

A key benefit to the consumer is the ability of a single, toll-free telephone call to reach one or more people capable of answering a question, resolving a problem or accepting an application. In North America, an enhanced 800 number provides single, free-call access to the company's call center. Once connected, delivery of a voice circuit, along with appropriate screen based information, is accomplished through automatic number identification (ANI). ANI is provided through arrangements with the public telecommunications carrier.

Using CTI software and communications links, the call center's switching system can deliver a voice circuit to the appropriate agent while giving instructions to an applications processor to provide appropriate computer screens. In some cases, the number dialed by the caller will determine the proper agent. For example, a company may publish two numbers in an advertisement; one for English and a different one for Spanish. Both numbers may be terminated at the same switching system; using a dialed number identification service (DNIS), the call will be delivered to the correct recipient.

In cases where a caller's problem cannot be resolved by the original agent CT technology makes it possible to accomplish a simultaneous transfer of both the voice circuit and the screen based information to perhaps a supervisor for special handling. Included in this transfer would be any changes the prior agent made in the record prior to the transfer. Administrative information such as call length, number of transfers and type of request is also transferred from station to station.

Customer service personnel must have access to customer profiles, transaction histories, billing histories and other account information to resolve service issues. Data may reside on a number of systems, some remote and some local. A distributed relational database management system (RDBMS) provides seamless access to required information regardless of the location or type of host system. Adherence to standards such as structured query language (SQL) is critical in this regard.

Many routine sessions can be scripted, meaning that an expert system software can literally put words in the mouth of agents. Such scripts could include prompting callers through an application process or taking an opportunity to cross-sell other financial or travel related services.

In the case where the customers require printed material, such as copies of receipts, a facsimile response sub-system can serve as a remote printer.

Implementation Issues

⇒ *Choosing long-distance and international carriers. to support virtual private network.*

⇒ *Selecting RDBMS package to support multiple sites and multiple hosts.*

⇒ *Deciding on distribution of centralized data versus synchronization of distributed data.*

⇒ *Developing a GUI based agent interface.*

Successful implementation relies on international carriers to create a seamless network among geographically dispersed facilities. Adequate network intelligence is required to make repeated transfers without tying up too many trunks on a switching system. From a distributed computing point of view, the GUI based workstation displays information, options and scripts as the product of a single database, masking agents from the complexity of the networking and distributed computing.

SOLUTION #5 EXPRESS PARCEL SERVICE

Business Issues

⇒ *Rapidly rising operating costs.*

⇒ *Inflexible, centralized host system.*

⇒ *Maintaining customer satisfaction.*

⇒ *Constant competitive pressure.*

The operating costs of express delivery services are rising rapidly with the impacts of inflation. Wages and benefits are subject to constant cycles of negotiation, which generally results in higher costs for the employer. The price of commercial vehicles has risen dramatically over the past ten years, along with the cost of their maintenance. And a host of other factors from taxes to communications costs are placing day-to-day pressure on profit margins. Unfortunately, many express parcel services also have expensive and inflexible mainframe support systems. This places them in the double jeopardy of rising costs and the inability to respond with more innovative and efficient methods of service delivery.

Naturally, the customer is only concerned with the delivery of quality service at the lowest possible price. If the service provider is too aggressive in curtailing its cost, which causes service to suffer, it risks the loss of existing customers. On the other hand, if the provider reacts to increasing costs by raising its rates, it is equally probable that existing customers will find other means to get their parcels from one point to another. Finally, anything less than a first-class reputation for service makes it difficult to attract new customers, which is essential to expanding the business and leveraging fixed costs.

Some progressive delivery services have, however, addressed this situation head on, treating it as an opportunity to justify a radical change in their way of doing business. These companies are applying state-of-the-art computing and communications technology to simultaneously reduce costs, improve basic service, and supply innovative new services to their clients. Of course, such improvements also increase competitive pressure in what is already an openly competitive environment.

Solution Profile

⇒ *Reduce paperwork to an absolute minimum.*

⇒ *Provide computer-assisted dispatch and tracking.*

⇒ *Improve information availability for customers.*

⇒ *Develop systems to improve management decision making.*

Great strides can be made to reduce the flow of paper through using integrated on-line entry and inquiry for everything from pickup orders to bills of lading to credit collection files. Whether the source of information required is existing host systems or new applications and databases, equipping call center em-

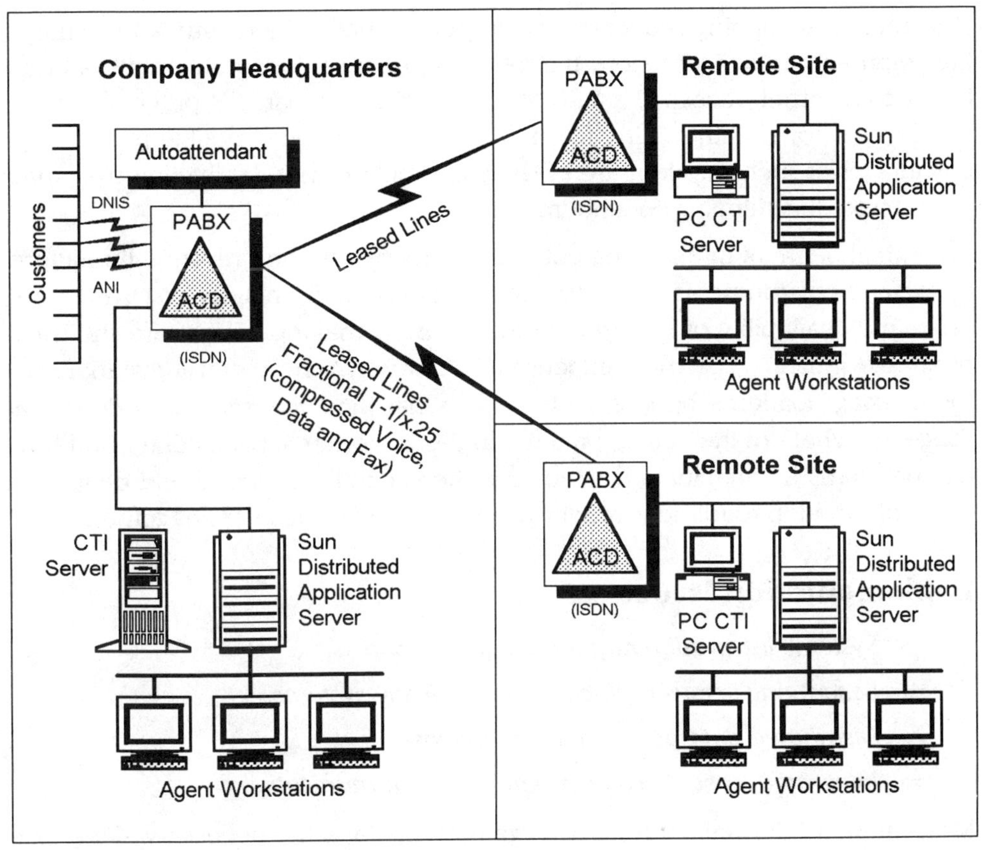

FIGURE 14.5 **The Express Parcel Service Solution**

ployees with high-end PCs or workstations and a common, graphical user interface (GUI) is a critical factor. Call center agents with appropriate information access can take a telephone order for a parcel pickup in as little as 25 percent of the time required by manual or dumb terminal methods, and that is further reduced when established customers are allowed to enter their own pickup orders directly into the service provider's systems.

On-line entry capabilities, computer assisted dispatch, and automated parcel tracking are a logical combination. With available technology, an automated process can be established using pagers to alert drivers of a pickup order, assign an electronic tracking identifier and provide it to the customer, determine the shipping sequence and resources, notify the receiving terminal of the shipment, track and record the parcel's progress during shipment, and allow the customer access to a VRU for a report on the parcel's progress. All of this can be done, with minimal human intervention and very little paper. The advantages of such a system are evident, particularly in the reduction of human error across the shipping continuum.

A final element in attacking both the cost structure and revenue limitations of existing environments is the use of monitoring and analysis systems to improve the availability of information for decision making. What are the most profitable routes? The most expensive? And why? How can individual customer usage patterns be segmented and what incentive programs does that suggest? What are the calling patterns in the customer service center and how can problems be anticipated? Many questions can be analyzed and dealt with if the information required is readily available and usable by management.

Implementation Issues

⇒ *Systems integration requires skilled resources.*

⇒ *Leveraging versus replacement of existing systems.*

⇒ *Distributed data access and management.*

⇒ *Integrated, cost-effective, multimedia communications.*

Most internal information systems organizations lack the necessary skills, attitude, and experience to fulfill the role of systems integrator in a re-systemization effort of this nature. For example, the thought process inherent in managing an existing, proprietary computer environment do not ordinarily lend themselves to the innovative, open systems thinking total customer service requires. Those thought processes include deep-seated organizational and cultural issues, which affect the decision to leverage or scrap an embedded base of expensive applications and computers. All facets of re-systemization need independent expertise, including specialists to review existing business processes and determine what traditional functions need to be automated and what can be eliminated.

Of the technical issues to be resolved, distributed data planning and the design of ubiquitous, cost-effective communications are among the most important. In a transportation organization with offices throughout the nation and around the world, universal access to timely and complete information will be the basis for reducing operating costs and improved customer service. Proven capabilities for implementing and managing fully distributed data are essential to success. Similarly, the skills to design and deliver an integrated, multimedia communications network that will be responsive to both customer and the company are required. In this solution we propose the elimination of existing mainframes and the implementation of an ISDN network.

Benefit Analysis

Although re-systemization efforts of this nature can be expensive, the efficiencies gained in key operations can easily justify the business case for making the investment. These efficiencies come from a combination of improved efficiency in key operations, reduced operating expense, and significant reductions in internal communications costs.

The most important benefit of all, however, is sustaining a competitive posture in a very dynamic industry. A reputation for excellent service can be enhanced by the astute implementation of technology. Moreover, creating differentiation over less progressive competitors, while saving operating costs, helps maintain existing customers and expand the business.

SOLUTION #6 INSURANCE CUSTOMER MANAGEMENT

Business Issues

- ⇒ *Provide quick quotes for prospects.*
- ⇒ *Resolve claims quickly.*
- ⇒ *Represent a wide range of products and services.*
- ⇒ *Offer 24 hour access.*
- ⇒ *Increase claims adjusters productivity.*

Insurance companies were among the first to deploy call center technology and customer management solutions. Property and casualty insurance carriers rely heavily on direct marketing programs to attract new customers and renew the existing base. Advertising campaigns often refer prospective clients to a single number for pricing, coverage or the location of the nearest agent.

Life insurance carriers must market investments in much the same way as an investment broker. Telephone based representatives require quick access to computer resources that generate annuity tables and furnish swift answers regarding coverage and premiums.

Activities after the sale are even more crucial to customer satisfaction and ultimate retention. Policyholders use the telephone as a substitute for trips to an agency's office. Customer management solutions can expedite these interactions by screening incoming calls, routing them to the proper agent or information resource and ensuring that all necessary information is available to the agent or the customer.

A recent addition to insurance claim processing systems is the advent of image processing and document distribution. If a client makes multiple calls regarding a single claim or policy, it is important for an adjuster or agent to be able to see a history of that claim, including a log of conversations between the client and the firm, and representations of documents or photos that accompanied the claim.

Solution Profile

⇒ *Integrated voice processing; autoattendant, interactive voice response, and speech recognition.*

⇒ *Intelligent call routing and agent supervision.*

⇒ *ANI screening.*

⇒ *DNIS screening.*

⇒ *CTI based coordinated call transfer/call diversion.*

⇒ *Distributed RDBMS, access to client files and other information.*

⇒ *Expert systems to script interaction with customers.*

⇒ *Facsimile response.*

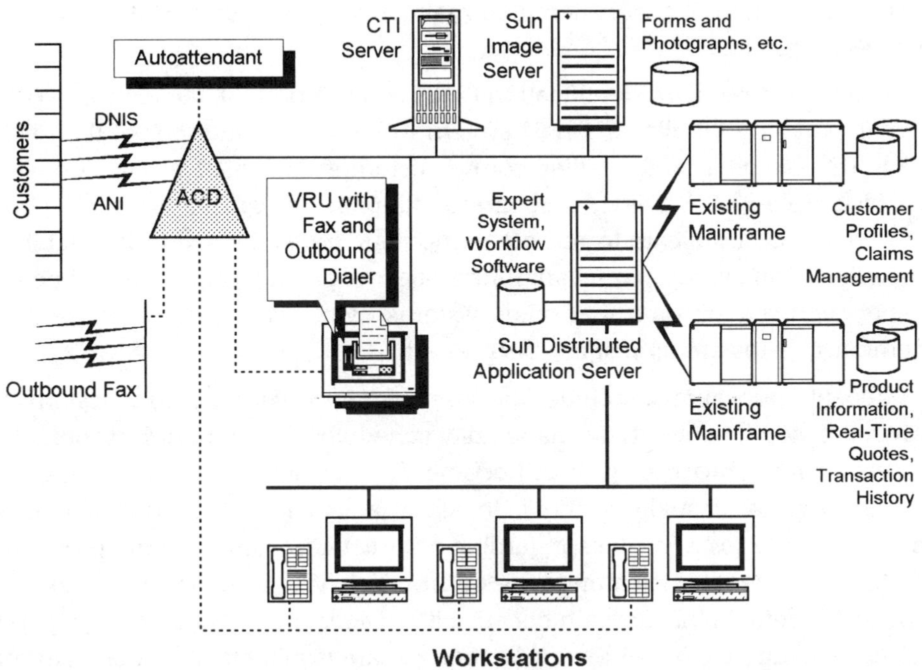

FIGURE 14.6 **Insurance Customer Management Solution**

⇒ *Distributed image processing.*

⇒ *Workflow software.*

The major objective of a high volume, telephone based customer management solution is to resolve a customer's query or problem in the minimum amount of time. This could begin with an autoattendant that greets the caller and prompts them to indicate the purpose of their call, or the extension number of the individual with whom they desire to speak. In some cases, speech recognition can be used to allow callers to simply say what they want. An alternative is to use dialed number identification service (DNIS) which allows the host system to route calls through an ACD based on the number originally dialed by the caller.

In either case, the objective is to reduce the amount of time it takes a caller to reach the person or resource that can solve the problem. In some instances the right person might be across the country, and the ACD is often put under the control of an external host system that maintains a database of station locations to expedite routing.

Automated number identification (ANI) is an important service for expediting call handling. It allows a host system to key off the number of the calling line to retrieve data on the caller from a customer database, transaction file or call history file. A CTI link, a separate talk-path through which a telephone switch can communicate to remote computers, transmits ANI to the database host which can serve up relevant information to agent workstations. Because agents often rely on data that other systems own, distributed database management systems are almost always present.

Recent innovations include the use of facsimile response systems to transmit copies of such things as annuity schedules or payment histories. Distributed image processing has become increasingly important, especially among property casualty carriers, to allow adjusters to call up all relevant documents, photos and notes regarding a particular claim. Claims processing resolution is often complex and time consuming. Workflow software has been especially useful in this domain because it provides an audit trail that all parties claims resolution can retrieve. Resources are available 24 hours per day through interactive voice response capabilities.

Implementation Issues

⇒ *System design and integration.*

⇒ *Software and hardware interoperability.*

⇒ *Security and privacy.*

⇒ *Interdepartmental and work group planning.*

Due to technological awareness sophisticated end users seem to be ahead of the systems integration community. Chief among the problems is determining the level of interoperability required among legacy computer systems and departmental DBMS systems.

Customer services managers must take special care to safeguard the system to prevent unauthorized access to corporate computer or telecommunications resources. Hackers have begun regarding call centers as ideal systems for break-ins. System administrators must install procedures to prevent intrusion or detect and trap intruders when they do break in.

Benefit Analysis

The primary benefit is the delivery of faster service. In the property and casualty department, customer management solutions technology can quickly screen calls and deliver them to the proper live or automated resources. Distributed database management can make certain that call recipients have access to all information required to solve problems. Agents make each call as productive as possible and are free to do more on behalf of their clients.

Agents can be prompted to offer a wider range of service to callers. Customers can take the best advantage of the full range of services offered by the agency. Twenty-four hour access provides better service to the company's clients. All these improvements elevate the agency's image with customers.

SOLUTION #7 RETAIL BANKING

Business Issues

⇒ *Focus on relationship and account retention.*

⇒ *Growth of span of control as banks consolidate and become regional.*

⇒ *De-emphasis on brick-and-mortar branches.*

⇒ *Differentiation through electronic services.*

⇒ *Unskilled telephone representatives fill the role of tellers and platform officers.*

⇒ *24 hour access to multiple services and information sources.*

⇒ *Special treatment of best customers.*

Retail banks are competing to maintain longer term relationships with their best customers. In response to changes in US banking regulations, regional super banks have grown through acquisition and consolidation with smaller competitors. In efforts to control costs, they are reducing the number of storefronts they operate in favor of automated teller machines an 24 hour telephone based services. Because the portfolio of products and services that commercial banks can offer is homogeneous, the quality and accessibility of customer service has become the key differentiator. Maintaining high levels of customer satisfaction, especially among attractive groups such as high net worth individuals, is a key business objective.

The design and development of robust customer service centers is crucial to the bank's success. They effectively extend the hours during which banking services are offered. They allow for relatively unskilled customer service representatives to perform the roles of experienced platform officers and tellers. They also act as integrated front ends to multiple bank back office systems, especially for acquisitive super banks still in the midst of years-long projects to consolidate back office systems.

Solution Profile

⇒ *Integrated voice response units.*

⇒ *CTI coordinated transfer of calls and screens.*

⇒ *Expert system script of query responses and cross-selling.*

⇒ *Application specific database integration.*

⇒ *Image retrieval.*

Twenty-four hour bank customer service call centers are among the most mature platforms in the customer management solution industry. Retail banks customarily publish a single telephone number to serve all branches in a particular state or region. In some cases, banks may publish different numbers in different departments; they use DNIS to route the calls to the proper call center, individual or voice response unit. ANI may be used to screen customers and expedite the delivery of their calls to specific bankers.

Virtually all cases employ some form of integrated voice response system to reduce the amount of time spent on line before reaching the required resource, either by prompting the customer to identify themselves and enter an extension number or by providing a menu of service or department options.

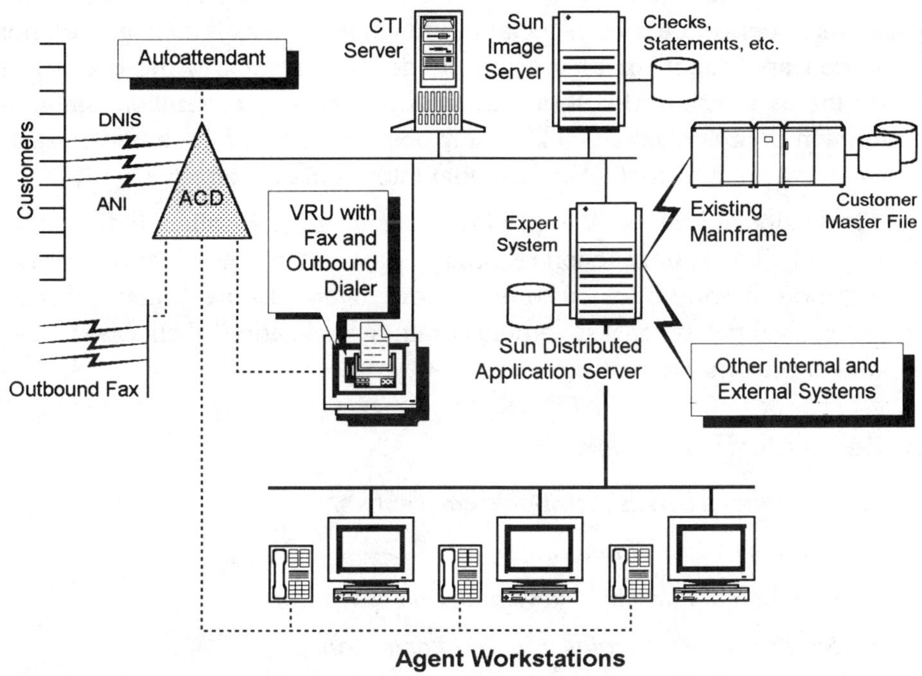

FIGURE 14.7 **Retail Bank Solution**

CTI links are critical for coordinating the delivery of voice circuits with the appropriate screen based information. They establish the communications link and protocols enabling an external host computer to gather the status messages and input from callers, interpret those messages and provide instructions both to the switch and applications servers.

Routine database queries such as requests for account balance and transaction history are handled by voice response units and require no input from service representatives. The VRUs, front ends to database servers, play back responses to queries using concatenated, digitized speech. Newer systems use text-to-speech technology which is less restrictive and easier to use.

At any point, callers may request to speak to a customer service representative. At that point the system will transfer both the voice circuit and relevant screens to the next available agent. Agents may also transfer a call to other personnel to perform a more specialized function such as initiating or completing a loan application or to provide needed expertise. In these cases, the screen that is transferred will display information on the duration, status and progression of an individual call. At any point, callers can be transferred back to their initial point of contact or any other intermediate stop.

It is possible for the bank to use facsimile response if callers request copies of canceled checks, transaction records or past statements. Internally it is becoming more common for customer service agents to use image processing technologies to pull up past statements or images of canceled checks to quickly respond to customer questions.

Implementation Issues

⇒ *Centralized versus distributed call centers.*

⇒ *Caller ID and validation.*

⇒ *Distributed database integration and synchronization.*

⇒ *Security, confidentiality and fraud detection.*

Call center managers constantly need to analyze the best way to accommodate additional traffic. They must also determine if there are special local needs that are best served by local agents.. In most cases, cost dictates the push towards centralization of call centers with WANs to interconnect disparate back office systems. Banks must also exercise caution to avoid fraudulent access and use of information and systems. Identification and validation of callers is the most efficient way of monitoring usage, but there is still exposure to fraud. For this reason banks may have one level of security for routine queries regarding account balance and transaction history and more stringent rules governing entitlement to order funds transfer and other transactional services. This additional security requirement is now met using special codes and speaker verification technology.

SOLUTION #8 HIGHWAY MANAGEMENT

Business Issues

⇒ *Traffic growth is outstripping highway capacity.*

⇒ *Average speeds are down dramatically.*

⇒ *Investment capital for new roads is insufficient.*

⇒ *Innovative improvements are a necessity.*

Few urban highways are not feeling the crunch of traffic at every rush hour. With an ever increasing population base and a more mobile society, the number of cars and trucks in and around the world's major cities is simply over-

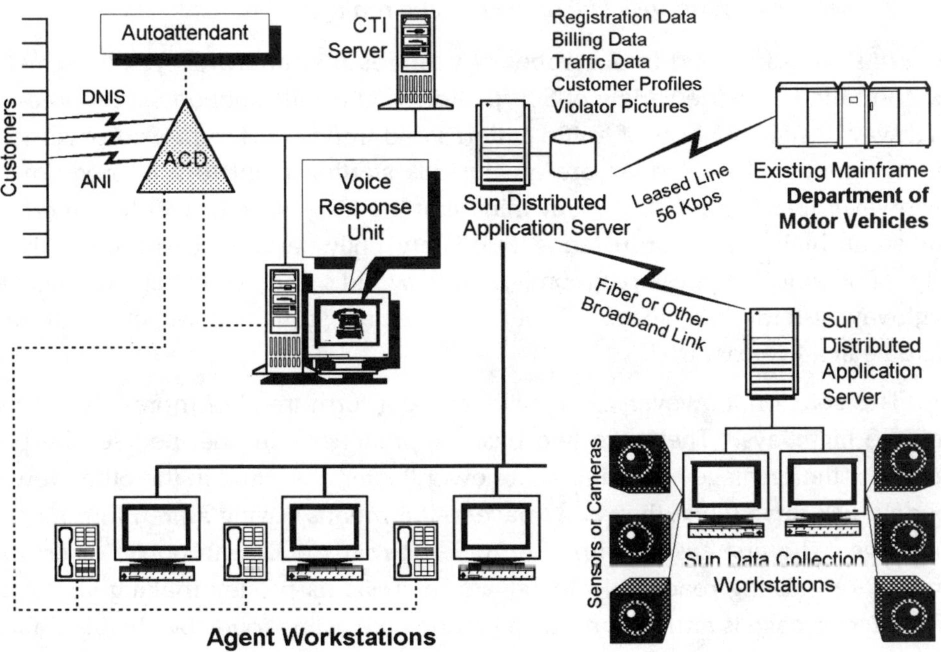

FIGURE 14.8 Highway Management Solution

whelming the highway systems. Once it might have been practical to simply build more roads, but where is the money to come from in today's economy, and who wants a new freeway through their neighborhood? There have been some bold moves to finance private toll roads in highly congested areas and that shows promise, but it is also a limited response and likely to stay that way. Some additional innovative ideas are needed to help gain the greatest value from existing roadways. One such idea, from the Pacific Northwest, is to tax the people who are contributing to the problem, namely commuters who drive at peak hours. The trick is to do that without toll booths, which only make congestion worse.

Solution Profile

⇒ *Better highway management is a starting point.*

⇒ *Smart highways offer some promise.*

⇒ *New toll roads can be part of the answer.*

⇒ *Selective taxation is still an increasing reality.*

Sun has been involved in a number of ventures that improve overall highway performance. A wide variety of Sun platforms currently support state-of-the-art highway design, analysis of traffic patterns and trends plus improved pavement management. These areas are an obvious starting point to any program of highway improvement. But many think that the long term answer lies in creating smart highways. For instance, Sun is currently partnering with the University of Illinois in developing one of the world's largest intelligent vehicle highway systems in Chicago. About 500 specially equipped vehicles will participate in a five year trial.

This solution, however, deals with the near term trend of more tolls for using the highways. There are two basic approaches. In one, people will pay more for the privilege of driving on a newer, faster road, and in the other, fewer people will drive when they don't have to if it means paying a premium for the privilege. The first case is aimed at new toll roads built with public or private money or existing roads sold to private interests as money making ventures. The second case is aimed at reducing traffic on existing roads by charging a toll for every user of the road during rush hour. Both of these cases, however, use the same technology to incrementally charge the user. All that is needed is instant, reliable vehicle identification.

Implementation Issues

⇒ *Two identification technologies are available.*

⇒ *Users need an easy way to register and pay.*

⇒ *Enforcement mechanisms must be adequate.*

In both Europe and the United States, some toll roads already use an optical scanning system that reads the bar code sticker on the windshield of a vehicle as it passes under the overhead sensor. If the sticker is missing or unreadable, a photograph or videotape record is taken of the vehicle after it passes the sensor point. This is an inexpensive system for both the user and the toll collector, but anyone who has used a supermarket with scanners knows it isn't foolproof. A more precise technology involves a smart card, about the size of a thick credit card, that contains an antenna, microchip and a lithium battery. Its advantage is superb accuracy, but the cost is about $30 US, which is relatively expensive from the consumer's perspective.

The challenge with either system is to make the registration and renewal process as painless as possible. This is done with IVR systems and on-line databases. An added advantage to this approach would be the ability to call 24 hours per day to check your billing status. Once registered, the authentication device could be mailed or even picked up at a variety of locations. For those people who attempt to avoid the tolls, an enforcement mechanism must be in place and rigidly employed. Tourists would also have to be included in the system through vehicle rental agencies and by way of registration points outside the toll zones of the affected highways. Debit cards could be issued in these cases with refunds for the unused portions.

Benefit Analysis

The principal economic benefit from such a system is the targeted taxation of the appropriate drivers, in exchange for the privilege of using a valuable and limited resource during high demand periods or as an alternative to more time consuming routes. Systems for automated toll collection have been working in

Europe for years; they operate in Denver, Colorado today and are currently being planned for suburban Los Angeles and the Washington, DC, area. They are economically sound, compared to other toll collection methods, particularly when set up on a debit system using prepaid accounts.

A considerable side benefit is the precise monitoring of traffic, by type that passes through a given point. This sort of analysis based on reliable statistics is an important element in better highway planning and traffic management. Finally, many toll roads in existence today cause terrible safety and congestion problems by the very nature of their antiquated collection systems. Even to slow down and throw coins in a basket contributes to erratic traffic.

SOLUTION #9 MUNICIPAL PARKING ENFORCEMENT

Business Issues

- ⇒ *Inefficient Parking enforcement wastes tickets.*
- ⇒ *Cities strapped for cash are sitting on unpaid tickets.*
- ⇒ *Repeat offenders flaunt the law and park anywhere.*
- ⇒ *Law abiding citizens are constantly inconvenienced.*

Anyone who has received a parking ticket is familiar with the cumbersome nature of nearly all municipal parking enforcement systems. The process generally starts with a violator receiving one page of a multi-page form. This form classifies the infraction and demands payment by a specific date. Unfortunately we all know what these forms look like. If the recipient wishes to contest the citation, he or she must schedule and attend a traffic court hearing. Even when the fine is paid, the person who wrote the citation is the only one of many tax supported employees who will handle every ticket during this complex processing cycle. The general lack of efficiency in handling large volumes of citations can directly offset both their revenue and enforcement value.

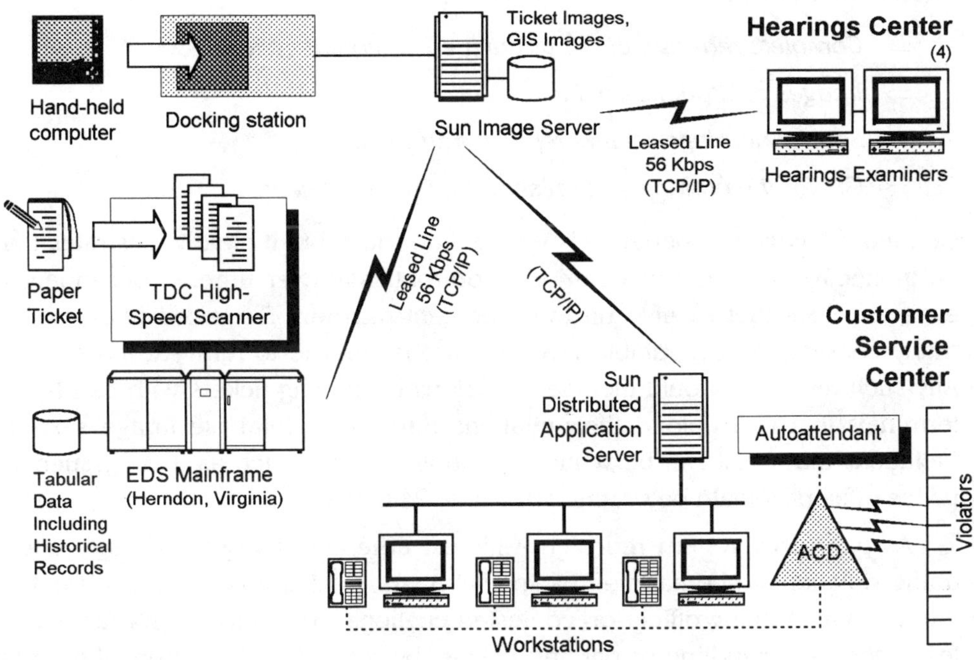

FIGURE 14.9 Parking Enforcement Solution - Chicago Example

The City of Chicago, Illinois, offers an extreme example of the declining value in enforcement. By the end of the 1980's, Chicago was adding 11,000 paper tickets a day to a backlog of 17 million unpaid citations. Only 10 percent of the violators voluntarily paid their fines and it took two years to get a court date. In a city needing cash, more than $420 million in potential revenues was uncollected.

A predictable result of such poor enforcement was a general disregard for parking restrictions and meters. A significant segment of the population simply ignored the law and parked wherever they liked for free. This created considerable hardship and hazards for pedestrians, busses, street level businesses and the general flow of personal and commercial traffic.

Solution Profile

⇒ *Complete redesign of the parking enforcement system.*

⇒ *Images of citations on line within 24 hours.*

⇒ *Special hearings centers replace traffic courts.*

⇒ *GIS for the parking system serves a variety of needs.*

Granted, Chicago's experience is worse than most, but it clearly illustrates that an ineffective system can deteriorate dramatically over time. Consequently, even for cities that receive up to 65 percent payment for levied fines, which many consider an acceptable level, it's probably time to reinvent the system. Any such redesign should include provisions for issuing tickets with hand-held terminals that can upload their citations directly to database image servers. Violations that must still be written on printed forms, such as those issued by police officers, should be scanned in within 24 hours.

As part of its system redesign, Chicago amended its ordinances to allow traffic violations to be handled by special hearings examiners, without the testimony of the issuing officer on contested citations. This allowed for faster and less expensive handling of parking tickets, by removing them from the badly overloaded traffic court system. The new hearings centers are located in selected areas of the city and people contesting one or more tickets can present their appeal at any convenient time during the assigned week. It is interesting to note that most never reach the hearing stage, however, because virtually any question can be answered by a telephone call to the system's customer service center.

Another key feature of any revamped system should be the ability to plot violations on a map of the municipality, using a GIS system. This is very helpful in developing recommendations for improved parking and traffic flow plus making enforcement decisions based on violation patterns. GIS systems for parking enforcement have even been used for resolving zoning issues and to assist the police in finding witnesses during investigations of crimes.

Implementation Issues

⇒ *Daily work activity must be correlated with historical records.*

⇒ *Workstations and mobile units are used to optimize enforcement.*

⇒ *Workstation output is provided to all hearing examiners.*

⇒ *GIS is a dynamic and key part of the overall system.*

⇒ *System development resources will depend on solutions scope.*

For the new system to be effective, all parking enforcement data, including inspection and repair records for meters should be on line. The objective should be to enter all new information on a daily basis, to data image servers, to insure its timely correlation with historic data. Dispatchers in the enforcement division can then use workstations to direct the activity of enforcement officers and towing companies. In large cities, optimum effectiveness is gained when mobile impoundment or enforcement crews are in constant radio contact with the dispatch center.

All available information related to any given citation must be given to the hearings examiners to avoid dismissing citations for unanswered questions. In Chicago, this is accomplished with Sun workstations capable of displaying all required images and text, on demand. This includes the ability to present meter inspection and repair records, citations plotted by date and time in any given area, and the actual image of any citation. Naturally, the plotting of citations is tied to the availability of GIS systems, but they are growing in popularity as elements of new parking systems.

In Figure 14.9 you will notice that an EDS mainframe was used. EDS, a major systems integrator was prime contractor on the Chicago system. It possesses all of the required skills for development of large complex systems involving both central and distributed processing. EDS is also doing all of the mainframe processing of tabular data for the City of Chicago as part of an outsourcing contract.

Benefit Analysis

- ⇒ *600 percent increase in collections in 18 months.*
- ⇒ *Parking meter revenues are up by almost 80 percent.*
- ⇒ *The traffic court is streamlined; police time is better used.*
- ⇒ *The new system costs the city $5 million less annually to operate.*
- ⇒ *Contractor has collected about 10 percent of the ten year backlog*

The benefits are the direct result of Chicago's strategic investment in reinventing its parking enforcement system. While most municipalities will not realize improvements of this magnitude, any benefits achieved should be directly related to the scope of the existing problem and the quality of the investment made. One important aspect of the investment here was the break with tradition by assigning a contractor the rights to collect all past due traffic citations.

SOLUTION #10 VERMONT STATE POLICE

Business Issues

- ⇒ *Inefficient paper based incident reporting system.*
- ⇒ *Staff absorbed in process and short on information.*
- ⇒ *Negative impact on law enforcement support for citizens.*

The Vermont State Police, with field offices in 12 cities, were using a manual incident and crimes data reporting system based on multipage forms. Police officers and their dispatchers were required to complete these forms, which were filed locally, dispatched to State Police Headquarters, and also forwarded to the state's central data processing center for database input. Five or six people would typically handle the forms from the field office to the computer center and as many people would handle the reports generated on their way back to the field. It was not unusual in this environment for an incident report to take 45 days to appear on computer records. For one year, the compiled data weighed 13 tons.

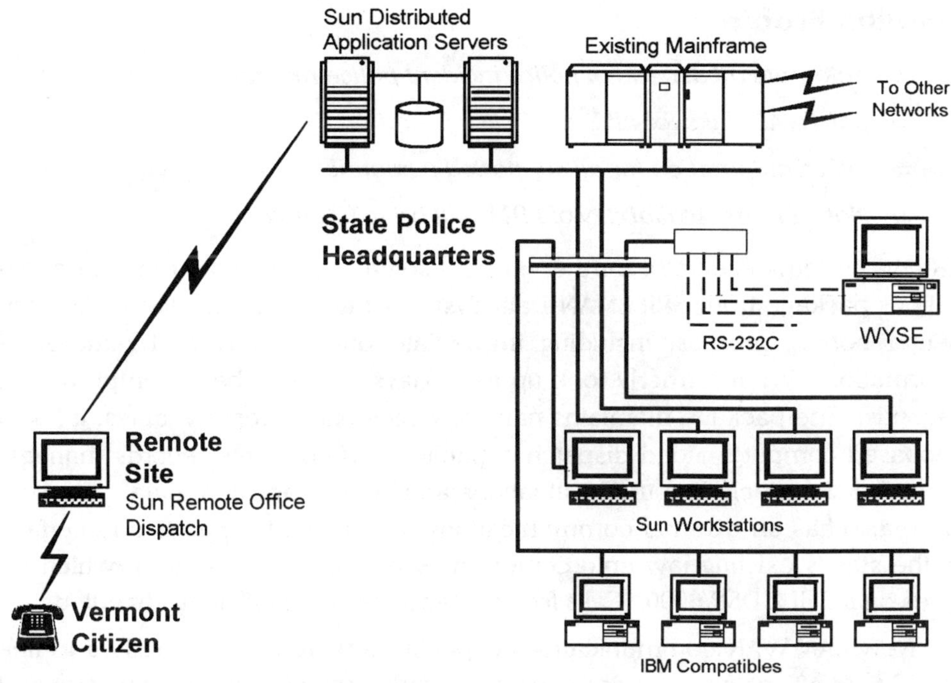

FIGURE **14.10 Vermont State Police Solution**

Obviously, despite the significant staff hours absorbed in paper processing, too little in the way of timely and easily accessible information was available to the police officers in performance of their daily duties. This invariably led to situations where the lack of information had a detrimental effect on the law enforcement support. An equally serious issue was the additional elements of risk that a lack of current and complete information can introduce in the already dangerous work of any police officer. Moreover, the combination of all these factors is likely to have a negative effect on the morale of any group of dedicated professionals.

Unfortunately, this set of issues is all too common in the law enforcement area, which suggests a rather broad scope of opportunity for similar systems. In fact, several other police organizations and affiliated interests in Vermont are already on line with similar, interconnected systems.

Solution Profile

⇒ *Automate all aspects of police incident police reporting.*

⇒ *LAN/WAN links for all State Police Locations.*

⇒ *Office automation for all supported personnel.*

⇒ *Note: Future availability of E911 will bring CTI links.*

ERI, the systems integrator for the Vermont solution, provided a modular applications package from SPILLMAN Data Systems that automates the entire incident reporting process, including immediate on-line entry and retrieval of information. What formerly took up to 45 days can now be accomplished in seconds. The package maintains names, vehicle, and property tables; it has a geobased computer aided dispatch capability and complete records management function, including traffic situations and vehicle accident data. The software also has an external communications module that supports an interface to the state's existing law enforcement message switching system, which resides on a BULL DSP/6000. Calls for assistance are now efficiently handled.

New LAN/WAN communications capabilities were also established to link the 12 field offices on one open access network. Any office may input data and every office can make an inquiry regarding any of the fully integrated cases on record. Further enhancements will even help police determine which types of vehicles to direct to a crime or accident scene by taking into account such factors as location, resource availability and situational variables. Some of those variables could be the additional information available from on-line connections to the Department of Motor Vehicles, the National Crime Information Center and other national databases.

This solution was designed to meet future national reporting requirements, by anticipating the automation of the cumbersome process of monthly statistical updates to the FBI's National Incident Based Reporting Systems. Additionally, more than 400 users of the new systems have received office automation training as part of the deployment package.

Implementation Issues

⇒ *Interface to existing Bull DSP/6000 platform.*

⇒ *Integration of existing Wyse terminals and PCs.*

⇒ *Sustained performance for 24 hour, 365 day service.*

There was a significant systems integration challenge in the design and development of this solution, which now has more than 500 users in the Vermont State Police and affiliated agencies. A similar or even greater challenge will be inherent in other police departments, where a variety of existing mainframe and minicomputer systems will require integration. Sun is better equipped than its competitors to deal with the heterogeneous connectivity required, particularly when working with experienced ISVs and system integrators.

The mission critical nature of the system is significant in such solutions. Dual ported, mirror disk platforms ensure sustained performance. CTI links will be required when a E911 system is installed. A system upgrade to accommodate image processing for fingerprints and mug shots is also planned.

Benefit Analysis

⇒ *Dramatic improvement in incident reporting.*

⇒ *Enhanced information access at all levels.*

⇒ *More rapid, better targeted law enforcement response.*

⇒ *More effective service to the citizens of Vermont.*

The quantum reduction of time and resources required for incident reporting is the most obvious benefit of the new system, but there are other important impacts from this modular approach to automating a public safety call center. For instance, more than 50 percent of Vermont's law enforcement activity is now being handled by one integrated computer aided dispatch/records management system, and that percentage is expected to rise.

Access to information has also been measurably improved from the field office level, where they no longer have to look through the paper files for an incident report not yet entered in the computer, to the on-line links to other state and national databases. That translates into a more rapid and better targeted law enforcement response.

|15|

Sun at Work in the Telephone Companies

Background

To better understand the workings of the telephone service provider business it is essential to review that industry's history. This will shed light on this highly regulated industry and provide insight into their current and future needs and directions. The monopolistic nature of this business has required the government to exercise stringent control of this industry. Its evolution has gradually given way to a more competitive market place.

Prior to the early 1980's, American Telephone and Telegraph, better known as AT&T was the provider of nearly all domestic telephone service. Their stronghold on local and long-distance service raised the ire of many; most notably Bill McGowan, founder of MCI. Individuals and organizations began petitioning the government through litigation to create a more competitive market for these services. The Justice Department in response to this pressure filed suit contending AT&T's control over the local phone service was hindering competition in the long-distance and telephone equipment markets. The result was the 1982, announcement by AT&T of changes to their organization followed by the 1984, modified judgment by Judge Harold Green which created seven Regional Bell Operating Companies (RBOCs) to offer service within geographically defined areas. These RBOCs were separated from long distance provider, AT&T, creating two distinct market segments; local exchange and interexchange service. AT&T retained its long-distance and manufacturing operations but was restricted from providing local telephone service. For the most part the RBOCs continue to enjoy their regional monopolies however that is giving way to change.

Shortly ensuing this 1984, decision AT&T realized competition for their long-distance business from MCI, Sprint and others. This competition has required all interconnect competitors to invest heavily in infrastructure building and upgrade their networks, forced innovation resulting in the introduction of new products and services, required them to spend significant money on advertising and other promotional activities while remaining price competitive. Today AT&T still carries over 70 percent of the long-distance traffic but their competitors continue to gain share. The effect of this quite heated competitive environment is readily witnessed via the massive rate oriented advertising campaigns that bombard businesses and consumers daily.

The government imposed separation of interconnect and local carrier business has required the long-distance carriers to rely on the RBOCs to provide connectivity to the local market for call completion. This requirement has resulted in a substantial revenue stream to the RBOCs from local access charges collected from the long-distance carriers. This fee arrangement did not go unnoticed. It served as a catalyst for the first real competition experienced locally by the RBOCs. Alternative-access providers entered the market to provide interconnects the needed access to the local markets. Their business was based on providing an alternative to the substantial fees charged by the RBOCs in "cherry-picked" cities. Metropolitan Fiber Services and Teleport Communications were among the first companies to enter this lucrative market. They installed fiber optic cables in major cities thus allowing the interconnects to bypass the RBOCs for this essential local access. These organizations have successfully siphoned off substantial traffic in major service areas to cause the RBOCs major concern. To date their targets have been business and large residential complexes where traffic is easily consolidated at the T-1 level.

Additional local competition has been provided by mobile and cellular carriers (McCaw Cellular, the largest in the industry, a recent acquisition of AT&T), cable companies and now the very interconnect companies which the RBOCs were separated from back in 1984. Joint marketing arrangements and consolidation of service providers has resulted in competition from RBOC sister companies offering non-land line based services in their geographically defined calling area. This competitive atmosphere has forced the RBOCs to become more aggressive in their pricing practices for local access plus look to their customer base for sources of additional revenues through new and improved services.

Defensive steps historically taken by the RBOCs include the introduction of *CLASS* (Custom Local Area Signaling Services) services such as call waiting, call forwarding and third party calling. Bolder steps now being taken include the implementation of enhanced services via *Advanced Intelligent Networks* (AIN). These AIN based services permit value added network products to be deployed. AIN services pose a substantial threat to competing services, and they are a source of non-tariffed revenue to the carriers.

AINs have a history dating back to 1965. Early versions were delivered via *stored program control* (SPC) on central office switches. Early features included Centrex offerings, speed-dialing, and three-way calling. Once effective, SPC services have become cumbersome both to implement and to enlarge due to the rapid growth in service lines, the need to tie feature utilization to billing systems, and the ever increasing range of enhanced services desired by RBOC customers. The first step towards resourceful deployment of AIN functionality utilized distributed data bases at network control points accessible by central office switches via *Common Channel Interoffice Signaling* (CCIS) network. Thus, the birth of distributed processing architecture in the network. Today new telecom services are not being added to the central office switch, but to computers that communicate with the switch. With AIN, the switch is replaced as the focal point of call processing activity by logic which is distributed throughout various network systems that function independently of each other.

The *client-server* model for distributed computing in the business world partitions software into client and server packages. The software operating on the user's desktop (or client) system provides only the user interface. Computer-intensive processing and database look-up functions are performed on a remote server.

As a result, AIN services will continually move from rigidly defined hardware/software implementations to new, flexibly defined, software-based services that can be upgraded without detrimentally affecting the fabric of communications switching which supports call routing and connectivity. Under the new client-server model, AIN services are delivered via the interaction of switches and computer systems that support AIN service logic. Such a switch recognizes the "triggering event" conditions for AIN service involvement, formulates an AIN service request, and responds to call processing instructions from the network server wherein the AIN service logic resides.

The Telecom Switch

The telecom switch located at the RBOC central offices and interexchange carrier points of presence is the foundation of the entire network. Without the switch there would be no telephone service. AT&T, Northern Telecom and Seimans/Rolm are a few of the major manufacturers of telecom intelligent network switches.

Advanced Intelligent Network service is essentially a collection of Service Logic Programs (SLP) that run on hardware configurations populating the AIN infrastructure. The switch is the hub of the architecture which identifies calls, requests call processing instructions and responds to the instructions received. *Call modeling* is the interface which permits the switch to understand and interact with the AIN service requests and respond to the AIN service logic. Standardization of this call modeling, accepted and strictly adhered to by all telecom switch manufacturers, is critical to network integrity. RBOCs and interconnect carriers have their central offices and points of presence populated with telecom switches from an assortment of manufacturers. Any service upgrade or modification requires all switches in the AIN to receive that upgrade and provide the same level of service of all other switches comprising the network. To ensure switch integrity Bellcore, the research arm of the RBOCs, created and published a call modeling standard for North America. In Europe and Asia, the CCITT and European Telecommunications Standard Institute (ETSI) have done the same.

The Bellcore North American standard for Advanced Intelligent Network call modeling is AIN 0.0, AIN 0.1 and AIN 0.2. Each successive AIN call model provides a greater breadth of Intelligent Network functionality.

The CCITT standard for call modeling is Compatibility Set 1 and 2, better known as CS1 and CS2. The ETSI has established its own Intelligent Network Application Protocol Standard; INAP.

Telecom Switch Evolution

Migrating from older, not as intelligent, switch technology to AIN is taking the switch manufacturers, RBOCs and Long Distance carriers years to accomplish. The switch manufacturers must add intelligence to their switch. The RBOCs and interconnects have financial constraints for their switch upgrade evolution speed. They have a tremendous capital investment in hardware which is tightly coupled with their current ongoing service offering. Potential service interruptions translate to lost dollars generated by their networks. The utmost of caution is exercised when dealing with network issues. In addition, have standard amortization schedules of approximately 30 years on hardware. These amortization schedules have been approved by the government regulatory bodies and are tightly tied to tariffed rates approved by these bodies. Changes to these schedules require approval by these governmental agencies.

Distributed Switches

Alternatives to the ideal AIN have been used by the RBOCs and interconnects as they make their route towards utopia. One approach is open architecture distributed switches which augment the abilities lower intelligence switches already in place. These switches intercept call handling and signaling information and facilitate the more advanced switching requirements. The key features of distributed switches include interface support for a variety of access technology. These interfaces include local analog loop start service, ISDN, T-1, E-1, HDSL, ADSL and SONET. Distributed switches also provide a generic interface between the switch and general purpose computer so the computer can then be used as an application platform or file server.

In essence distributed switches act as a virtual destination for the central office switch when certain conditions occur.

The Simplified Message Desk Interface (SMDI)

SMDI has been used to connect voice mail and voice messaging systems to the telecom switch in lieu of a true AIN. This Bellcore standard alleviates network compatibility problems. SMDI consists of an RS-232 data link and multiple lines arranged in a hunt group. It interfaces with virtually any voice store and forward system on the market.

The Cellular Standard IS-41

The potential for call control and switch compatibility problems exists in the cellular telephone intelligent network. These switch configurations are newer in comparison to their land line counterparts operated by the RBOC therefore newer technology is readily deployed. The real problem arises as these networks expand regionally and nationally through merger and acquisition the potential for conflicts between switch technology deployed escalates substantially. Thus the need for the IS-41 standard. Transported on SS7 protocol, IS-41 enables a cellular system's home location register (HLR) to automatically communicate the new service area of the subscriber's calling profile by establishing a visited location register (VLR). This is a high-speed method of making it possible for a subscriber to access all the same features from any location their cellular company services area. In addition IS-41 automatically tracks the subscriber's location and automatically notifies the home system so calls can be delivered without user interaction.

The Advanced Intelligent Network (AIN)

To better understand the local and long-distance telephone service offerings we need first be familiar with the many hardware and software components and how they work together to provide telephone based services. (Figure 15.1 represents a typical AIN configuration.) To provide a better understanding of this technology each component of the network is tersely described, Bellcore standard network connectivity and component specifications are highlighted and a call is traced through the AIN.

AINs are composed of Network Elements, which transport services, and Network Systems, the software-based services which control and actually perform the services. Network Elements include *Service Switching Points* (SSPs), *Signaling Transfer Points* (STPs), and non-SSP switches. Network Systems contain *Service Control Points* (SPCs), *Adjunct Processors,* and *Intelligent Peripherals* (IPs).

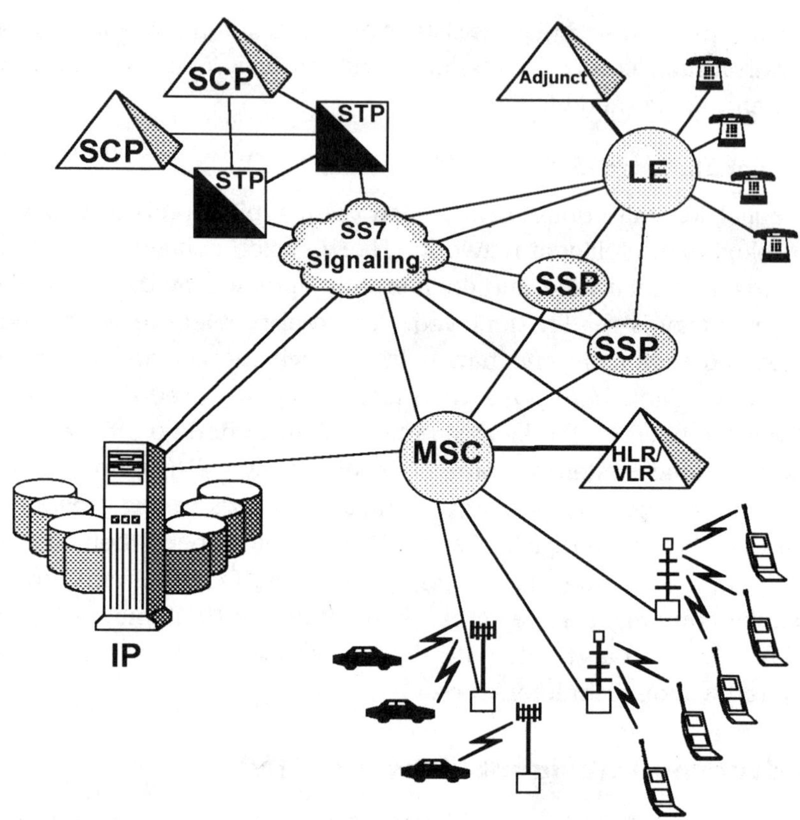

FIGURE **15.1** **Advanced Intelligent Network**

Other systems in the network architecture include the Service Management Systems (SMS), a workstation that manages and deploys services in the SCP network. Also attached to the SMS, is a Service Creation Environment (SCE), a platform providing graphical tools to create, simulate and test services before they are deployed by the SMS. Other computers attached to the network, but which fall outside the traditional AIN definition, are Operation Support Systems (OSS), that provide specific operations capabilities such as network administration, surveillance, testing, traffic management, billing and data collection.

Figure 15.1 graphically details the typical Advanced Intelligent Network configuration.

Network Elements

The Service Switching Point (SSP), an integral part of the AIN switching network, permits interaction between the callers and their desired service(s) which are located in the SCP. The SCP communicates with one or more Adjunct Processors via Ethernet, and one or more SCPs via SS7, to obtain call processing information and instructions. The SSP communicates with one or more IPs via PRI-ISDN to provide appropriate call processing and enhanced services.

The Signaling Transfer Point (STP) provides essential SS7 network links to SCPs. The STPs are generally used redundantly to ensure service reliability.

Network Systems

All control functions for AIN calls are handled by an *Adjunct Processor* or *Service Control Point.* Most AP or SCP are high capacity, high availability computers that supply information to switching systems required call handling. They delineate requiring special treatment. Both Adjunct Processors and Service Control Points are stand-alone Network Systems containing service logic (i.e., application programs and service logic programs) and data bases used to support various AIN services.

The Adjunct Processor is tightly coupled to the AIN switching system via a high-speed Local Area Network (LAN). This high-speed connectivity used between the adjunct and the AIN switching system because the Adjunct Processor best supports services requiring quick response to users actions. Adjuncts support services associated with the Local Exchange. These include call set up and CLASS services (e.g., automatic callback, call forwarding, etc.)

Service Control Point provides Service Logic Programs in response to external messages received from an AIN switch and internal messages that originate within the SCP. The SS7 network allows the SCPs to be fully interconnected with AIN switching systems through one or more of the signaling transfer points. SCPs are particularly well suited to support network service capabilities such as those required for 800 and 888 toll free services, area number calling or follow me number person locators.

Both the Service Control Point and the Adjunct Processor are highly sophisticated data base systems. For the most part the Adjunct Processor is associated with the local exchange while the SCP is typically associated with call setup for long distance services provided by the interexchange carriers.

Intelligent Peripheral computers contain communications processing hardware and software which control, manage and provide resources such as text-to-speech, audio announcements, speech recognition, Fax, messaging services and digit collection. The AIN switch passes message parameters to instruct the IP to perform specific functions. When the IP has completed the requested functions, it returns status and data, collected from the call which includes the person who called, to AIN service logic via the AIN switch data.

Most intelligent peripherals today are vehicles to deliver voice messaging, voice response, Fax and Voice Dialing services. In the future, they will evolve into highly sophisticated platforms that can provide advanced AIN services such as imaging and video-on-demand. Intelligent Peripherals communicate with SSTs via PRI-ISDN and with SCPs and Adjunct Processors with SS7.

Signal System 7 (SS7). SS7 is the glue that makes it all work. This standardized signaling protocol was developed by Bellcore (the research arm of the RBOCs) and maintained in accordance to their specifications. It could easily be said that the Advanced Intelligent Network is an architecture built around the SS7 standard. This out-of-band signaling system facilitates all basic network functions which include call set-up, connection and tear-down upon completion. For example, when a caller dials a number the local central office switch immediately sends instructions to search for the best available voice path. While locating the appropriate open transmission path SS7 automatically check the line end to end for integrity, sets all switching mechanisms to reach the desired destination and opened the appropriate account for billing purposes. All this is performed in less than 1/20th of a second. Pretty amazing. If a malfunction occurs an alternative path is instantaneously identified and the call is transferred to the operational line. The efforts of SS7 are assisted by connectivity to service- providing data bases.

Earlier information demonstrates that SS7 network connects SCPs with SSPs and Adjunct Processors with Intelligent Peripherals. Some networks use SS7 to connect Adjunct Processors to the network instead of using a high-speed LAN.

How All This Works

When placing a call from you home or office or from anywhere for that matter the number dialed as well as the number you are calling from contains information for the call completion. When the call is placed the number dialed in communicated via SS7 technology necessary information to connect the callers voice channel to that of the dialed party. The initial digits determine the call will be local, long-distance, toll-free or pay-per-call. The last set of digits determine the local switch and the exact circuit that will handle the final interconnection.

All local AIN configured switches are programmed to recognize calls requiring special processing. These are identified by the switch through software which recognizes a triggering event. When it happens these trigger conditions are communicated through the SSP as a query to an SCP data base requesting call handling information. The SCP contains information detailing the action(s) to be taken as defined by the trigger.

For example, when an 800 and 888 toll free number is issued a record is created in a SCP data base rather than a physical circuit location established. This same situation transpires when you utilize call forwarding capabilities on your phone. When activating call forwarding service you dial in the appropriate code with your DTMF tone keypad then the forwarding number. Incoming calls are then routed to a SCP data base which automatically transfers the call and dials the forwarding number

Hardware for the Telecom

Hardware used for all network level services in the telecom must adhere to stringent standards for ruggedness, reliability and redundancy. This equipment requires seven days per week twenty-four hours per day capability complying for the most part to the "Five 9's" standard. The Five 9's is telco jargon for fault tolerant systems which must be operating error free 99.999% of the time. That translates to five minutes an fifteen seconds total down time per year.

Bellcore's NEBS Standard

Bellcore periodically publishes Technical References under the umbrella of its Network Equipment Building System (NEBS) standard. NEBS standards detail the "minimum generic requirements" appropriate for all new equipment systems used in RBOC central offices and other telephone buildings. Many of these standards have been adopted and adhered to by interexchange, cellular and international carriers as well.

Periodic release of these Technical References reflect the experiences of the RBOCs when incorporating and maintaining technology in their networks. Each RBOC determines how stringently they will adhere to NEBS standards. When issuing a request for proposal, submitting a purchase order or awarding an equipment contract the RBOC may include some or all the NEBS requirements which must be met.

NEBS standards address issues such as the electrical environment and electrical safety, space planning, test methods for temperature and humidity, fire resistance, earthquake protection, acoustic noise, clearing heights for equipment and cabling, floor loading limits for equipment, cabling and lighting specifications, requirements for totally floor-supported equipment, detail and use of standard floor plans plus compatibility issues for cable pathways.

NEBS specifications for computer based technology consider serviceability; specifically while the network continues operation. Reflecting this criterion is the "hot swappable" or "hot pluggable" nature of the components. Components of subsystems and subsystems themselves can be replaced, upgraded and serviced without disturbing system operation. These levels of operational requirements have been established due to the nature of RBOCs and interexchange carrier business. Even minor equipment failures could mean millions of dollars of lost revenue, damaged reputation and lawsuits from their customer base. Therefore, equipment manufacturers who design equipment for telco environments created Fault Tolerant and High Availability systems.

Fault Tolerant Systems

Fault Tolerant systems are designed for availability 99.999% of the time. To accomplish this manufacturers have designed systems which contain specialized hardware which detect any hardware fault and instantaneously switch to a redundant hardware component. These redundant components are usually housed within the same chassis as the fails component. This instantaneous switching is virtually seamless. This architecture is generally built upon two or more central processors executing identical instruction streams in tight syn-

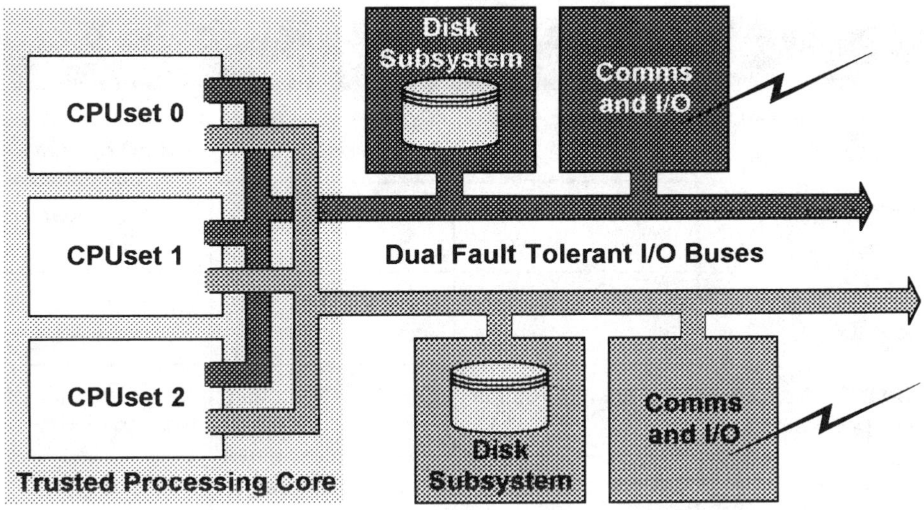

FIGURE 15.2 **Fault Tolerant Architecture (Source: Integrated MICRO Products)**

chronization. This "trusted" processing system forms a self-checking, self-correcting core unit. In addition to fault checking and bad block mapping, the system verifies each read and write operation with a process called end-to-end checksumming. During any write operation a checksum is generated and written to both disks simultaneously.

The systems "trusted core" consists of CPU sets running in lock step combined with memory subsystems. Each CPU set normally drives two fault tolerant I/O buses with each bus having the capability of accommodating several I/O modules. Figure 15.2 provides a graphical presentation of the above detailed fault tolerant architecture.

Disk drives used in each system are configured as mirrored pairs. When a new drive is installed in the system mirroring software automatically copies all data without interrupting normal system operation. These drives are "hot swappable" to secure operation even with a drive failure.

High Availability Systems

High Availability provides a high degree of system reliability through the use of networked system clusters at a lower cost than fault tolerant architectures.

TABLE 15.1 System Reliability Matrix (adapted from Motorola Computer Group)

System Reliability Matrix			
	Standard Computer	High Availability	Fault Tolerant
Architecture	Simplex	Partial Redundancy	Full Redundancy, Self-Checking
Impact of Failure on Data	Possible Loss	Possible Loss	No Loss
Impact of Failure on Application	Service Interruption	Service Interruption	No Interruption of Service
Recovery	Maintenance Call	Maintenance Call or by User	Automatic
Repair	Maintenance Call	Maintenance Call or by User	By User

These system configurations have a cluster manager which tracks the status of all the servers which are cluster components in the network. When the cluster manager detects a server failure it initiates a reconfiguration process which makes the data services of the failed server available as transparently as possible. Under normal conditions one server in a cluster configuration sits idle as a standby for one to three servers in operation. This idle server duty is rotated among the other servers in the cluster in accordance with a predetermined schedule. All active servers in the configuration concurrently work on the same data simultaneously.

The basic difference between the fault tolerant and high availability system approach is displayed in Table 15.1. Fault tolerant systems cost more but have 99.999% system operation levels. High availability systems provide minimal interruption.

Sun Microsystems and the Telco Business

The telecom market customer list of Sun Microsystems reads like a *Who's Who* in telecommunications. Approximately ten percent of Sun's business is with communications carriers. Some systems purchased are deployed as workstations and client-servers for non-AIN network related use. Others are integral components to AIN networks. To date all AIN installed system have been utilized as Adjunct Processors and Service Control Point systems. Now utilizing the technology from Linkon Corporation and Newbridge Microsystems, Sun Microsystems will provide Intelligent Peripheral technology featuring a wide array of communications processing capabilities for deployment in Advanced Intelligent Networks.

In the early 1990's, Sun experienced a substantial growth in the use of high-end server level technology by the telecom industry. These system were deployed to reduce cost by replacing more expensive mainframe architectures, modernize the network, improve flexibility and responsiveness in their systems, accelerate development and deployment schedules plus improve the integration of telecom services. Sun's client-server and networked computing approach is particularly well suited to the distributed configurations of telecom network technology. This has assisted Sun's penetration of this high-growth market.

Systems deployed by the telecom industry are used for Business Support Systems, Operational Support Systems, Intelligent Network Systems and would soon be used for Intelligent Peripherals which provide value added services. Details on these uses are as follows.

Business Support Systems

Business Support Systems handle billing, provisioning, directory assistance and customer management. All areas are critical to the telecom's operation yet do not actually process network traffic for call completion. Software for these business areas and hardware integration is provided by third party vendors.

Operational Support Systems

Network data collection, Local Area Network (LAN) surveillance, traffic analysis, switch and component maintenance plus facilities management. Operational support systems by definition support all operations of the carriers business. Some of the software operating on these systems is supplied by third party vendors while a substantial portion is written and supported by internal telecom MIS groups.

Intelligent Network Systems

Service Control Point systems (SCP), Service Maintenance Systems (SMS), Service Creation Environment (SCE), Adjunct Processors, Private Virtual Networks, Fraud Abatement Systems and 800, 888 and 900 Call Service Systems, Home Location and Visitor Location Register are all AIN components discussed in detail earlier in this chapter.

Intelligent Peripherals

Intelligent Peripherals are used to deploy enhanced functionality for AIN network use. These servers provide value added services through which the RBOCs generate non-tariffed fee based revenue. Early implementation of Intelligent Peripherals by the telcos were primarily used to offer voice mail and voice messaging services. These services were soon followed by Fax store and foreword systems. With the increased awareness of the Internet, E-mail is in the telcos sights as a service offering. Unified messaging, Voice + Fax + E-mail in the same mailbox, will soon be seen. Voice Dialing trials have been conducted by many of the RBOCs and wireless carriers to provide hands free operation for mobile users and a convenience offering for household and business use. These tests have been well received. The carriers are currently establishing service introduction schedules. Video on demand will be an Intelligent Peripheral offering some time in the future. Many trials have been conducted however the carriers are still trying to determine how they will actually make any money with this service.

With the exception if video, all enhanced functionality features are available for use today through the deployment of Linkon Corporation (New York, NY) technology in combination with Sun Microsystems Servers.

Middleware Technology from EBS

SS7 and other AIN signaling technology is currently available for use in the Sun system architecture through EBS of Shelton, Connecticut (203) 925-6100. EBS provides a family of fully open system telecom platforms, operating on Sun-

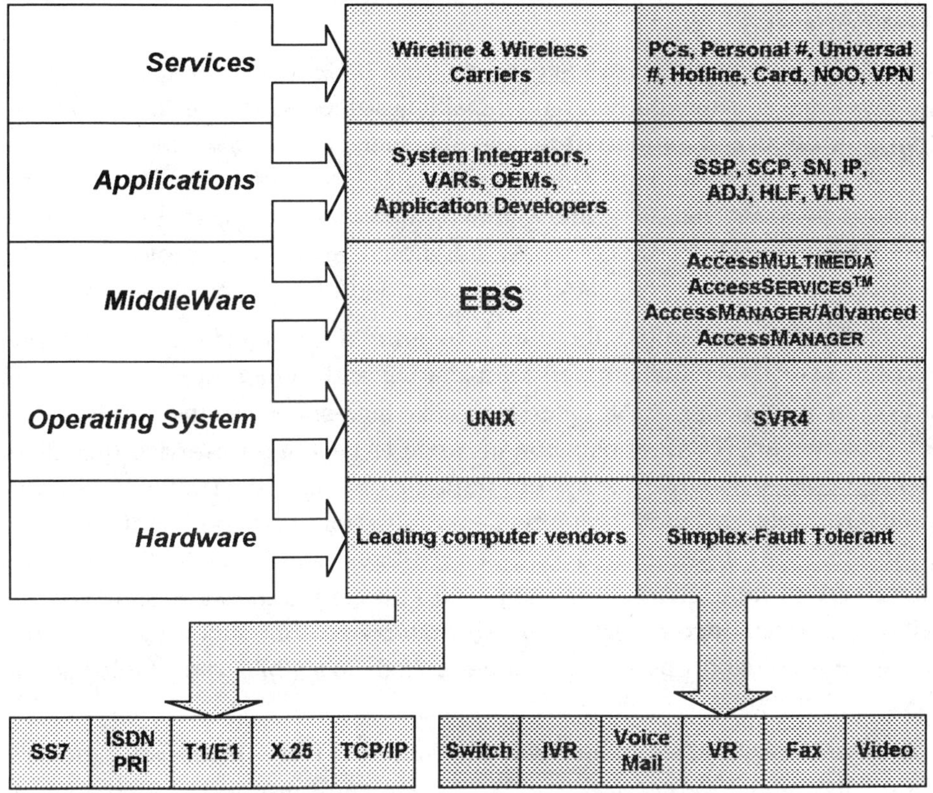

FIGURE 15.3 **Where EBS "middleware" technology fits in.**

Soft's Solaris, which feature common APIs which are used as building blocks in an advanced intelligent network. This building block approach is flexible and permits AIN technology to be deployed in phases.

The EBS technology is considered "middleware" as it is software which resides in the layer between the application software and the operating environment. It permits signaling protocol to interact with the application software and hardware providing the application programmer a standardized interface without his need to be familiar with the nuances of the signaling protocol. Figure 15.3 diagrams where the EBS "middleware" technology resides.

The three levels of middleware available from EBS are AccessMANAGER™, AccessMANAGER/Advanced™ and AccessSERVICES™.

AccessMANAGER™

AccessMANAGER™ with an API accessible from C/C++, and all the SS7 signaling control and transaction handling capabilities needed for the intelligent network applications. It supports multiple signaling points on a single platform, using an optional Intelligent Network Emulator, to provide the functions required for message handling and signaling network management, even with different variants of the protocol stack. Additional layers of the protocol stack—SCCP, TCAP, TUP, ISUP and OMAP—can be added on an as needed basis.

With TCAP, for example, there are powerful SCP enhancements for multiple instances of TCAP that support load sharing and , when needed, seamless hand-off of incoming TCAP queries. This high-speed enable/disable load sharing allows any TCAP application instance to be put into service quickly or shutdown gracefully while the system remains operational. Traffic flows transparently from one instance to another, even after the load sharing of the previous instance has been turned off. This allows TCAP application updates without service interruption, adds application level redundancy, and provides additional performance on symmetrical multi-processing platforms. This fully open system solution operates on simplex, high availability and fault tolerant UNIX based systems.

AccessMANAGER™ complies with all ANSI, CCITT and Japanese standards. Any country-specific or network-specific variant can be provided within an average of 90 days.

AccessMANAGER/Advanced™

AccessMANAGER/Advanced™ extends the built-in API of AccessMANAGER™ with advanced APIs and library interfaces that are crucial to the success of services, such as PCS, AIN 0.1, AIN 0.2, AIN 1, CS1 and CS2 services. It supports a load sharing client server architecture which allows applications to run transparently across heterogeneous network nodes.

Advanced features, such as rules-based applications start-up and heartbeat capabilities, are provided by an Application Manager™. By defining the start-up order of every application at any point in time, then utilizing heartbeat messaging to allow the exchange of state information between the applications and the Application Manager™, it is ensured that every application feature or service will start or restart at the point it was interrupted, whether triggered by a preplanned service configuration or a totally unanticipated event, such as a power outage.

This highly flexible real-time call control and configuration capability and on-demand monitoring of the application status is built around the Access-MANAGER™ architecture, which provides high performance Streams processing elements and EBS-designed open system software to couple what are ordinarily two distinct and separate subsystems—the SS7 messaging subsystem and the IPC messaging subsystem—and implement the functions needed by both subsystems. The result is a unique "soft switch" environment where every application can register as a named object—including alarms, status and log facilities—and interact.

The log feature provides standards-defined capabilities for sequencing and forwarding log and error messages to specific destinations in a hierarchical progression. This allows users to immediately view the triggering events—in sequential cause-and-effect context, to simplify the problem analysis.

AccessMANAGER/Advanced™ simplifies service deployment with a rich layer of APIs, tool kits, network management interfaces such as SNMP and TMN, a user-friendly object-oriented Graphical User Interface available in Motif version, and an intelligent network emulator, enabling the creation of multiple signaling points on a single CPU.

For example, message-streams can be diverted from a single SS7 port to a simulation process, enabling powerful de-bugging capabilities, such as on-line inject and trace simulation, for the distribution and re-routing of messages from process to others. By making the creation of loopback capabilities fully transparent, AccessMANAGER/Advanced™ allows users to quickly create and support highly customized facilities, such as custom logging facilities that link customer-specific and carrier specific call billing and monitoring processes.

A trace feature enables on-demand monitoring of the status of any application or service without slow down or disruption. Library calls can be invoked to automate this process, or trace activation calls can be made to access this shared memory information.

AccessSERVICES™

AccessSERVICES™ completes the EBS trilogy with call control, totally eliminating application dependency on signaling, and allowing the introduction of a diverse array of resources without affecting the application. It is built around a three-layered architecture using resources, server and application layers to provide complete call control. AccessSERVICES™ extends the AccessMANAGER/Advanced™ and AccessMANAGER™ middleware software platforms.

AccessSERVICES™ is compatible with AIN 0.1, 0.2 and 1.0 call modules plus CCITT CS1 and CS2 call modules. It also provides higher granularity of points-in-calls, more triggering detection points for full customization of the call modules, and the capability to introduce new service dynamically while the system is processing calls. It gives developers the complete decision flexibility required to create new intelligent network-based services. For example, the developer can choose to offer new services by creating local service logic programs on the AccessSERVICES™ platform. Or the developer may choose to use the agents to interface the service logic programs that reside on other intelligent network nodes, such as Service Control Points or Adjunct Processors. These agents can be created by the developer, or EBS can provide off-the-shelf agents such as AIN 0.1, AIN 0.2, CS1, 1129,1129+ and 1129 CORE.

AccessSERVICES™ seamlessly converts signaling events to call events, eliminating the application's dependency on different signaling protocols, simplifying applications development and allowing the introduction of a diverse array of switch and media resources without affecting the application. It is designed for ease in creating AIN network nodes such as intelligent peripherals, enhanced services platforms, service nodes, protocol gateways, service switching points, mobile switching centers and adjunct processors.

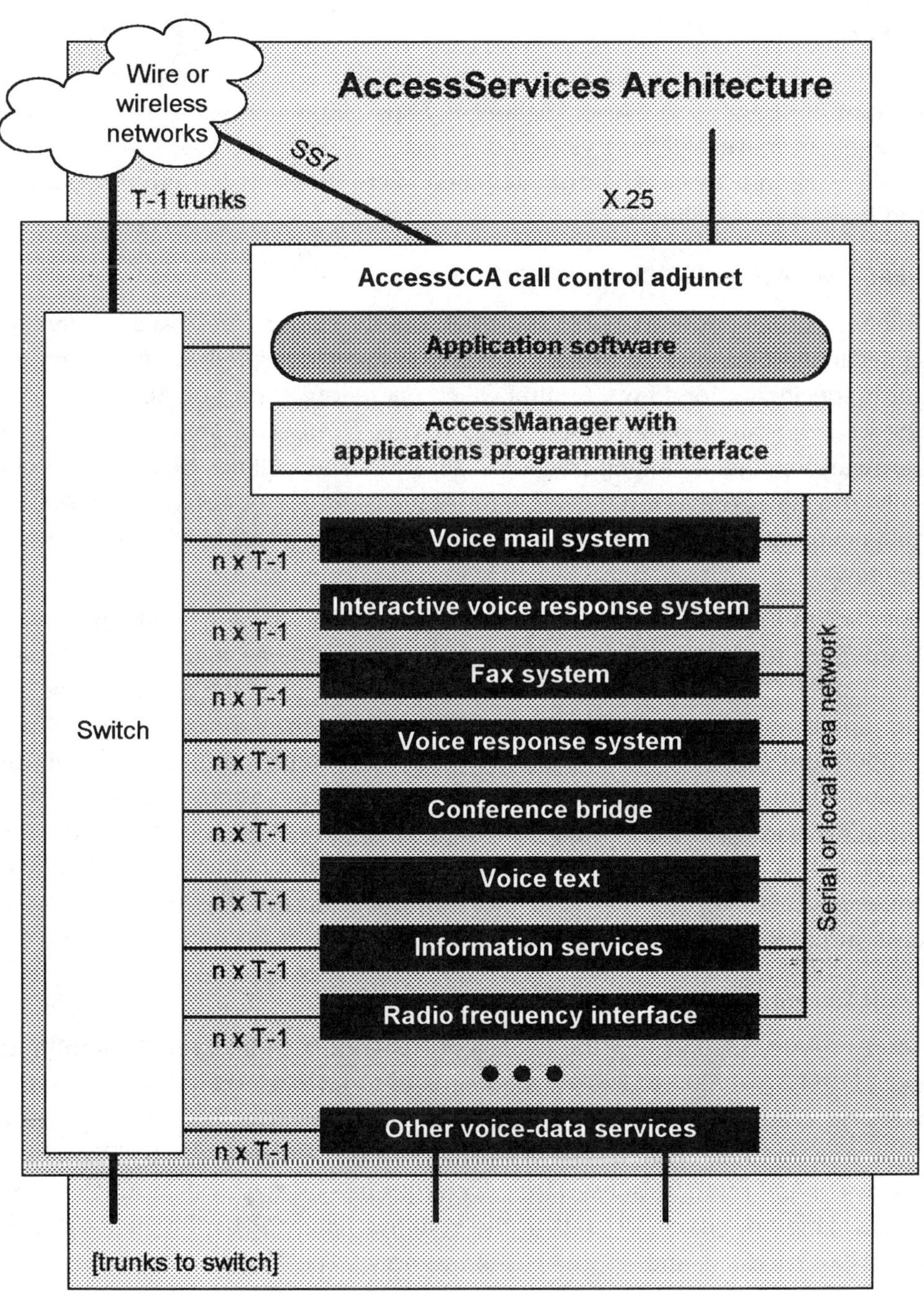

FIGURE 15.4 **AccessSERVICES Architecture** (Source: EBS, Inc.)

Hardware

Sun Microsystems does not currently manufacture a NEBS-compliant or fault-tolerant computer. However, a line of Sbus NEBS compliant computers is available from Texas Microsystems of Houston, Texas. An Sbus Fault Tolerant SPARC system is available from Integrated Micro Products of Durham, England and Bannockburn, Illinois.

NEBS Compliant RISC Based SPARCstation

The Texas Microsystems (713) 514-8200, Model 9632 SPARCstation Telecommunications System is a SPARC-based computer designed to Bellcore standards specifically for 48 volt Central Office operation. It runs on Solaris or SUN OS 4.1.2 operating systems and offers a vast number of software and Sbus hardware options including Datakit, X.25, ISDN and others. Designed specifically as a "telco" platform it is ideal for Adjunct Processor applications, service enhancement or any AIN application requiring up to 28 MIPS of UNIX processing power.

The model 9632 features a microprocessor based, battery backed Alarm/Communication/Control unit that is removable as a module for minimum MTTR. This unit monitors all operational aspects of the computer including: input voltage, output voltage, CPU operation, fan operation failure, temperature (high and low) and circuit breaker status. When a fault is detected, a time stamped ASCII message is transmitted to a remote management site, a dry contact closure is initiated and a front panel LED is illuminated. This action will occur even under total power failure as the unit is internally battery backed. This alarm unit is user programmable and allows messages to be customized for a variety of applications.

Fault Tolerant SPARC Computer

The FT-SPARC platform for the Intelligent Network from Integrated Micro Products, (708) 374-1700 addresses the needs of today's telecommunications network providing cost effective distributed intelligence. The FT-SPARC system is built around a fully trusted core consisting of two CPU sets. Each CPU set is a processor and memory subsystem which runs in lock step with, and is capable of checking on, other CPU sets in the system. Each CPU set contains one or two Mbus modules and up to 1 GB of memory. The Mbus modules may con-

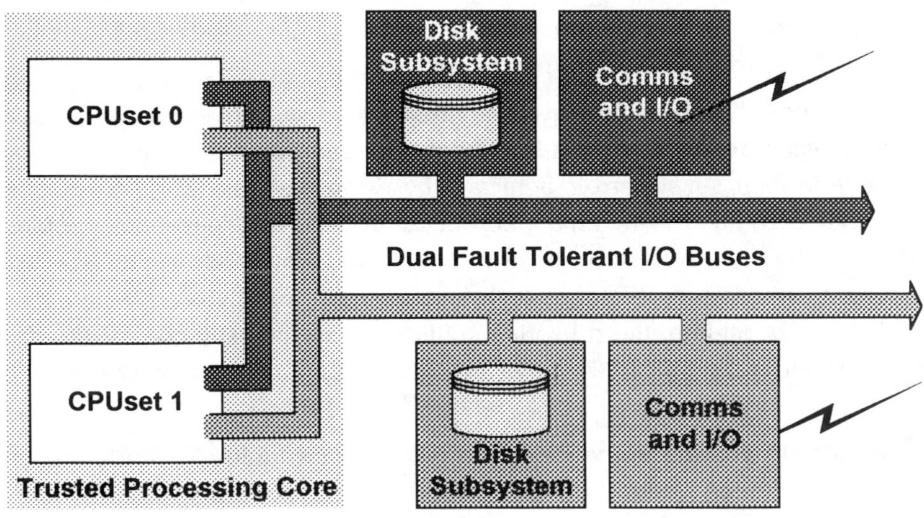

FIGURE 15.5 **Fault Tolerant Architecture**

tain one or two SuperSPARC processors, providing up to four processors per CPU set. The CPU sets in turn drive two fault-tolerant input-output (FTIO) buses.

Up to eight I/O modules can be plugged into each FTIO bus. An I/O module can be either a native FTIO bus device controller or a communications subsystem. It can also be a module containing a bridge from the FTIO bus to a standard bus, such as VME or Sbus, and a standard bus I/O card. This allows the FT-SPARC system to take full advantage of the vast array of existing I/O controllers available on standard buses. External I/O can be accomplished via a rear panel connector or a Comms Chassis accepts plug-in boards based on a variety of physical interface standards.

Block I/O in the FT-SPARC is implemented by using an IOSet, which features built in SCSI devices and an Ethernet connection.

The multiple CPU set design of the FT-SPARC processing subsystem forms the self-checking and self-correcting core of the fault-tolerant unit. By combining the CPU sets, which execute identical instruction streams in tight synchronism, the FT-SPARC system offers transparent fault tolerance. When the I/O operation occurs, the CPU sets verify that they are operating in harmony, a process which is built into the hardware as part of the standard I/O bus cycle and requires virtually no system overhead.

Based on SCSI-II technology, the FT-SPARC disk subsystem modules includes a disk, power supply and SCSI controller in a hot replaceable unit. Fault tolerance in their subsystem is achieved by using the disk drives as mirrored pairs, with each half relying on a different controller. This mirroring is enforced entirely by the disk subsystem driver software, and disk reads are scheduled to the least busy drive, providing a high level of read throughput. When a new disk is brought on-line, the mirroring software automatically copies data from the other half of the pair to the new disk while maintaining normal system operation.

In order to guarantee the detection of hardware faults, FT-SPARC runs an additional protocol within the reliable CPU set core, a process called end to end checksumming. During any write operation, a checksum is generated within the device driver and written to both disks. In the reverse direction (i.e. a read operation), the checksum is verified for correctness within the device driver. The FT-SPARC disk subsystem runs continuous latent fault checking and bad-block mapping.

The FT-SPARC system exhibits a very high tolerance to power spikes and brownouts. A redundant, uninterruptible power subsystem allows the FT-SPARC to withstand component faults and external power failures. The standard system provides two on-line replaceable power modules. Each power module uses an external 90-240V AC supply to charge internal batteries, which in turn supply a nominal, uninterrupted and clean 28V DC current to the dual redundant system power buses. Optional 48V DC power modules are available for Central Office applications.

If external power fails, the batteries will support the entire system until power is restored or a backup generator is placed on-line. In the event of prolonged failure, power monitoring software will initiate an orderly shutdown of the system.

FT-SPARC supports a rich set of local and wide area communications, including standard asynchronous serial I/O, Ethernet, Solaris networking support, X.25, and SS7. Fault -tolerant communications are implemented using dual controllers on separate I/O buses, with automatic fault detection and recovery.

Should one controller fail, the other is brought on-line without interruption to applications.

Because FT-SPARC's fault-tolerant design features are separated from the operating system, the machine functions as a standard UNIX workstation—the Sun SPARCstation running Solaris. This means application software can run unmodified on the fault-tolerant platform. By making the system's fault tolerance transparent, IMP ensures adherence to industry standards as well as the lowest possible system cost.

Unlike conventional UNIX systems, FT-SPARC is designed to operate non-stop during service failures and network repairs. Automatic diagnostics and self-checking device driver software report hardware failures to an intelligent knowledge base, the FT-SPARC Configuration Management System (CMS). The CMS responds to hardware failures by reconfiguring the system to operate without the failed module. It notifies the user automatically and can "phone home" to report the event to a designated service center. In addition each module carries a record of its service history in EEPROM. When modules are replaced the CMS oversees the testing and re-integration of the spare part.

|16|

What The Telcos Should Do

The RBOCs monopolistic grip on their local market has been gradually slipping from their hands for the past several years. Inertia and government decree have assisted the efforts of more nimble competitors to siphon-off some of the RBOC's most profitable business by providing better service at a cheaper price.

In some instances the hands of the RBOCs have been tied due to tariff regulations and restrictions on certain business practices by the government. However, the RBOC businesses least regulated, generating non-tariff revenue, have been subjected to the most blistering attack. With the government gradually removing the restrictions which prohibit non-RBOC entities from providing local area dial tone, the RBOCs dormant market place is becoming their battleground.

Patrick's Opinion

My recommendations to the RBOCs may challenge traditional beliefs, require legislative change and run counter to the views of management. These recommendations are based on my knowledge (sum total of ones' experience) of the RBOC business based on first hand experience as a residential and business customer, a vendor, a foreword thinker and a practical businessman.

These views are my own. They do not represent the thoughts or feelings of my family, friends, acquaintances, business associates and those of you who thought you knew me.

The catalyst for these recommendations is my philosophy that a business should do what it does best and limit its activities to that. A businesses should continually strive to perform a service or deliver a product better than they have in the past, increasing productivity while maintaining reasonable prices and profits. A business should keep its core product competitive, enhance the product to extend its life, and guard each customer with a vengeance treating them as if they were the only one they had.

Viewing an RBOC in light of my espoused philosophy brings a smirk to a sane man's face. This however need change. The RBOCs have several unique market opportunities which are theirs to miss. These opportunities are certainly within their grasp. They need to stand up immediately and embrace them.

Like the RBOCs, AT&T and IBM once had a near monopoly in their market. Competition and changed government regulations have significantly altered their market presence forcing a reshaping of their marketing and business approach to better participate in their rapidly changing market.

Traditional Telco Network and Enhanced Services

The "core" business of a telco is providing network based services and billing. These services must remain the primary product focus of all carriers going foreword. They must maximize the usefulness of their networks placing control in the hands of their customers and teaching them how to best use this capability. They need to continually expand the bandwidth available thus retaining their position as the network utility of choice. Their offering must span from analog POTS lines through the digital domain of T-1 and ISDN to Frame Relay and ATM. The telcos are the plumbers installing, managing, maintaining and expanding the "pipes" that permit the flow of communications.

The telcos need to implement new pricing strategies which favor their customers and provide a reason for continuing a business relationship with the telco.

Basic Service

All telco networks, regardless of line type (POTS, T-1, ISDN, Leased Line ...), need to feature a packaged basic service priced at a flat monthly fee. This basic service price should include all line, service and maintenance fees. Time, distance and volume priced local service is yesterday's dream in today's market. If a telco is trying to increase the use of a particular network service (i.e. ISDN) price incentives, such as lower fees for the basic service package, should be instituted for a limited time or on a permanent basis.

Premium Service

Custom Local Area Signaling Services (CLASS) such as call waiting, conference calling, *69 service, automated callback and call forwarding should be classified as premium services. Obviously, premium service would be priced at a flat monthly fee in addition to basic service.

Platinum Service

Platinum service would include voice activated dialing, voice messaging, fax messaging, E-mail messaging, unified messaging and other enhanced services. These services need be positioned differently than earlier generation premium services.

For example, fax and E-mail messaging should be sold as add-on services, positioned similar to a virtual telephone system in a developing country, for individuals lacking access to fax and modem capabilities. Individuals and businesses possessing fax and E-mail capabilities could use these services as a resource to handle over-flow. The service would self-activate any time their machine was busy or inoperable. Fax and E-mail messages could be reviewed by the subscriber using their household telephone and the text-to-speech capabilities of the service. If desired these messages could be then transmitted to the appropriate equipment at a later date.

Fax and E-mail messaging should be priced at a flat fee plus additional storage charges for large users.

Long Distance Service

It's amazing how history tends to repeat itself. The long distance carriers want to provide local dial tone while the RBOCs want to become long distance carriers. This consolidated service offering was once available from AT&T just over ten years ago.

As this commingling of services becomes a reality the long-distance carriers will target the densely populated metropolitan markets with wireline services. They are already in the local markets with wireless service.

The RBOCs are a natural for offering long-distance service. They currently control local access and the pricing for the provision of that service to long-distance carriers. This access has been effectively targeted by local bypass providers offering service to commercial accounts.

A recent *Wall Street Journal* article discussed the rumored merger of Bell Atlantic and NYNEX. The most revealing bit of information in the article was that 30 percent of all long distance traffic originates and terminates within the combined Bell Atlantic and NYNEX calling areas. The significance of this information lies in the fact that this thirty percent is probably worth more than the combined revenue from the provision of local service in that same calling area.

The RBOCs need to form a cartel for the provision of long distance service. This cartel could provide domestic long-distance service at a cheaper rate than the service currently available through the traditional long-distance carriers and their resellers. RBOC Cartel long-distance service packaged with local basic, premium and platinum service could provide an attractive offering to the customer and a profitable service for the RBOCs.

New Network Services

To remain competitive the telcos need offer new network based services. Services such as ISDN, Frame Relay and ATM are at the forefront of this new service offering. The carriers need offer these services with the historic degree of security, reliability and management capabilities they have been known for with their current service offering. Some new service offerings will require the RBOCs to add value to their networks. Much of this value will require the expertise of third parties. This expertise can be obtained through joint venture and joint marketing arrangements.

Internet and Intranet

The incredible market coverage of the telcos, combined with explosive growth of the Internet, presents an immediately opportunity for the telcos to become the dominate on-ramp to the Internet. With the number of computer households and business on the rise there will continue to be an expanding need for this service. Larger companies and universities direct connect to the Internet. They represent a small portion of Internet users. The rest of us, already telco customers, need the telco for the connection even when using an Internet service provider.

Sun Microsystems, clearly the leading supplier of Internet servers and a leader in browser technology, would be the ideal joint venture or joint marketing candidate for the telcos. By installing Sun equipment at central office locations the telcos could immediately become Internet service providers. In addition, a joint marketing campaign established between Sun and the telcos could make the Netra™ product available at a very attractive price to their customers.

Telcos not wishing to become Internet service providers could become service bureaus providing individuals, businesses and organizations "turn key" access to the Internet. This service offering would parallel those of traditional audiotex service bureaus providing billing and processing services for information providers.

Intranet, using TCP/IP technology, is the inter-company use of Internet type technology. It can be LAN or WAN based. LAN based Intranet installations do not provide business to carriers. However, once they become WAN based, then telco participation is necessary.

Internet Messaging

As discussed in Chapter 5, Internet messaging is an essential future service offering by the carriers. Currently, there are numerous of Internet telephone companies beginning/planning to provide voice based telephone service. Their technology uses real-time, highly compressed voice, transported as data over the Internet, for decompression at the other end of the circuit to make conversation possible. This technology need be full-duplex to permit an interactive conversation.

VE-mail

Network latency problems accompanied by the requirement for the calling party and the called party to be logged on the Internet at the same time makes these services questionable for less than casual use. However, this service is quite an interesting option for traditional E-mail use. The real universal use of this technology is VE-mail (vocalized E-mail).

VE-mail uses the same technology as Internet telephone services. The only difference being the receiving party need not be on-line at the same time as the sender. Like standard E-mail, the vocalized version is sent to the standard E-mail address of the intended recipient.

Groupware

Move over Microsoft, the carriers have the capability to provide software access over their networks which minimizes the need for consumers and businesses to own the software. Groupware provides the capability for a multi-location corporations to have numerous individuals, in separate locations, share resources while seamlessly collaborating with one another. They merely access the software and information as if it were available on their LAN.

Carriers can make these kinds of applications possible, without the customers having to develop the network infrastructure themselves.

It's quite interesting that Groupware is built on an ideology Sun Microsystems has espoused since the mid-1980's, "the network is the computer". Their distributed processing approach to computing is legendary. Hard to believe they were that far ahead of the time.

August 1995, Lotus Development Corp., now IBM, announced the commercial availability of AT&T Network Notes service. AT&T Network Notes is Lotus Notes, a groupware tool that many organizations have been using for years to enhance and streamline their internal business processes, available on the AT&T network. These tools permit corporations to take the same tool, widely used within the corporation, outside for use at other locations, by strategic partners and remote located employees.

Groupware is a new growing trend in value added network services available through carriers. This trend will continue as telcos and others bring added value to their network through joint venture and joint marketing arrangements.

One of the most difficult things about using groupware between businesses is managing it. These new public notes services put all of the maintenance, administration, security, troubleshooting and cross-organizational billing in the hands of the carrier providing the service.

Companies can now spend more doing business rather worrying how they are going to do the work.

Other Services

Directory Assistance, Operator Services and Telemarketing

Traditional directory assistance and operator services positions have been automated some but not for all services. Typically, with directory assistance, a caller asks an operator for the telephone number of an individual or business. The caller then listens while a text-to-speech computer voice gives the actual number.

The entry of credit card numbers has been automated to eliminate the need of operators to handle them. However, operator assisted calls still require human intervention to complete. The operator listens to the request to make a person-to-person or collect call, keys the information into a computer, stays on line to speak to the called party and only end their involvement when the party has been reached or the charges have been accepted. These activities are costly and do not generally offer value added, revenue producing service, even when priced on a per use transaction basis.

The RBOCs need to reduce costs and turn all directory and operator assistance centers into revenue producing business units. This can be done by identifying revenue generating work that uses the existing embedded base of Sun SPARCworkstations and people.

Using voice recognition technology, live answers can be virtually eliminated. Calls will be answered by voice response units and customers responding in natural speech. For example, to place a collect call, the calling party is asked to state his or her name. The response, "Will Chafin", is recorded. When the called party answers, she hears the IVR ask: "Will you accept a collect call from...." then the recorded voice of the calling party says "Will Chafin". The called party either accepts the charges and the system completes the call or the called party rejects the call and it is terminated. There must, however, always be a clear path to an operator for emergencies and special needs, such as those involving children and the physically challenged.

Voice technology is so sophisticated today that the user may not realize that they are not talking with a live operator. The system will handle a variety of foreign languages. Today, callers to directory assistance in some areas receive the requested number and an option to be connected to that number for an additional charge. This service will expand to areas where it is not currently available as the telcos continue their effort to reduce operating expenses while increasing revenue.

The real revenue opportunity lies in leveraging the system's capabilities of directory and operator assistance centers. They currently handle thousands of calls efficiently and effectively 24 hours per day, 365 days per year. The proliferation of 800 numbers for sales of products and services ranging from dicing/slicing machines to mutual funds has created a growing market for centralized service bureaus to handle all the calls. The RBOCs existing operator centers are in a perfect position to move into that high-volume telemarketing sales business.

Although fully automated service bureaus handle product sales, some products and services require live agents for effective selling. This is the business Sun Microsystems should help the telephone companies set up. The existing operator services workstations could display the appropriate product service based on DNIS and simultaneous connectivity to remote databases. Orders will be placed and the client systems will be updated, along with the collection of billing and statistical data.

The system requires the same response time and fault resistance as the directory and operator assistance platforms used today. Billing for completed calls and for sales calls handled will require additional project time. Operator service units already bill for incoming directory assistance calls and for operator handled completions. Where possible, RBOCs must leverage existing capabilities. The operators will require formal sales training.

An interesting product differentiator for their telemarketing service offering is the capability of the telco to direct bill purchasers located in their calling area, for their product purchased, as part of their monthly telephone service invoice.

Substantial savings will occur by the elimination of direct operator handling of directory assistance, collect, third-party billed and person-to-person calls. One RBOC calculated that saving one second of operation time per call is worth approximately $1 million per year.

Data Archiving

There exists a huge need for secured data archiving. As the use and compilation of data increases with our growing information society this need will expand accordingly. The telcos with their reputation for continuous, twenty-four hour per day, seven day per week service would seem quite capable of providing secure data archive services. They are already connected via analog and digital circuits to potential users. Offering and billing for the service would seem quite feasible with a modicum of work. Sun SPARCservers with disk arrays configured in a high-availability system architecture would provide sufficient processing and storage capability to offer such a service. If the telcos used large bandwidth high-speed networks they could quickly download and upload substantial amounts of data.

Subscribers to this service would permit the telco on-line access to their databases during off-hours to perform the scheduled download. This would be done daily or weekly depending on the needs of the subscriber. Uploads of the information would be provide on an as needed basis.

Network Consulting and Outsourcing

As corporations continue to re-engineer, refocus and downsize their operations they will look to the outside for assistance at an increasing rate. This assistance will be in specialty areas where they lack sufficient expertise and manpower, and tasks which are more cost effectively performed by a third party. With the growth in communications technologies it is reasonable to assume that some organizations will need assistance with internal and external communications problems. The telcos, with their years of experience in managing telephone and data networks, possess the expertise to handle these assignments. They can assist a corporation from a consulting stage all the way through an outsourcing arrangement where the telco handles that aspect of the customers business.

Telco Billing

The billing coverage of the telcos is legendary. With approximately 97 percent penetration of their geographically defined calling area they clearly have the most comprehensive market coverage in existence.

Their highly sophisticated billing system permits monthly invoicing for all services used by each customer. This automated system, in place for several years, possesses flexible billing capabilities. In addition they have highly experienced customer service representatives that accommodate inquiries and adjust for errors when need be.

In addition to billing for local service, the telcos also bill for long distance service provided by other carriers, calls made to toll numbers (900, 976, 970, 550 and 540) where service is provided by third parties, and toll free (800 and 888) calls where the service is owned by a customer.

Their billing coverage and experience makes them a prime candidate for expansion into providing billing services for other organizations. For example, the telco could bill for cable television, direct TV, pay-per-view, gas, electric, prepaid telephone cards and even Internet service as part of their monthly statement.

As stated earlier, an additional feature of their telemarketing services the telcos could invoice for purchases made by their telephone service customers, as part of their monthly statement. This would provide an efficient alternative to using their credit card as a payment vehicle.

Theories abound about cybercash, e-cash and other encrypted secured transaction and invoicing capabilities needed for consummating Internet purchases. The telcos are a natural for the provision of this service. Whether the telco was the Internet service provider, or acting as a service bureau for Internet service providers, they could easily act as an invoicing and payment vehicle for purchases made and services offered.

The billing areas of many telcos currently use distributed processing technology provided by Sun Microsystems. Sun technology works well with the distributed nature of their calling area, and provides a near fail-safe redundant design which will insure reliability.

Package Pricing and Service Use Incentives

The RBOCs need to upgrade their marketing skill-set. Current customer account growth is critical to profitability. Customers need to be continually cross-sold and up-sold services. Packaged pricing strategies need be tested and utilized which encourage additional service use, and reward customers with credit vouchers towards their bill. A more personalized relationship need be developed between the RBOC and the customer. Relationship managers need be established to assist customer satisfaction and monitor account profitability.

Quantitative marketing analysis has determined that customers incented to use additional services through free service trial introductions and packaged pricing techniques are more likely to use additional services than those merely sent a statement stuffer containing service and pricing information.

The telcos are entering a new era of product and service offering. Every day new "like" products will appear. If the efforts of the providers of these like products go unchecked the telco runs the risk of losing that customer for that particular service. Statistics document it is cheaper to keep an existing customer than to gain the business of a new one. Oh well, enough said. The choice is theirs.

This is all just one man's opinion.

|17|

Complete Computer Telephony Solutions, Applications Generators & C Tool Kits

Currently Available for the Sun Platform

System integrators (SIs), value added resellers (VARs) and the MIS department of many companies are finding it quite easy to deliver powerful computer telephony solutions which are tightly integrated with their distributed Sun architecture. Many of these solutions are high availability systems, interacting with their enterprise databases to provide customer management solutions never before available on a Sun.

There is a fine assortment of tried and tested computer telephony software currently operating with Solaris on a SPARCserver™ or SPARCworkstation™, performing a wide variety of communications processing functionality. These systems are either custom assembled by SIs and VARs, purchased already configured from Access Worldwide Channel Integration, or self assembled by innovative organizations. There is considerable latitude; the way you decide to go is up to you. As we make our way through this technology you will soon discern that these are powerful solutions available on a truly powerful platform.

SOFTWARE SOLUTIONS

Aum Tech

Aum Tech, Incorporated provides a suite of Companion software that allows you to provide a variety of voice processing solutions. The types of solutions available to any Companion installation are interactive voice response, voice mail, fax, automated attendant and many more.

Aum Tech's product offering includes Telecom Server, TCP/IP Server and SNA Sever software. All software is available for the Solaris operating system.

Telecom Server Software System

The Telecom Server is the front-end interface to a range of telephony-based Servers. Stand-alone Telecom Server applications can provide basic voice processing such as audiotex and voice mail. However, the Telecom Server can communicate with other server applications to provide much greater functionality. For example, a Telecom Server application coupled with a TCP/IP Information Server that accesses remote information is an Interactive Voice Response application.

The Telecom Server architecture is open. The application is activated by calling the system. The application answers the call and, with pre-recorded speech prompts, asks the identity of the function to perform. The caller enters a menu selection with DTMF or by spoken word. The input is converted by the Linkon board into digital form, and is then interpreted by the application. The application either takes the caller to another menu or performs a requested action. The application terminated when the caller hangs up or requests a transfer to another operator.

The Telecom Server is a voice processing resource. The Linkon boards and software allow the Telecom Server to perform advanced voice processing functions. The digital signals from the Linkon board reach the Telecom Sever via voice processing software. The software is critical to an application in that it performs special functions such as speech recognition, fax, and DTMF tone detection.

Aum Tech's Companion software capitalizes on Linkon's Universal Port technology. The universal port function lets the Telecom Server software to dynamically support voice, fax, speech recognition and text-to-speech. Such universal application access simplifies the engineering and administration of application interfaces. System engineers no longer need to dedicate a fixed percentage of Telecom Server resources to a specific application. In other words, they do not need to dedicate each incoming voice channel to either voice, fax, speech recognition or text-to-speech application. The full range of universal port functions can be used in a single application without adding extra boards. As a result, engineers need only determine the number of channels required to handle the aggregate volume of calls to a certain application. They do not have to engineer all the machine slots and network access points on the basis of call types.

The Telecom Server API's incorporate a number of capabilities that allow application designers to create advanced applications quickly and easily. Some of the voice processing functions available with the Telecom Server are:

⇒ DTMF detection and generation

⇒ Alpha/numeric input from a DTMF tone phone

⇒ Voice input from a DTMF tone or rotary phone

⇒ Menu and logic branching

⇒ Interruptible and non-interruptible announcements

TCP/IP Server System Software

A host machine supporting the TCP/IP network communications protocol is accessible to applications via the TCP/IP Server. This server manages the communications interfaces to the TCP/IP host-based information. The TCP/IP Server establishes log-on sessions and performs various service functions on the TCP/IP host's native environment. The environment provides access to information systems and SQL information databases either through low-level programs or through SQL-based 4GLs.

The TCP/IP Server applications act as servers which process requests from client applications. For example, when a client application, excluding the Telecom Server, translates a transaction request, it may decide that the transaction should be processed by the TCP/IP Server. The TCP/IP Server application can handle the request itself or forward the request to a TCP/IP host through a terminal session or a program-to-program session. It then extracts the response data and formulates a message to be sent back through RPCs to the Telecom Server application. After sending the response to the appropriate client, a TCP/IP Server application will clear its transaction screen and wait for the next transaction request to arrive.

The TCP/IP Server links with TCP/IP-compatible hosts by means of an Ethernet card that supports TCP/IP.

TCP/IP Server API's incorporate a number of capabilities that allow application designers to create advanced applications quickly and easily. Some of the service features available with TCP/IP include:

⇒ **Communications Management.** TCP/IP communications between the application and the TCP/IP host can take place as a telenet terminal session. If the TCP/IP emulation style changes, TCP/IP Service API's shield the application developer from the emulation specific API calls, and instead provide a consistent developer level API for any application function that works with the emulation.

⇒ **Emulation.** The TCP/IP Server provides program interface emulation's, full functional screen interface emulations and emulation key simulation. These capabilities allow a developer to perform basic screen scraping as well as placing information on a screen. Virtual terminal activity is established with the ability to emulate control sequences that the host can understand. Because program interface emulation is also available, TCP/IP applications would not be required to create their own SQL queries to access information. Instead, they could use information access programs that are already available on the host. This feature is significant in that terminal programs are common in industry. By including the TCP/IP Server in its environment, a business can leverage the utility of its computer resources.

⇒ **SQL Database Access.** DBMS (Database Management Systems) software products generally provide a data collection facility. Most DBMS support queries, report generation and data manipulation. In some cases, information may reside within a DBMS and may be accessible via SQL calls. The TCP/IP Server System Software has the capability to support applications that generate SQL providing another alternative to accessing information. The DBMS software can reside on a separate machine or may co-reside with the TCP/IP Server.

⇒ **Multiple Host Access.** The TCP/IP Server is not limited to single host access. Each application running on the TCP/IP Server can go against a different host machine, with a different database or application, to serve a transaction request.

SNA Server System Software

A host machine supporting IBM's SNA network communications protocol is accessible to applications via the SNA Server. A SNA Server manages the communications interface to SNA host-based information. Typically, the SNA host is an IBM mainframe or compatible with IBM's System Network Architecture (SNA). The SNA Server establishes log-on sessions and performs various service functions on the SNA host's native environment. The environment, which could be CICS, MVS/TSO, and/or IMS, typically provides access to complex information systems and large information databases.

SNA Server applications act as servers which process requests from client applications. For example, when a client application, executed on the Telecom Server, translates a transaction request, it may decide that the transaction should be processed by an SNA Server. The SNA Server application can handle the request itself or forward the request to an SNA host through a terminal session. The application then extracts the response data and formulates a message to be sent back through RPCs to the Telecom Server application. After sending its response to the appropriate client, an SNA Server application may clear its transaction screen and wait for the next transaction request to arrive.

SNA Server API's incorporate a number of capabilities which allow application designers to create advanced applications quickly and easily. Some of the SNA Server features available with the SNA Server are:

⇒ **Communications Management.** SNA communications between the application and the SNA host can take place using HLLAPI SDLC API's. If the SNA emulation style changes, SNA Server API's shield the application developer from the emulation-specific API calls, and instead provide a consistent developer-level API for any application function that works with the emulation.

⇒ **Emulation.** The SNA Server can provide program interface emulation, full function screen interface emulations and emulation key simulation. These capabilities allow a developer to perform basic screen scraping as well as placing information on a screen. Virtual terminal activity is established with the ability to simulate control sequences that the host can understand. Because program interface emulation is also available, SNA applications would not be required to create their own SQL queries to access information. Instead, they could use information access programs which are already available on the host.

CallStream Communications

CallStream Communications, Inc., provides a full line of time tested computer telephony solutions featuring the most advanced, flexible, open architecture messaging solutions. CallStream technology provides the ability to voice enable your existing applications or create your own specialized interactive voice response (IVR) application. You can create telephone front-ends to corporate databases, fax on demand and audiotex applications that incorporate voice recognition and text-to-speech technologies. You can empower your employees with a Windows based interface to voice, fax and E-Mail messages. With CallStream software it is easy to interface with telephone company central office switching systems and provide tens of thousands of users with advanced messaging features.

The CallStream has product line includes VoiceStream Messaging™, VoiceStream IVR™ and FaxStream™ plus an applications generator which will be covered in that section of this chapter.

VoiceStream Messaging™

VoiceStream Messaging is a high performance system which incorporates a full featured suite of call processing and messaging capabilities. These capabilities include voice and fax messaging, automated attendant, pager support, off premises message delivery, group messaging, multilingual operation and custom audiotex applications.

System users receive a mailbox for voice and fax which permits them to send and receive messages, perform call handling and access other mailbox functions. Standard, company, system administrator and application mailboxes can be set up. VoiceStream Messaging™ can be integrated with any other CallStream products to provide additional functionality when needed.

The system administration is performed through a full-screen menu that can be accessed on the system console, through remote terminals, over LANs or via modem. Most administrative functions can be performed while the system is operational.

System reporting information includes number of calls per port per day, total calls per hour, total call minutes per hour, total busy minutes per hour, number of messages sent per hour, number of mailbox accesses per hour, number of extensions dialed, number of extensions answered, number of extensions not answered and caller left a message, number of extensions not answered and caller pressed "O" for assistance and number of extensions not answered and caller tried another extension. All system statistics can be converted to Comma Separated Value (CSV) format files, for inclusion into spreadsheet programs or text format suitable for inclusion into database applications.

VoiceStream Messaging™ can be connected to analog lines or digitally via T-1, E-1 and ISDN.

VoiceStream IVR™

Voice Stream IVR is an interactive voice response run-time and development platform that provides the ability to easily create highly customized IVR applications, without programming. In addition to audiotex functions, VoiceStream IVR provides telephony access to SQL-compatible and Dbase-compatible databases.

Fax reception, creation and transmission are standard features which can be used for one and two call fax-on-demand applications. There exists programmable access to voice mailbox functions and features plus application access to UNIX commands and executables for additional application control.

System administration is performed through a full-screen menuing system that can be accessed on the system console, through remote terminals, over local area networks and via modem. Most system functions can be performed while the system is live and functional.

VoiceStream IVR™ provides a system configuration utility to customize the system parameters for individualized operation. System and applications prompts can be recorded as needed. System back-ups can be scheduled and applications restored.

VoiceStream IVR™ requires a minimum of 7 Mb of free RAM for a basic four port system and 1 Mb RAM for each additional four ports. The system files require 40 Mb of hard drive for a basic configuration with one language and 10 Mb for each additional language.

The system requires 10 Mb per hour of recorded speech, using standard OKIDATA 3 KB/sec compression. Higher and lower speech compression algorithms are available with Linkon technology.

VoiceStream IVR™ has comprehensive statistics and reporting features. It provides data detailing number of calls per port per day, total calls per hour and total call minutes per hour. As is the case with other CallStream products all data can be converted to Comma Separated Value format for inclusion into spreadsheet programs, or text format, suitable for inclusion into database applications. With VoiceStream IVR™ Developer's Tool kit custom reports can be created.

FaxStream™

FaxStream™ is easy to use fax-on-demand software which can be quickly implemented on a Sun SPARC platform. Routine inquiries and requests for information are easily handled 24 hours per day 7 days per week; whenever the caller desires the information. With just one telephone call the customer, prospect or anyone else can look up and access information of interest. Voice prompts guide callers in the retrieval of documents using reference numbers or catalogues. FaxStream™ then transmits the requested information by fax/modem through a one call or two call method.

The one call method allows callers using fax telephone to request and receive information during the same telephone call. After document selection the caller is prompted to press the "start" button on the fax telephone.

The two call method allows callers using any common telephone to look up and select the required information. After the caller identifies the telephone number of the destination fax machine, the caller simply hangs up and in seconds information arrives.

FaxStream™ requires a minimum of 6 Mb of unused RAM for a basic four port system and 1 Mb RAM for each additional four ports. Thirty megabits of hard disk storage are required for the system software and 35 Mb disk capacity for 2.5 hours of speech storage.

Linkon

Linkon Corporation, in addition to their DSP based communications processing technology, (see chapter 13) Linkon provides an application suite which works with TeraVox™ and ProVox™, the company's "C" tool kits (discussed later in this chapter).

This application suite provides developers a base structure from which they can create customized applications which meet their unique specifications. The suite is composed of LinkFax™, Linkon's fax-on demand application and LinkMail™ a full featured integrated messaging system for voice, fax and E-mail.

LinkMail™

LinkMail is Linkon's messaging system which integrates voice, fax and E-mail messages in a single virtual mailbox. LinkMail™, written in C, is available in three forms. Level One works with voice messaging only. Level Two supports voice and fax. Level Three supports voice, fax and E-mail (modem) interfaces. All messages are treated equally, no matter whether voice, fax or E-mail; truly Unified Messaging. Most companies take the vanilla LinkMail™ and add their own flavoring, integrating the result with their product or as their product. Source code is what Linkon sells; there are no royalties, licenses or other fees.

Mailbox owners can send messages to each other and receive messages from anyone. Messages can be replied to, forwarded, saved or deleted. Each message has statistics associated with it; these statistics can be heard, analogous to reading the envelope of a letter.

Distribution lists can be used for sending or forwarding messages. Each mailbox can have up to 999 such lists.

LinkMail™ has three prompt levels available; terse, standard and verbose. Mailbox owners may switch among these as often as desired.

Multiple language support is a standard feature. Each mailbox owner selects the language to hear for prompts. Mailbox owners may record their own greetings or use the default greeting. Each mailbox may have one normal greeting and two alternates.

Personal Identification Numbers protect the privacy of each mailbox. The PIN can be changed an unlimited number of times by the mailbox owner or an administrator.

Once LinkMail™ is purchased, you are free to do with it what you want (except resell the source code; of course).

LinkMail™ Noteworthy Features:

⇒ LinkMail™ permits callers to interrupt prompts to make selections at any time.

⇒ Certain keys, especially * (cancel) and # (entry finished) work identical everywhere

⇒ Messages are played in order received

⇒ Messages marked by mailbox holder for deletion are not deleted until call is ended

⇒ LinkMail works on many phone lines simultaneously

⇒ There can be a different language on each phone line

⇒ Notification of the arrival of messages is automatic

⇒ Forwarded messages can be forwarded again

⇒ For conservation of resources, copies of messages delivered to multiple destinations do not contain audio; they just point to the original message. A messages is removed from LinkMail™ when it has been deleted by all recipients.

⇒ Sending mail to oneself is allowed

⇒ Owners can exit without hearing all new messages. Unheard messages will remain classified as new until they are played.

⇒ When recording a message it can be played an unlimited number of times before it is accepted, re-recorded or deleted.

⇒ Messages not explicitly deleted are considered saved

⇒ Audio and other data can be imported into LinkMail as part of the message.

⇒ In the unlikely event of trouble, error messages are sent to the standard error device in human-readable format, with meaningful text. Error messages are spoken over the phone to the caller as well.

⇒ All LinkMail™ data are stored in a speech file. There are no external databases.

LinkFax™

LinkFax™ is Linkon's software for fax-on-demand and the complete integration of fax services. It is written in C and available in source code form. Most companies use the fax-on demand capabilities of LinkFax™ to provide information to clients and potential customers. As is the case with LinkMail™, many organizations start with the vanilla LinkFax™, add their own flavoring, integrate the result with their products and sell the result.

Callers are greeted with a welcome message and asked to choose a category of data. Within a chosen category can be choices of sub-categories or individual items to be sent. There is no limit to the depth of categories, subcategories, sub-sub-categories, et cetera.

Speaker independent speech recognition has been integrated into LinkFax™. Callers may use DTMF tone, speech recognition or both.

Information selected may be faxed back during the same call (one call faxing) or sent in a return phone call (two call faxing).

LinkFax™ Noteworthy Features

⇒ Speaker independent speech recognition has been completely integrated into LinkFax™.

⇒ Callers select one call or two call faxing.

⇒ Available items are announced by name.

- ⇒ Callers are not required to remember "magic numbers" to obtain information.
- ⇒ An unlimited number of fax objects may be selected for transmission.
- ⇒ Callers receive verbal reports of all items selected at any time.
- ⇒ LinkFax™ prompts can be interrupted with DTMF tone.
- ⇒ Certain keys, "0" (states what was selected), "*" (finished with the category) and "#" (finished selecting items) perform identically everywhere.
- ⇒ If the caller selects all items within a category, LinkFax™ ascends the tree structure automatically.
- ⇒ Items are sent in the order selected.
- ⇒ LinkFax™ can be active on many phone lines simultaneously.

Trigon Business Alliance

Trigon Business Alliance has created the industry's only application which allows complex engineering and schematic diagrams to be sent or received automatically from widely used computer-aided design (CAD) software to any location equipped with standard fax technology.

FaxDraw™

Their FaxDraw™ software enables organizations with libraries of centralized CAD drawings to make them available to service or other personnel via secure telephone retrieval. FaxDraw™ version 1.0 provides fax on demand capabilities for delivering high-resolution drawings to standard fax machines. The software operates on Sun SPARCworkstations™ with AutoDesk's AutoCAD design software and Linkon Corporation's Faxpeak™, a fully integrated fax card.

The FaxDraw™ software provides a secure interface that allows users to select drawings from an AutoCAD screen and fax them to remote locations. When using a DTMF tone phone, users may also call a FaxDraw™ server and retrieve drawings from remote locations via fax modem. These inbound FaxDraw™ requests are subject to the multiple layers of security available in the Solaris™ operating system. Various levels of security can be customized, as needed, for a particular user's business.

FaxDraw™ supports A and B (8.5" X 11" to 11" X 17") size portrait and landscape style tiling output for faxing and then assembling B, C (17" X 22"), D (22" X 34") and E (34" X 44") size drawings to meet end user field work print size requirements. This fax application requires a minimum of 32 Mb memory, 40 Mb disk storage and 32 Mb swap space be available on the SPARC™ system. The FaxDraw™ software runs with AutoCAD release 12 or newer operating on SunOS or Solaris™ 2.4 or more recent.

APPLICATIONS GENERATORS

CallStream Communications

VoiceStream Plus/X™

CallStream Communications Inc., in addition to the applications software detailed above, CallStream has developed **VoiceStream Plus/X™**. VoiceStream Plus/X™ is an advanced interactive applications generator that combines the collective power of easy to build voice scripting and fax retrieval, with databases and other libraries of information, in order to create client-server applications across open system networks. With VoiceStream Plus/X™ interactive voice and fax response applications can be designed and built effortlessly, with minimal training in a "Top-Down" approach.

VoiceStream Plus/X™ features many enhanced technologies such as multiline speaker independent speech recognition, text-to-speech, interactive voice and fax all available

In every port any time it is needed for easy integration into an application.

It has a Tool Kit composed of "nodes" which are the building blocks of the application. For examples, Message and Record nodes are used to speak or record phrases, while the FaxQ and FaxSend nodes are used to schedule and fax information residing on the network. Nodes are manipulated within the workspace of the display. Other nodes allow you to access the UNIX shell for external command execution, and access to database. The addition of speech recognition, voice commands and text-to-speech for ASCII data sources such as databases and E-mail greatly increases the ability of telephones to act as data terminals into corporate data residing open system networks.

Graphical icons permit users to drag and drop nodes which makes the creation of applications quite easy. Node setup dialog gives detailed control over the actions to be taken by the nodes. The system uses these nodes to build interactive decision trees or "voice menus". Input to the system can be through DTMF or speech recognition while output from VoiceStream Plus/X™ can be spoken or faxed. This process permits applications to be quickly built. The nodes in the tree can speak a phrase, number or obtain a value from a database, prompt input from a telephone, create fax response or save information to disk. All applications are developed in a top down approach.

VoiceStream Plus/X™ has the ability to execute proven external commands and programs locally and remotely via LAN. By using its SYSCAL node, programming languages can be used to build custom interfaces to existing databases and applications; this greatly extends this software's' capability.

Voicetek Corporation

Generations

Voicetek Corporation developed a sophisticated applications generator called Generations™. It is a single, integrated software environment that is used for the design, documentation and deployment of a wide variety of interactive voice response applications, from the most simple announcement-type application to highly advanced voice system integrated with a database. With an open, modular design, Generations™ is designed to support:

⇒ an object-oriented user interface for application development

⇒ multiple heterogeneous computer operating platforms in traditional and fault-tolerant environment

⇒ a cross-platform development infrastructure that facilitates the development of portable applications that can be targeted to the largest possible user base

⇒ a multimedia environment where such technologies as fax, speech recognition and text-to-speech can be integrated within a voice application to enhance the efficiency of the application

⇒ a client-server environment that allows users to tie together a variety of hardware and system software types across a network

⇒ access to host databases residing locally or at remote sites

⇒ modular architecture for easy growth form a single system, minimum configuration of a few ports to multi-node systems linked over a LAN/WAN topology

The flexibility of Generations™ is the result of an underlying object-oriented structure that generates a unique set of reusable objects referred to as cells. These cells were designed to include the fundamental actions common to all voice processing applications. The development of an application is a matter of simply selecting the appropriate cells and building the call flow for the solution you need. The administrative facilities are also provided to control the applications, modify prompts and report system statistics.

Applications are built, therefore, by linking a series of cells together, and connecting them to model the actions required in a given call flow. Each cell type is defined by set of parameters that tell the cell exactly how it's suppose to act. These parameters default to the most commonly used value, but if it doesn't meet the requirements of the application, the parameter can be changed within a specified range. Such parameters might include identifying the number of prompts to play or the amount of time allowed for DTMF input.

In all voice processing applications, data such as DTMF input must be collected, stored and manipulated for a short period of time so that it can be used by different cells. Generations™ provides an elegantly simple mechanism for this purpose called buffers. The data storage contained provides application specific data storage locations. There are system buffers defined by Generations™, and user buffers definable by the developer. These buffers can handle information like ANI digits, DID digits, current messages or mailbox Ids.

Generations™ provides user access to functions through five basic interfaces: application management, prompt management, system management, data access and reports and statistics.

Through these interfaces, users totally control the development and operation of the voice application, using either a character-based interface or the more flexible graphical user interface if your terminal supports X Windows. These interfaces provide the developer all the needed tools to create any kind of voice application.

Application Management

Offered under a development license, Application Management, consists of a character or object-oriented graphical application editor that is used to create and edit applications. The cell types mentioned earlier are the standard building blocks used to design an application, which controls the activity on a port and performs numerous actions from answering a telephone call to hanging it up. If programmers want to access special databases or technologies within the application not yet provided within the palette of cells offered, Generations™ provides an envelope that allows the seamless integration of C code via the User Functions. Applications can be created, edited and loaded on-line without interrupting existing services.

Offering the first object-oriented interface for the development of voice applications, Voicetek built the editor to reflect varying levels of programming skills and leveraged the familiarity users would have with the Motif-style look, including icons, pull-down and pop-up menus, and point-and-click flexibility in moving objects on the screen. By building these elements within the interface, Generations™ empowers users to have total control of their application.

Prompt Management

A major factor in all voice applications is the creation and loading of prompts. With Generations™' Prompt Management interface, prompts can be reviewed, recorded, installed, deleted, backed up to removable media, restored and distributed over LAN or WAN. They can be loaded on-line over the telephone, through a microphone, or from a tape; and the process can be semi or fully automatic. Users can record individual prompts or lists of prompts, or record with DTMF prompt numbers. The prompts will be phased in automatically while applications remain on-line.

The optional Prompt Manager is an interactive voice editing tool that allows you to retrieve a prompt from the VRU or a back-up file on the application processor, examine and modify its contents, and either distribute the editor voice prompt to the VRU(s) or save it on the application processor.

System Management

Based on the same flexibility in the system management and administration as exists for the development of applications, Generations™ allows users to load, run or unload an application. What's important to understand is that users can bring on a new revision of an application or move an application to another port while the system is on-line. If a caller happens to be on-line at the time, the changes on that port will take effect after the caller hangs up.

Generations™ on the application processor is connected to a voice response unit either by RS-232-C or Ethernet. Generations™ can support multiple VRUs to expand to virtually unlimited port and storage capability by networking systems and clusters of systems together. Expansions depend only on the application processor the systems are connected to. This capability is especially critical in environments that handle heavy voice processing traffic, such as in telephone companies, large service bureaus or large corporate applications.

If a system includes systems remotely networked, administration can be centralized. Through the graphical Systems Monitor, disk and port status can be viewed easily and overall activity monitored. In addition, with 100% up-time an important requirement for almost all applications, Generations™ supports major alarm handling should communications be lost between the application processor and the VRU. When an alarm is set off, the transaction log file within Generations™ displays a message, and the user will see alarm notification on the System Monitor.

Statistics and Reports

Understanding the need to track statistical information for such activities as billing and to justify services, Generations™ offers a rich choice of standard reports that can be selected through the Statistics and Reports interface. Reports can be viewed and run without interrupting the operation of the application, and they can be sent to a printer for a hard copy print-out. Reports available are call detail, cell usage, trunk usage, subscriber information and transaction log. Custom reports can be generated exporting underlying statistics.

Data Access

Interfacing with databases is critical to the smooth access and input of information within all IVR applications. Interfaces can be built easily between Generations™ and SQL relational databases through the SQL Easy option, or specific databases meant to handle functions like billing through a User Function. Generations™ provides an SQL cell type that allows users to delete, insert, select and update data.

Generations™ also provides the facility to access any data either through its gateway products or through a User Function cell. To integrate communications with a host within an application, developers simply choose one of three cells that allow information to be entered into an interactive terminal session or extracted from it.

Generations™ also supports two internal databases, which, through its Database Editor, allows developers to create Message and Information Databases. The Message Database, important in voice messaging applications, consists of mailboxes that are associated with a number, usually the phone number of the subscriber who will access the box for the message deposited in it. More than one message database can be supported within Generations™ to accommodate multiple applications. Messages can be retrieved either first in, first out (FIFO), or last in, first out (LIFO).

The Information Database provides a facility to store different kinds of data. Each Information Database consists of a set of records, which, in turn, consists of a series of fields. Each field may have a field description to help track the kinds of data being stored. These fields are the same for every record within the particular Information Database created.

Multimedia Support

In an age when information is delivered through several different kinds of media, it is imperative that a development and deployment software like Generations™ supports existing and emerging technologies. Voice/fax allows developers to easily integrate fax capabilities within a voice application. In order entry applications, for example, a copy of an order can be faxed to the caller, or a marketing department can automatically distribute by fax datasheets on the company's products.

Similarly, voice recognition provides developers with the tool to integrate voice recognition within an application to support rotary-phone users or diverse ethnic neighborhoods where multiple languages are spoken.

C Tool Kits

Linkon Corporation

Linkon Corporation has developed two C tool kits for communications processing. These tool kits have been used by many developers to create their applications. Developers have reported applications development time savings of 50 to 75%.

TeraVox™

TeraVox™ is a fully featured application development environment capable of supporting any type of application requiring interactive communications. The API is a set of powerful, high-level C programming language libraries of complex computer telephony functions. TeraVox™ provides an optimized file management system for voice data storage. Some of TeraVox™'s unique features grew directly out the experience of operating a large service bureau supporting 1,500 telephone lines, and thousands of hours of message storage. The API is completely scaleable from the small 4-port systems to the very large multi-node systems required by large corporations.

With TeraVox™, the normal UNIX file system for all voice and database input/output is bypassed. A completely new application-specific file manager is provided which is custom tailored to the desired audio response environment. This file manager differs from standard file systems in a number of ways. All control structures are memory resident for high performance. Control structures are replicated and checksummed on file storage. Raw disk I/O is used so that the UNIX disk caches are not disturbed, and "elevator" seeking is performed to reduce disk arm movement. The TeraVox™ file manager allows a developer to create and maintain lists of objects like mailboxes, or messages in a mailbox. These kinds of features and many others give TeraVox™ robust, high performance voice storage transfer rates. Typical read/write rates are 700 KB to 1 MB per second.

TeraVox™ has over 25 command line utilities for system administration and reporting. Standard formation captured includes call counts by application by hour, system call counts by hour, system call statistics and event logging.

TeraVox™ contains a C function library that contains over 90 entries, through which user applications communicate with the TeraVox™ system tasks. In this way, all application development is performed using standard UNIX tools. C programmers have virtually no learning curve and can begin developing voice applications almost immediately. The TeraVox™ library is simple enough to provide the productivity increases usually associated with applications generators, but powerful enough to create truly innovative voice applications.

An Inter-Task Communications (ITC) facility tightly couples TeraVox™ drivers and system tasks. The TeraVox™ ITC is streams based at the task level, providing for deep queuing of commands.

Special extensions to the UNIX kernel are supplied in the form of device drivers, eliminating the need for custom kernel objects. These extensions supply high resolution timers and some address conversion services required by TeraVox™ system tasks.

ProVox™

ProVox™ is an open client-server implementation of C programming language libraries for complex computer telephony functions. "Open" is defined by the fact that it can use any UNIX database for file management, and processes run both on the telecommunications gateway as well as on the server. Built on TeraVox™, ProVox™ is designed for applications that need to work with very large databases and complex file management. ProVox™ was developed under the direction of a major telco, and was designed for WAN telecommunications.

SUMMARY

Sufficient CTI software is available for the Sun Platform to meet the needs of most developers. The range of this software is from off-the-shelf applications, through easy to use application generators to C tool kits for the more adventurous. If need be, a developer can use Linkon's DDI (Driver Development Interface) to create a custom application without the overhead of an application generator.

Those seeking Sun computer telephony technology without the application development and integration issues can purchase a pre-configured TeleFusion server from Access Worldwide Channel Integration (see chapter 12).

|18|

Java™

Java™, as referred to by Scott McNealy, CEO of Sun Microsystems Inc., is the company's applications building technology that you "write once, run anywhere". It is the basis of Hot Java™, combined with Netra™ which is taking the Internet by storm. It is so good that it will soon be seen in many other places.

The Java™ programming language and environment is designed to solve a number of problems inherent to modern programming practices. It is a language that started as part of a much larger project at Sun. That project intended to an advanced software for use in consumer electronics. Most consumer electronics devices are small, reliable, portable and are built upon technology which requires the use of real-time embedded systems. When the project started the system architects intended to use C++ but encountered numerous problems. Early out these were thought to be just compiler technology problems, but as additional development time was put into the project the problems continued. Finally, it was determined the problems would be best solved by changing the language and that is exactly what was done.

Java™ is a dynamic object-oriented, multithreaded high performance programming language which is architecture neutral and portable. It is robust and secure and works well in distributed environments.

> *Blocks of indented text in this chapter will be used for examples of the Java™ issues covered. To facilitate understanding of this incredible language we are using a fictional software company called RunSoft Inc., which produces tutorial programs for phys-*

ics. RunSoft software was designed to interact with the user, providing not only text and illustrations in the manner of a traditional textbook, but also providing a set of software lab benches on which experiments can be set up and their behavior simulated. For example, the most basic software version allows students to put together levers and pulleys to see how they interact.

Java™ Is Simple

The Java™ architects' goal was to build a system that could be programmed easily without a lot of esoteric training and leveraged today's standard programming practices. Most programmers working today use C, and most programmers using object-oriented programming use C++. However, even though C++ was found unsuitable, Java was designed as closely to C++ as possible in order to make the system more comprehensible.

Java™ omits rarely used, poorly understood, confusing features of C++ that experience has proven bring more pain than benefit. The omitted features primarily consists of operator overloading, multiple inheritance and extensive automatic coercions. However, the Java™ language does have method overloading.

The engineers added auto garbage collection thereby simplifying the task of Java™ programming but at the same time making the system somewhat more complicated. A good example of a common source of complexity in many C and C++ applications is storage management; the allocation and freeing of memory. By having automatic garbage collection the Java™ language not only makes the programming task easier, it also dramatically cuts down on bugs.

To further simplify Java™ the engineers kept it small. The goal was to enable the construction of software which could run stand-alone in small machines. The size of the basic interpreter and class support is about 40K bytes; adding the basic standard libraries and thread support, essentially a self-contained microkernel, adds an additional 175K.

The people at RunSoft wanted to spend their time thinking about levers and pulleys, but instead spend considerable time on mundane programming tasks. Their core expertise was teaching not programming, however, programming is what they did. One of the more complicated of these programming tasks was figuring out where memory was being wasted across their 20,000 lines of code.

Java™ Is Object-Oriented

Object-oriented is unfortunately one of the most overused buzzwords in the industry. Object-oriented design is very powerful because it facilitates software reusability and the clean definition of interface.

Simply stated, object-oriented design is a technique that focuses the design on data (objects) and the interface to it. To use carpentry as an analogy, an object-oriented carpenter would be mostly concerned with the chair he was building, and secondarily with the tools used to make it. A non-object-oriented carpenter would think primarily of his tools. Object-oriented design is also the mechanism for defining how modules "plug and play".

The object-oriented facilities of Java™ are essentially those of C++, with extensions from Objective C for more dynamic method resolution.

The folks at RunSoft had many things in their simulation, among them, ropes and elastic bands. In their initial C version of the product they ended up with a pretty big system because they had to write separate software for describing ropes versus elastic bands. When they rewrote their application using object-oriented style, they found they could define one basic object that represented the common aspects of ropes and elastic bands, and then ropes and elastic bands were defined as variations, or subclasses, of the basic type. When it came time to add chains, it was easy because they could build on what had been written before, rather than writing a whole new object simulation.

Java™ Is Distributed

Java has an extensive library of routines for coping easily with TCP/IP protocols like HTTP and FTP. Java™ applications can open and access objects across the Net via URLs with the same ease that programmers accustom to when accessing a local file system.

The people at RunSoft initially built their programs for CD ROM. They however, had some ideas for creating interactive learning games which they wanted to test for their next product. For example, they wanted to allow students on different computers to cooperate in building a machine to be simulated. All the networking systems they had seen were complicated and required special esoteric software specialists. So, they gave up.

Java™ Is Robust

Java™ is intended for writing programs that must be reliable in a variety of ways. Java™ puts substantial emphasis on early checking for possible problems, later dynamic (runtime) checking and eliminating situations that are error prone.

One of the advantages of a strongly typed language like C++ is that it allows extensive compile-time checking from C, which is relatively lax; particularly method/procedure declarations. Java™ requires declarations and does not support C-style implicit declarations.

The linker understands the type of system and repeats many of the type checks done by the compiler to guard against version mismatch problems.

The single biggest difference between Java™ and C/C++ is that Java™ has a pointer model that eliminates the possibility of overwriting memory and corrupting data. Instead of pointer arithmetic, Java™ has true arrays. This allows subscript checking to be performed. In addition, it is not possible to turn an arbitrary integer into a pointer by casting.

The people at RunSoft had their application basically working in C pretty quickly, but their schedule continued to slip because of all the small bugs that kept slipping through. Tremendous problems occurred with memory corruption, versions out-of-sync and interface mismatches. What they gained because C let them pull strange tricks in their code, they paid for in quality assurance time. They also had to reissue their software after the first release because of the bugs which got through undetected.

Java™ does not eliminate all quality assurance issues but it does make it significantly easier.

Very dynamic languages like Lisp, TCL and Smalltalk are often used for prototyping. A major reason for their success is attributed to the fact that they are quite robust. You need not worry about freeing or corrupting memory. Programmers can be relatively fearless about dealing with memory because they don't have to worry about it getting corrupted. Java™ has this property and programmers find it quite liberating.

Dynamic languages are good for prototyping because they allow you to defer implementation details during the initial work of design. Java™ has exactly the opposite property; it forces you to make choices explicitly. With these choices comes assistance. You can write method invocations and if you get something wrong, you are informed about it at compile time. You don't have to worry about method invocation error. You can also get a lot of flexibility by using interfaces instead of classes.

Java™ Is Secure

Java™ is intended to be used in networked/distributed environments. Due to this capability much emphasis is placed on security. Java™ enables the construction of virus-free, tamper-free systems. The authentication techniques are based on public-key encryption.

There is a strong correlation between "robust" and "secure". For example, the changes to the semantics of pointers make it impossible for applications to forge access to data structures or to access private data in objects that they do have access to. This closes the door to most activities of virus.

> *Someone wrote an interesting "patch" to the PC version of RunSoft's system, and posted this patch to one of the major bulletin boards. Since it was easily available and added some interesting features to the system. Many people downloaded it. It hadn't been checked out by the people at RunSoft, but it seemed to work. Until the following April 1st, when thousands of folks discovered nude pictures popping up in their children's lessons. Needless to say, even though they were in no way responsible for the incident, the employees of RunSoft had a substantial amount of damage to control.*

Java™ Is Architecture Neutral

Java™ was designed to support applications on networks. In general, networks are composed of a variety of CPU and operating system architectures. To enable a Java™ application to execute anywhere on the network, the compiler generates an architecture neutral object file format. The compiled code is executable on many processors, given the presence of Java runtime system.

This is useful for networks and single system software distribution. In the current PC market, applications writers need to produce versions compatible with Intel based systems as well as with Apple Macintosh. With the PC market diversifying into many CPU architectures, and Apple moving off the 68000 towards the PowerPC, porting software to all platforms becomes unfeasible. With Java™, the same version of the application runs on all platforms.

The Java™ compiler does this by generating bytecode instructions which have nothing to do with a particular computer architecture. Rather they are designed to be both easy to interpret on any machine and easily translated into native machine code on the fly.

> *RunSoft is a small company that started out producing applications for the PC market. After all, that is the largest market. After a while, the company grew large enough that they could, and needed to, do a port to the Macintosh. As it would turn out the port was a larger effort than the RunSoft engineers ever anticipated. After spending considerable time and money the project was abandoned. Needless to say a port to the PowerPC Macintosh and MIPS NT machines was totally out of the question. RunSoft couldn't catch the wave when it was happening and a competitor jumped in and grabbed the opportunity.*

Java™ Is Portable

Java's™ architecture neutral design is a large part of its' portability. There is, however, much more to it than that. Unlike C and C++, there are no "implementation dependent" aspects of the specification. The size of the primitive data types are specified, as is the behavior of arithmetic on them. For example, "int" always means a signed two's complement 32 bit integer, and "float" always means a 32-bit IEEE 754 floating point number. Making these choices is feasible because essentially all interesting CPU's share these characteristics.

The libraries that are part of the system define portable interfaces. For example, there is an abstract Window class and implementations of it for UNIX, Windows and Macintosh.

The Java™ system itself is quite portable. The new compiler is written in Java™ and the runtime is written in ANSI C with a clean portable boundary. The portability boundary is essentially POSIX.

Java™ Is Interpreted

The Java™ interpreter can execute Java™ bytecodes directly on any machine to which the interpreter has been ported. Since linking is a more incremental and lightweight process, the development process can be much more rapid and exploratory.

As part of the bytecode stream, more compile-time information is carried over and available at runtime. This is what the linker's type checks are based on, and what the RPC protocol derivation is based on. It also makes programs more amenable to debugging.

> *The programmers at RunSoft spent considerable time waiting for programs to compile and link. They also spent a lot of time tracking down senseless bugs because some changed source files didn't get compiled, despite using a fancy "make" facility, which caused version mismatches. They had to track down procedures that were declared inconsistently in various parts of their program. Additional months were lost in the schedule.*

Java™ Is High Performance

While the performance of interpreted bytecodes is usually more than adequate, there are situations where higher performance is required. The bytecodes can be translated on the fly at runtime into machine code for the particular CPU the application is running on. Those accustom to the normal design of a compiler and loader will find this somewhat like putting the final machine code generator in the dynamic loader.

The bytecode format was designed with generating machine code in mind, so the actual process of generating machine code is generally simple. Reasonably good code is produced; it does automatic register allocation and the compiler does some optimization when it produces the bytecodes.

In interpret code Java™ is getting about 300,000 method calls per second on a Sun Microsystems SPARCstation™ 10. The performance of bytecodes converted to machine code is almost indistinguishable from native C or C++.

> *When RunSoft was starting up, they did a prototype in Smalltalk. This impressed the investors enough to get them funded, but really didn't help them produce the product. In order for them to make their simulations fast enough and the system small enough, the code had to be rewritten in C.*

Java™ Is Multithreaded

There are many things happening at the same time around the world. Multithreading is a method of building applications with multiple threads. Unfortunately, writing programs with many things happening at once can be much more difficult than writing in conventional single-threaded C and C++ style.

Java has a sophisticated set of synchronization primitives that are based on the widely used monitor and condition variable paradigm that was introduced by C. A. R. Hoare. By integrating these concepts into the language they become much easier to use and are more robust. Much of the style of this integration came from Xerox's Cedar/Mesa system.

Other benefits of multithreading are better interactive responsiveness and real-time behavior. This is limited, however, by the underlying platform. Standalone Java™ runtime environments have good real-time behavior. Running on top of other systems like UNIX, Windows, Macintosh or Windows NT limits the real-time responsiveness to that of the underlying system.

> *Many things were happening at once in their simulations. Ropes were being pulled, wheels were turning, levers were rocking and input from the user was being tracked. Because RunSoft had to write this in a singlethread form, all things that happened at the same time, even though they had nothing to do with each other, had to be manually intermixed. Using an*

"event loop" made things a lot cleaner, but it was still a mess. The system became fragile and hard to understand. They were pulling in data from all over the Net. Originally they were doing it one chunk at a time. This serialized network communication was very slow. When they converted to a multithreaded style, it was trivial to overlap all of their network communication.

Java™ Is Dynamic

In many ways Java is a more dynamic language than C or C++. It was designed to adapt to an evolving environment.

For example, one major problem with using C++ in a production environment is a side effect of the way that code is always implemented. If company A produces a class library, a library of plug and play components, and company B buys it and uses it in their product, then A changes it's library and distributes a new release, B will almost certainly have to recompile and redistribute their own software. In an environment where the end user gets A and B's software independently problems can result.

Further, if A distributes an upgrade to its libraries then all of the software from B will break. It is possible to avoid this problem in C++, but it is extraordinarily difficult and effectively means not using any of the language's 00 features directly.

RunSoft built their product using the object-oriented graphics library from 3DPC Inc. 3DPC released a new version of the graphics library which several computer manufacturers bundled with their new machines. Customers of RunSoft that bought these new machines discovered, to their dismay, that their old software no longer worked.

By making these interconnections between modules later, Java™ completely avoids these problems and makes the use of the object-oriented paradigm much more straightforward. Libraries can freely add new methods and instance variables without any effect on their clients.

Java™ understands interfaces; a concept borrowed from Objective C which is similar to a class. An interface is simply a specification of a set of methods that an object responds to. It does not include any instance variables or implementations. Interfaces, unlike classes, can be multiply-inherited. They can be used in a more flexible way than the usual rigid class inheritance structure.

Classes have a runtime representation. There is a class named Class, instances of which contain runtime class definitions. If, in a C or C++ program, you have a pointer to an object but you don't know what type of pointer it is, there is no way to find out. However, in Java™ finding out based on the runtime type information is straightforward. Because casts are checked at both compile-time and runtime, you can trust a cast in Java™. On the other hand C and C++, the compiler just trusts that you're doing the right thing.

It is also possible to look up the definition of class given a string containing its name. This means that you can compute a data type name and have it easily dynamically linked into the running system.

To expand their revenue stream, the employees of RunSoft wanted to architect their product so that new aftermarket plug-in modules can be added to extend the system. This was possible on the PC, but just barely. They had to hire a couple of new programmers because it was so complicated. This also added problems when debugging.

Java™ Is Successful

As of January 1996, Java™ has been licensed by several big-name companies, from Microsoft to Netscape Communications to IBM and Toshiba. It is literally an overnight success. Already programmers have used Java™ to create some 600 "applets" for everything imaginable. And because Java™ will be built into the next release of Netscape's Web browser, millions of personal computer owners will be users in no time at all.

In summary, the Java™ language provides a powerful addition to the tools that programmers have at their disposal. Java™ makes programming easier because it is object-oriented and has automatic garbage collection. In addition, because compiled Java™ code is architecture-neutral, Java™ applications are ideal for a diverse environment like the Internet.

|19|

The Future of Computer Telephony

In the past few years we have witnessed dramatic changes in the way we think, the way we work, the way we shop and the way we communicate. As a result, the world is shrinking, corporations are downsizing, the SOHO and telecommuter markets are exploding while our personal arsenal of resources has expanded globally.

Without leaving the comfort of our home or office we can accomplish many tasks that formerly required personal contact, transportation and many other inconveniences. We have decentralized and distributed everything.

Not long ago we waited days for the corporate MIS department to generate a report. We retrieved telephone messages from a secretary who would also complete our typing when time permitted. We reviewed our mail and manually forwarded a response. We returned our phone calls from our office, home or a pay phone inbetween. Boy, that all seems pretty strange today.

Computers and telephone systems began as separate centralized functions. Technological developments have increased the utility of each. Both are rapidly racing towards distributed pieces of intelligence along networks that have become strikingly similar. In fact, in some instances the same network is now used for both concurrently.

There is an interesting phenomenon driving this rapid evolution. It seems that almost annually the computer industry is introducing new systems with increased processing power and speed. The competitive environment of the computer market has fueled these advances, and provided sufficient incentive for price decreases as a product differentiator. This has exponentially increased the number of systems sold. These new systems have found their way into businesses and households alike. Included in these purchasers are the carriers.

During this same time period the carriers have continued to upgrade their networks with intelligent digital switches and cabling capable of handling higher bandwidth digital transmissions. They have integrated computer technology into their network and used an advanced signaling protocol (SS7) to manage the pieces. The carriers have begun the implementation of their Advanced Intelligent Network (AIN) which will provide enhanced functionality and network management capabilities (see Chapter 15). These enhancements will expand their service offering and provide network control to their larger users.

Businesses have continued to decentralized much of their internal data sourcing through the installation of internal LANs and external WANs. This has increased the utility of the data and exponentially increased the productivity of their workers. In addition, many have begun the integration of these data networks with voice messaging, fax and E-mail capabilities to centralize the corporate resources at the workstation level. This effort was first seem in the corporate call centers, and is now moving to the desktop of the managers and others. It is truly an exciting time.

The SOHO market and early adopter households are installing stand-alone and networked computers, at an increasing rate, many of which possess the capability of accessing external data sources and telephone networks for voice, fax and E-mail messaging. These individuals look to the telcos for this connectivity and service.

The Internet, the new ubiquitous business and entertainment resource, is yet another conduit in this increasingly networked society.

Intelligence is being added to copy machines, heating/cooling systems, corporate and household lighting systems, televisions and VCR's, microwaves, stoves, elevators and other areas too numerous to mention. Processing power is cheap and easily managed. Our homes and offices are getting smarter all the time.

Cellular and other mobile technologies are freeing us to roam freely about our planet while never being out of contact. We have single number service which follows us to wherever we might go, beepers, messaging centers and personal communicators which provide constant contact capabilities for those so inclined. One technology crossing the path of the other, managed by the intelligence we have permitted in our network. Today you can literally take communications capabilities as far as you would like.

Worldwide Client-Server

When you really think about it, the world is evolving into a highly distributed client-server network topology providing communications and interaction capabilities with anyone, anywhere as long as they have access to the network. The network is ubiquitous and further defined each day it is used.

Client-server technology puts the user in control. It is highly flexible and easily adapts to changing environments. It makes existing systems more useful and offers a tremendous performance growth path. In this structure everyone is capable of becoming clients and servers on the network and can vacillate between each role as needed.

The only difference between the networks used by businesses and those used by households will be the size and the type of components. Both will provide networked technology where access is centralized and available to all. The smart home will be no different than a smart business form a network access perspective. The physical facility of each will be centrally controlled by intelligent devices connected to the network and managed by the individual(s) controlling that particular network segment.

The future of computer telephony will be further expansion of the worldwide client-server. New exciting capabilities will be regularly added as our appetite and capabilities expand. Adherence to standards will permit this unification and growth. With technology the world is at our finger tips.

|20|

Where To Go for Help

Depending on your knowledge level and need their are specific places you can turn to for help. System integrators and value added resellers needing advice and desiring to purchase product can contact:

Access Worldwide Channel Integration
1426 Pearl Street
Boulder, CO 80302
(800)-730-6462

Sun Microsystems Inc.
Contact your local Sun Business Office, or call:
(800) USA 4 SUN

Original Equipment manufacturers seeking detailed product and pricing information should contact:

Linkon Corporation
140 Sherman Street
Fairfield, CT 06430
(203) 319-3175

Companies needing assistance with all or part of their computer telephony strategy, from design through implementation, may contact:

Millennium Partners
66 Pearl Street
Suite 504
New York, NY 10004
(212) 613-3137

Computer Telephony Glossary

This glossary is excerpted from Newton's Telecom Dictionary, which is available from Flatiron Publishing Inc. 800-542-7279 or 212-691-8215.

23B+D
An easy way of saying the ISDN Primary Rate Interface circuit. 23B+D has 23 64 Kbps (kilobits per second) paths for carrying voice, data, video or other information and one 64 Kbps channel for carrying out-of-band signaling information. ISDN PRI bears a remarkable similarity to today's T-1 line, except that T-1 can carry 24 voice channels. In ISDN 23B+D, the one D channel is out-of-band signaling. In T-1, signaling is handled in-band using robbed bit signaling. Increasingly, 23B+D is the preferred way of getting T-1 service since the out of band signaling is richer (delivers more information -- like ANI and DNIS) and is more reliable than the in-band signaling on the older T-1.

2B+D
A shortened way of saying ISDN's Basic Rate Interface interface, namely two bearer channels and one data channel. A single ISDN circuit divided into two 64 Kbps digital channels for voice or data and one 16Kbps channel for low speed data (up to 9,600 baud) and signaling. Either or both of the 64 Kbps channels may be used for voice or data. In ISDN 2B+D is known as the Basic Rate Interface. In ISDN, 2B+D is carried on one or two pairs of wires (depending on the interface) -- the same wire pairs that today bring a single voice circuit into your home or office. See ISDN.

700 SERVICE
An "area code" you dial, as in 1-700-XXX-XXXX. This "area code" has been reserved for the long distance companies to do so as they will. So dial 1-700- and then certain "office codes" (the next three digits), you'll get AT&T and a service it calls Easyreach. It will give a service allowing your calls to follow you. Dial other 700 office codes, get MCI and you'll be able to make intralata "long distance" phone calls, such as those from Manhattan to Westchester County -- for presumably cheaper than with the local phone company. 700 service is still evolving. Each carrier has the right to create whatever service it wants with its 700 numbers.

800 SERVICE
Eight-hundred service. A generic and common (and not trademarked) term for AT&T's, MCI's, US Sprint's and the Bell operating companies' IN-WATS service. All these IN-WATS services have 800 and 888 as their area code. Dialing an 800-number is free to the person making the call. The call is billed to the person or company being called. The telephone company suppliers of 800 services use various ways to configure and bill their 800-services. One way: you can buy an 800 line which will ring on your normal phone line. You'll only pay per call, but you won't receive any incoming call if you're making an outgoing one. You can even terminate an 800 number on your cellular phone or your home number. You might pay a flat monthly rate plus "so-much" (i.e. timed usage) per call. That timed usage may include some calculation for the distance the incoming call traveled. 800-Service is now available for calls from Canada and many countries overseas Europe, though many of those companies pay a normal toll call to reach an American "toll-free" 800-number.

800 Service works like this: You're somewhere in North America. You dial 1-800 or 1-888 and seven digits. Your local central office sees the "1" and recognizes the call as long distance. It ships that call to a bigger central office (or perhaps processes the call itself). At that central office it's processed, a machine will recognize the 800 or 888 "area code" and examine the next six digits. Those six digits will tell which long distance carrier to ship the call to locally. That carrier then holds the call, while it sends a data communications query to a computer containing a big "translation" database. That computer will look up the 800 number and send back information on how the carrier should route the call. There are many ways the carrier can route the call. The simplest is to say "Dial the following number, i.e. 212-691-8215 and connect the call. Or it might say Connect it to your dedicated T-1 number 12345 in New York City.

As a real-life example, Telecom Library, publishers of this book, has an 800 number, namely 800-LIBRARY (or 800-542-7279). When you call that number, MCI routes that number to the first available circuit on the dedicated T-1 line which we have rented from MCI's POP (Point Of Presence) to our New York City office.

Because 800 long distance service is essentially a database lookup and translation telephone service, there are endless "800 services" you can create. You can change the routing instructions based on time of day, day of week, number calling etc. In May of 1993 the FCC mandated that all 800 numbers will become "portable." That means that customers can take their 800 telephone number from one long distance company to another, and still keep the same number.

888 SERVICE
North America is running out of 800 numbers. So it is adopting a new prefix
-- 888. At the time of writing, it looked like the first 888 number would come in around April, 1996. That 888 prefix will have all the characteristics of today's 800 Service.

900 SERVICE
A generic and common (and not trademarked) term for AT&T's, MCI's, Sprint's and other long distance companies' 900 services. All these services have "900" as their "area code." Dialing a 900-number is free to the company or person receiving the call, but costs money to the person making the call. Here's the story: 900 service was introduced as the industry's "information service" area code. You'd dial 1-900-WEATHER, for example, and punch in some touchtones in response to prompts and you could hear the weather in Sydney, Australia or Paris, France, wherever you might be planning your next vacation. For this service, you'd be charged perhaps 75 to 95 cents a minute. And you'd get the bill as part of your normal monthly phone bill. That was the original idea. Then some people got the idea that 900 would make a wonderful porn number and they started advertising "Call 900-666-3333 and speak with Diana. She really wants you." And they started charging $5 a minute. When huge 900-call bills started appearing on people's bills, there was an outcry from many subscribers who wouldn't pay the bills. Some children called on their parents' phones. Employees made calls from work and the company's accountants went nuts. So the industry retreated from 900 porn. Then someone thought -- "Why not sell things through an 900 number?" We could sell a set of ginzu knives for just calling this 900 number. No messing with credit cards or checks. The bill goes straight on your phone bill. At about the same time someone thought that 900 numbers would be great for running sweepstakes. "Call up, register your name for a free trip with a racing car team to the Australian Indianapolis 500. Three lucky people will be chosen. The call will cost you only $2.75." So 900 services became a new type of gambling.

The long distance companies providing 900 services reacted predictably to some of the newer services. They clamped down on who they would sign up, which service and/or product you could, or could not sell. And, rather than charging "a piece of the action" as they did in the beginning, the long distance companies began to charge for them as if they were normal long distance calls: charge a set-up fee, a fee for carrying the call, a fee for collecting the money and a fee for the possibility of bad debts. There are variations on these themes.

A & B BITS
Bits used in digital environments to convey signaling information. A bit value of one generally corresponds to loop current flowing in an analog environment. A bit zero corresponds to no loop current, i.e. to no connection. Other signals are made by changing bit values; for example a flash-hook is set by briefly setting the A bit to zero.

A & B SIGNALING
Procedure used in most T-1 transmission links where one bit, robbed from each of the 24 subchannels in every sixth frame, is used for carrying dialing and controlling information. A type of in-band signaling used in T-1 transmission. A and B signaling reduces the available user bandwidth from 1.544 Mbps to 1.536 Mbps.

A/D CONVERTER
Analog to Digital converter, or digitizer. It is a device which converts analog signals (such as sound or voice from microphone), to digital data so that the signal can be processed by digital circuit such as a digital signal processor.

ABCD SIGNALING BITS
These are bits robbed from bytes in each DS-0 or T-1 channel in particular subframes and used to carry in band all status information such as E&M signaling states.

ACCESS CODE
A series of digits or characters which must be dialed, typed or entered in some way to get use of something. That "something" might be the programming of a telephone system, a long distance company, an electronic mail service, a private corporate network, a mainframe computer, a local area network. Once the user dials the main number for the service he must then enter his assigned Access Code to get permission use the system. An Access Code becomes an Account Code when it is used for identifying the caller and doing the billing. Access Code may also mean the digit, or digits, a user must dial to be connected to an outgoing trunk. For example, the user picks up his phone and dials "9" for a local line, dials "8" for long distance, dials "76" for the tie line to Chicago, etc. In programming a phone system such as Northern Telecom's Norstar, there are Access Codes to begin Startup, Configuration programming, and Administration programming.

ADJUNCT PROCESSOR
A computer outside a telephone switching system that "talks" to the switch and gives it switching commands. An adjunct processor might include a database of customers and their recent buying activities. If the database shows that a customer lives in Indiana, the call from the customer might be switched to the group of agents handling Indiana customers. Adjunct processors might be concerned with energy management, building security etc.

ADPCM
Adaptive Differential Pulse Code Modulation. A speech coding method which calculates the difference between two consecutive speech samples in standard PCM coded telecom voice signals. This calculation is encoded using an adaptive filter and therefore, is transmitted at a lower rate than the standard 64 Kbps technique. Typically, ADPCM allows an analog voice conversation to be carried within a 32Kbit digital channel; 3 or 4 bits are used to describe each sample, which represents the difference between two adjacent samples. Sampling is done 8,000 times a second. In short, ADPCM, which many voice processing makers use, allows encoding of voice signals in half the space PCM allows.

ADSI
Analog Display Services Interface. ADSI is a Bellcore standard defining a protocol on the flow of information between something (a switch, a server, a voice mail system, a service bureau) and a subscriber's telephone, PC, data terminal or other

communicating device with a screen. The simple idea of ADSI is to add words to, and therefore a modicum of simplicity of use to a system that usually uses only touchtones. Imagine a normal voice mail system. You call it. It answers with a voice menu. Push 1 to listen to your messages, 2 to erase them, 3 to store them, 4 to forward them, etc. It's confusing. You have to remember which is which. ADSI is designed to solve that. It's designed to send to your phone's screen the choices in words that you're hearing. You then have the choice of responding to what you hear or what you see. Your response is the same -- a touchtone button. ADSI's signaling is DTMF and standard Bell 202 modem signals from the service to your 202-modem equipped phone. From the phone to the service it's only touchtone. With ADSI, you don't hear the modem signaling because every time the service gets ready to send you information, it first sends a "mute" tone. ADSI works on every phone line in the world.

For ADSI to work visually, you'll need a special ADSI-equipped phone or a piece of ADSI software in your PC. The nice feature of ADSI is that the standard is so flexible, it can work on cheap phones with a small display and more expensive phones with a bigger display and on a PC with a real big display. These three diagrams show a little of the basic concepts behind ADSI -- Information Page Mapping and Softkey Mapping.

ADVISORY TONES
Signals such as dial tone, busy, ringing, fast-busy, call-waiting, camp-on and all the other tones your telephone system uses to tell you that something is happening or about to happen.

AGENT
1. This term comes from the huge telephone call-in reservation centers which the airlines, hotels and car rental services run. An agent is the person who answers your call, takes your order or answers your question. Agents are also called Telephone Sales Representatives or Communicators. The term "agent" was first used in the airline business. It came from gate or counter ticket agent.

2. An "Agent" is the person or persons you have legally authorized to order your telephone service and equipment from telephone companies.

AGGREGATE BANDWIDTH
The total bandwidth of channel carrying a multiplexed bit stream.

AIN
Advanced Intelligent Network. A term promoted by Bellcore (Bell Communications Research Inc.), adopted by Bellcore's owners, the regional Bell holding companies, and by AT&T and virtually every other phone company to indicate the architecture of their networks for the 1990s and beyond. While every phone company has a different interpretation of what their AIN is, there seems to be two consistent threads. First, the network can affect (i.e. change) the routing of calls within it from moment to moment based on some criteria other than the

normal, old-time criteria of simply finding a path through the network for the call. Second, the originator or the ultimate receiver of the call can somehow inject intelligence into the network and affect the flow of his call (either outbound or inbound). The concept of AIN is simple. Before calls are sent to their final destination, the network queries a database. "What should I do at this very moment with this phone call?" Depending on the response, depends the disposition of the call. That database may belong to the phone company. Or it may belong to the customer. It makes no difference, so long as they're connected. And various carriers (phone companies) have proposed and implemented various ways of joining these databases. Initial AIN services tend to be focused on inbound 800 toll-free calls. Although no two phone companies seem to have the same idea as to what an Advanced Intelligent Network is, (some call it just an Intelligent Network), it generally includes three basic elements:

1. Signal Control Points. SCPs. Computers that hold databases in which customer-specific information used by the network to route calls is stored.

2. Signal Switching Points. SSPs. Digital telephone switches, which can talk to SCPs and ask them for customer-specific instructions as to how the call should be completed.

3. Signal Transfer Points. STPs. Packet switches that shuttle messages between SSPs and SCPs.

All three communicate via out of band signaling, typically using Signaling System 7 (SS7) protocol. The AIN has increased in complexity, as carriers have added voice response equipment that can prompt callers to enter further instructions as to how they'd like their call handled. Despite the differences between AIN networks, all work fundamentally the same, according to Mark Langner of TeleChoice, Verona, NJ: The SS7 identifies that a call requires intelligent network processing. The SSP creates a query to find out how this call should be handled. The query is passed via out-of-band signaling through STPs to an SCP. That interprets the query based on the criteria in its database and information provided by the SSP. Once the SSP has determined how the calls is to be handled, it returns a message through STPs to the SSP. This message instructs the SSP how the call should be handled in the network. According to Langner, the number of actions that could take place at the SCP are truly infinite. The call could be translated into a different number for completion. It could be routed to a user's private network for on-net handling. It could be sent to a voice response unit in the carrier network, where a message is played to the caller. Or it could even be blocked, preventing completion of the call.

ALTERNATE ROUTING
A feature used with long distance calls that permits the telephone system (typically a PBX) to send calls over different (alternate) phone lines. It might do this because of congestion of the primary phone lines the calls would normally be sent over. Alternate routing is often confused with LEAST COST ROUTING, in

which the telephone system chooses the least expensive way (available at that time) to route that call. Least Cost Routing typically works with so-called "look-up" tables in the memory of the PBX. These tables are put into the PBX by the user. The PBX does not automatically know how to route each call. It must be told by the user. That "telling" might be as simple as saying "all 312 area codes will go via the AT&T FX line." Or it might be as complex as actually listing which exchanges in the 312 area code go by which method. LEAST COST ROUTING tells the calls to go over the lines which are perceived by the user to be the least cost way of getting the call from point A to point B. Alternate routing happens when the least cost routes get congested and alternate routes (typically more expensive) are found from the look-up tables in the PBX's memory.

ANALOG
Comes from the word "analogous," which means "similar to." In telephone transmission, the signal being transmitted -- voice, video, or image -- is "analogous" to the original signal. In other words, if you speak into a microphone and see your voice on an oscilloscope and you take the same voice as it is transmitted on the phone line and ran that signal into the oscilloscope, the two signals would look essentially the same. The only difference is that the electrically transmitted signal (the one over the phone line) is at a higher frequency. In correct English usage, "analog" is meaningless as a word by itself. But in telecommunications, analog means telephone transmission and/or switching which is not digital. Outside the telecom industry, analog is often called linear and covers the physical world of time, temperature, pressure, sound, which are represented by time-variant electrical characteristics, such as frequency and voltage.

ANALOG DIGITAL CONVERTER
An A/D Converter. Pronounced: "A to D Converter." A device which converts an analog signal to a digital signal.

ANI
Automatic Number Identification. A phone call arrives at your home or office. Somewhere in that phone call is a series of digits which tell you the phone number of the phone calling you. These digits may arrive in analog or digital form. They may arrive as touchtone digits inside the phone call or in a digital form on the same circuit or on a separate circuit. You will need some equipment to decipher the digits AND to do "something" with them. That "something" might be throwing them into a database and bringing your customer's record up on a screen in front of your telephone agent as he answers the call. "Good morning, Mr. Smith." Some large users say they could save as much as 30 seconds on the average IN-WATS call if they knew the phone number of the person calling them. They would avoid asking regular customers for routine identification information since it would all be there in the database. ANI is touted as one of ISDN's most compelling advantages -- but it is really an advantage of Signaling System 7 (and therefore distinct from ISDN) and you don't need ISDN to get ANI. In the US, there are various types of "ANI." There's the ANI you get from a long distance phone

company, which may arrive over the D channel of an ISDN PRI circuit or on a dedicated single line before the first ring. In contrast, the signaling for Caller ID, as delivered by a local phone company, is delivered between the first and second rings. In Canada, caller ID for both local and long distance is delivered in the same technical way
-- between the first and second rings. In the US, there are no accepted standards, as yet. In November of 1995, local phone companies in the US are scheduled to deliver both local and long distance ANI exactly as they do today in Canada -- between the first and second rings. Thus normal dial-up users, who subscribe to caller ID (and usually pay a few extra dollars a month for the privilege) will be able to figure who's calling before, or as they pick up the incoming call. At one stage, ANI was not available in many states. But those restrictions are disappearing. There are some people who believe ANI is long distance and delivered by long distance phone companies; and Caller ID is local and delivered by local phone companies. And these same people believe the technologies of delivery are different. In reality, ANI and Caller ID are rapidly becoming synonymous.

AREA CODE
A three-digit code designating a "toll" center in the United States, Canada and Mexico. Until January, 1995 the first digit of an area code was any number from 2 through 9. The second digit was always a "1" or "0." In January 1995, North America (i.e. the US and Canada) adopted the North American Numbering Plan (NANP) and second digits could be any number. This dramatically increased the number of possible area codes -- from 152 to 792 and the number of phone numbers to more than six billion.

ASR
1. Automatic Speech Recognition.

2. Automatic Send-Receive teletype or telex machine. Such a machine, if left on and loaded with paper, will receive incoming messages and print them, even when nobody is present.

ASYNCHRONOUS TRANSFER MODE
ATM is the technology selected by the Consultative Committee on International Telephone & Telegraph (CCITT) International standards organization in 1988 (now called the ITU-T) to realize a Broadband Integrated Services Digital Network (B-ISDN). It is a fast, cell-switched technology based on a fixed-length 53-byte cell. All broadband transmissions (whether audio, data, imaging or video) are divided into a series of cells and routed across an ATM network consisting of links connected by ATM switches. Each ATM link comprises a constant stream of ATM cell slots into which transmissions are placed or left idle, if unused. The most significant benefit of ATM is its uniform handling of services, allowing one network to meet the needs of many broadband services. ATM accomplishes this because its cell-switching technology combines the best advantages of both circuit-switching (for constant bit rate services such as voice and image) and packet-switching (for

variable bit rate services such as data and full motion video) technologies. The result is the bandwidth guarantee of circuit switching combined with the high efficiency of packet switching.

AUDIOTEX
A generic term for interactive voice response equipment and services. Audiotex is to voice what on-line data processing is to data terminals. The idea is you call a phone number. A machine answers, presenting you with several options, "Push 1 for information on Plays, Push 2 for information on movies, Push 3 for information on Museums." If you push 2, the machine may come back, "Push 1 for movies on the south side of town, Push 2 for movies on the north side of town, etc."

AUTOMATIC CALL DISTRIBUTOR
ACD. A specialized phone system designed originally for handling many incoming calls, now increasingly used by companies also making outgoing calls. You receive and make lots of phone calls typically to customers. You need an ACD. Once used only by airlines, rent-a-car companies, mail order companies, hotels, etc., it is now used by any company that has many incoming calls (e.g. order taking, dispatching of service technicians, taxis, railroads, help desks answering technical questions, etc.). There are very few large companies today that don't have at least one ACD. Many smaller companies, like the company that publishes this dictionary, also has one.

An ACD performs four functions. 1. It will recognize and answer an incoming call. 2. It will look in its database for instructions on what to do with that call. 3. Based on these instructions, it will send the call to a recording that "somebody will be with you soon, please don't hang up!" or to a voice response unit (VRU). 4. It will send the call to an agent as soon as that operator has completed his/her previous call, and/or the caller has heard the canned message.

The term Automatic Call Distributor comes from distributing the incoming calls in some logical pattern to a group of operators. That pattern might be Uniform (to distribute the work uniformly) or it may be Top-down (the same agents in the same order get the calls and are kept busy. The ones on the top are kept busier than the ones on the bottom). Or it may be Specialty Routing, where the calls are routed to answerers who are most likely to be able to help the caller the most. Distributing calls logically is the function most people associate with an ACD, though it's not the most important.

The management information which the ACD produces is much more valuable. This information is of three sorts: 1. The arrival of incoming calls (when, how many, which lines, from where, etc.) 2. How many callers were put on hold, asked to wait and didn't. This is called information on ABANDONED CALLS. This information is very important for staffing, buying lines from the phone company, figuring what level of service to provide to the customer and what different levels of service (how long for people to answer the phone) might cost. And 3. Information on the origination of the call. That information will typically include

ANI (Automatic Number Identification -- picking up the calling number and DNIS (Direct Number Identification Service) picking up the called number. Knowing the ANI allows the ACD and its associated computer to look up the caller's record and thus offer the caller much faster service. Knowing the DNIS may allow the ACD to route the caller to particular agent or keep track of the success of various advertising campaigns. Ad agencies will routinely run the same ad in different towns using different 800 phone numbers. Picking up which number was called identifies which TV station the ad ran on.

B CHANNEL
A "bearer" channel is a fundamental component of ISDN interfaces. It carries 64,000 bits per seconds in both directions, is circuit switched and is able to carry either voice or data. Whether it does or not depends on how your local telephone company has tariffed its ISDN service.

BANDWIDTH
The range of electrical frequencies a device can handle. The amount of bandwidth a channel is capable of carrying tells you what kinds of communications can be carried on it. A wideband circuit, for example, can carry a TV channel. The capacity to move information. A person master hardware, software, manufacturing and marketing-and plays the oboe or some other semi-obscure musical instrument-is "high bandwidth." The term is believed to have originated in Redmond, WA in the headquarters of MIcrosoft. People there who are super-intelligent and have generally broad capabilities are said to have "high bandwidth."

BIT RATE
The number of bits of data transmitted over a phone line per second. You can usually figure how many characters per second you will be transmitting
-- in asynchronous communications -- if you divide the bit rate by ten. For example, if you are transmitting at 1200 bits per second, you will be transmitting 120 characters per second. In real life, it's never this simple, however. The total bits transmitted will depend on re-transmissions, which depends on the noise of the line, etc.

BOC
Bell Operating Company. The local Bell operating telephone company. These days there are 22 Bell Operating Companies. They are organized into (i.e. owned by) seven Regional Bell holding companies, also called RBOCs, pronounced "R-bocks," or RHCs.

BRI
Basic Rate Interface. There are two subscriber "interfaces" in ISDN. This one and PRI (Primary Rate Interface). In BRI, you get two bearer B-channels at 64 kilobits per second and a data D-channel at 16 kilobits per second. The bearer B-channels are designed for PCM voice, slow-scan video conferencing, group 4 facsimile machines, or whatever you can squeeze into 64,000 bits per second full

duplex. The data (or D) channel is for bringing in information about incoming calls and taking out information about outgoing calls. It is also for access to slow-speed data networks, like videotex, packet switched networks, etc.

BROADBAND
A transmission facility that has a bandwidth (capacity) greater than a voice grade line of 3 kHz. (Some say that to be "broadband" it should be 20 kHz.) Such a broadband facility -- typically coaxial cable -- may carry numerous voice, video and data channels simultaneously. Each "channel" will take up a different frequency on the cable. There'll be "guardbands" (empty spaces) between the channels to make sure each channel doesn't interfere with its neighbor. A coaxial CATV cable is the "classic" broadband channel. Simultaneously it carries many TV channels. Broadband cables are used in some office LANs. But more common are the baseband variety which have the capacity for one channel only. Everything on that cable to be transmitted or received must use that one channel. That one channel is very fast, so each device needs only to use that high speed channel for only a little of the time. The problem is getting onto the channel.

BUS NETWORK
All communications devices share a common path. Typically in a bus network, a "conversation" from each device is sampled quickly and interleaved using time division multiplexing. Bus networks are very high-speed -- millions of bits per second -- forms of transmission (e.g. on a local area network) and switching. They often form the major switching and transmission backbone of a modern PBX. The printed circuit cards which connect to each trunk and each line are plugged into the PBX's high-speed "backbone" -- i.e. the bus network.

BUS SPEED
The speed at which the computer's CPU (central processing unit) communicates with other elements of the computer. For example, the speed at which data moves between the CPU and your serial ports.

BUSY OUT
To cause a line to return a busy signal to a caller. Busying out lines going into a computer is useful when the computer is not available, i.e. during maintenance periods. This way callers do not get connected to modems with no computers to talk to. This is also known as "taking the phone off the hook." In a voice phone system with trunks that rotary (or hunt) on, sometimes busying one or more broken trunks out helps calls rotary on to trunks that are still working. This way, someone doesn't end up on your third trunk with endless ringing, while your 4th, 5th, 6th etc. trunks are free, leaving them wondering where you are and wondering why you're not getting any calls.

CABLE MODEM
Modem designed for use on TV coaxial cable circuit. The idea is simple: Put the cable modem on a home cable TV line, provide the subscriber a circuit on which

he can dial up Prodigy, CompuServe, America On Line, the Internet, etc. And charge a little extra for the privilege.

CADENCE
In voice processing, cadence is used to refer to the pattern of tones and silence intervals generated by a given audio signal. Examples are busy and ringing tones. A typical cadence pattern is the US ringing tone, which is one second of tone followed by three seconds of silence. Some other countries, such as the UK, use a double ring, which is two short tones within about a second, followed by a little over two seconds of silence.

CALL CENTER
A place where calls are answered and calls are made. A call center will typically have lots of people (also called agents), an automatic call distributor, a computer for order-entry and lookup on customers' orders. A Call Center could also have a predictive dialer for making lots of calls quickly. The term "call center" is broadening. It now includes help desks and service lines. For more information on Call Centers, please read CALL CENTER Magazine. 212-691-8215.

CALL COMPLETION RATE
The ratio of successfully completed calls to the total number of attempted calls. This ratio is typically expressed as either a percentage or a decimal fraction.

CALL CONTROL
Call control is the term used by the telephone industry to describe the setting up, monitoring, and tearing down of telephone calls. There are two ways of doing call control. A person or a computer can do it via the desktop telephone or a computer attached to that telephone, or the computer attached to the desktop phone line (i.e. without the actual phone being there). That's called First Party Call Control. Third-party call control controls the call through a connection directly to the switch (PBX). Generally third-party call control also refers to the control of other functions that relate to the switch at large, such as ACD queuing, etc.

CALL DETAIL RECORDING
CDR. A feature of a telephone system which allows the system to collect and record information on outgoing and incoming phone calls -- who made/received them, where they went/where they came from, what time of day they happened, how long they took, etc. Sometimes the data is collected by the phone system; sometimes it is pumped out of the phone system as the calls are made. Which ever way, the information must be recorded elsewhere
-- dumped right into a printer or into a PC with call accounting software.

CALL DURATION
The time from when the call is actually begun (i.e. answered) to the instant either party hangs up. Call Duration is an important concept for traffic engineering.

CALL ESTABLISHMENT

The process by which a call connection is created.

CALL FORWARDING
A service available in many central offices, and a feature of many PBXs and some hybrid PBX/key systems, which allows an incoming call to be sent elsewhere. There are many variations on call forwarding: Call forwarding busy. Call forwarding don't answer. Call forwarding all calls, etc.

CALL PROGRESS
The status of the telephone line; ringing, busy ring/no answer, voice mail answering, telephone company intercept, etc.

CALL PROGRESS ANALYSIS
As the call progresses several things happen. Someone dial or touchtones digits. The phone rings. There might be a busy or operator intercept. An answering machine may answer. A fax machine may answer. Call progress analysis is figuring out which is occurring as the call progresses. This analysis is critical if you're trying to build an automated system, like an interactive voice response system.

CALL PROGRESS TONE
A tone sent from the telephone switch to tell the caller of the progress of the call. Examples of the common ones are dial tone, busy tone, ringback tone, error tone, re-order, etc. Some phone systems provide additional tones, such as confirmation, splash tone, or a reminder tone to indicate that a feature is in use, such as confirmation, hold reminder, hold, intercept tones.

CALL QUEUING
Incoming or outgoing calls may be queued pending an answer. The idea of call queuing is to save money.

CALL RELEASE TIME
The time it takes from sending equipment a signal to close down the call to the time a "free condition" appears and the system is ready for another call.

CALL SETUP
The first six PICs (Point In Call) of the Originating BCSM (Basic Call State Model), or the first four PICs of the Terminating BCSM. Definition from Bellcore in reference to Advanced Intelligent Network.

CALL TRANSFER
Allows you to transfer a call from your phone to someone else's. On some phones you do this by punching in a bunch of numbers. Some you do it by hitting the "transfer" button and then the number you want to send the call to. The fewer buttons and numbers you have to punch, the easier it will be for your people.

CALL WAITING

Call Waiting is a feature of phone systems that lets you know someone is trying to call you. You're speaking on the phone. A call comes in for you. You might hear a beep in your ear or see a light on your phone turn on. Or you might hear a beep and see a message come across the screen of your phone. When you hear the beep, you can, if you wish, put the present call on hold and answer the new one. Or you can ignore the new one, hoping it will go away, and perhaps send it to your attendant/operator, or voice mail. Call Waiting can be done manually by your telephone system operator. Or it can be a service which you buy as a monthly from your local phone company.

A major problem with call waiting is if you're on a data call from your PC, the call waiting "beep" will often cause your modem to hang up, thus destroying your data call. There are two solutions to this, the obvious one being turn off call waiting. Some phone systems will allow you to turn it off. The less obvious one is modify your modem's initialization string. Here's how. In all Hayes and Hayes-compatible modems, there's a S10 register. It tells the modem how long before it hangs up after losing carrier. In Hayes modems, the S10 register is set for 1.4 seconds. The typical call waiting tone is 1.5 seconds. Solution, increase the S10 register to six seconds (to be sure). Use your communications software. Go into terminal mode, then type: ATS10=60. You must put this command in every time you power up, because the Hayes 1200 modem (and others) have volatile memory. But the Hayes 2400 and higher speed asynchronous modems have non-volatile memory. They remember the six seconds after they've been switched off. The command to write this to memory is ATS10=60&W. The "&W" means write it to memory.

CALLBACK

1. A feature of some voice and data telephone systems. You dial someone. Their phone or computer is busy. You hit a button or code for "call-back." When their phone becomes free, the phone system will call you and them simultaneously. You can only use this call-back feature on things internally in your phone system -- calling other people, calling long distance lines (which might be busy), calling the dictation pool, etc.

2. A quick way of referring to international callback, which works thus: Calling the United States from many countries abroad is far more expensive than calling those countries from the United States. A new business called International Callback has started. It works like this. You're overseas. You dial a number in the United States. You let it ring once. It won't answer. You hang up. You wait a few seconds. The number you dialed in the U.S. knows it was you calling. There is a piece of equipment on that number that "hears" it ring and knows it's you since no one else has that number. That was your special signal that you want to make a call. A switch attached to that line then calls you instantly. When you answer (overseas, obviously) it conferences you with another phone line in the United States and gives you U.S. dial tone. You can then touchtone from overseas your American number, just as if you would, were you in the U.S. There are huge savings. U.S. international callback operators can offer as much as 50% savings on calls from South America, where international calling rates are very high. The

process of international callback is being automated with software and dialing devices. International callback is also helping to bring down the high cost of calling the U.S. from overseas. A company called kallback in Seattle, WA has received a service mark from the U.S. Patent and Trademark Office for the words "callback" and "kallback" and sends letters to and threatens law suits against companies who use "their" words.

CALLER ID
Your phone rings. A name pops upon on your phone's screen. It's the name of the person calling you. Or it may be just the caller's phone number. It's called Caller ID and the information about name and/or calling phone number is passed to your phone by your telephone company's central switch. There are basically two forms of "caller ID" -- one provided by your local phone company and one provided by your long distance company (chiefly on 800 calls). Caller ID, generic term, is a term most commonly applied to the service your local phone company provides, usually called CLASS. In CLASS, the information about who's calling and/or their phone number is passed to your phone between the first and second ring signaling an incoming call.

CENTREX
Centrex is a business telephone service offered by a local telephone company from a local central office. Centrex is basically single line telephone service delivered to individual desks (the same as you get at your house) with features, i.e. "bells and whistles," added. Those "bells and whistles" include intercom, call forwarding, call transfer, toll restrict, least cost routing and call hold (on single line phones).

Think about your home phone. You can often get "Custom Calling" features. These features are typically fourfold: Call forwarding, Call Waiting, Call Conferencing and Speed Calling. Centrex is basically Custom Calling, but instead of four features, it has 19 features. Like Custom Calling, Centrex features are provided by the local phone company's central office.

Phone companies peddle Centrex is leased to businesses as a substitute for that business buying or leasing its own on-premises telephone system -- its own PBX, key system or ACD. Before Divestiture in 1984, Centrex was presumed dead. AT&T was, at that time, intent on becoming a major PBX and key system supplier. Then Divestiture came, and the operating phone companies recognized they were no longer part of AT&T, no longer had factories to support, but did have a huge number of Centrex installations providing large monthly revenues. As a result, the local operating companies have injected new life into Centrex, making the service more attractive in features, price, service and attitude.

CHANNEL
1. Typically what you rent from the telephone company. A voice-grade transmission facility with defined frequency response, gain and bandwidth. Also,

a path of communication, either electrical or electromagnetic, between two or more points. Also called a circuit, facility, line, link or path.

2. An SCSA term. A transmission path on the SCbus or SCxbus Data Bus that transmits data between two end points.

3. A channel of a GPS (Global Positioning System) receiver consists of the circuitry necessary to tune the signal from a single GPS satellite.

CHANNEL BANK
A multiplexer. A device which puts many slow speed voice or data conversations onto one high-speed link and controls the flow of those "conversations." Typically the device that sits between a digital circuit
-- say a T-1 -- and a couple of dozen voice grade lines coming out of a PBX. One side of the channel bank will be connections for terminating two pairs of wires or a coaxial cable -- those bringing the T-1 carrier in. On the other side are connections for terminating multiple tip and ring single line analog phone lines or several digital data streams. Sometimes you need channel banks. Sometimes, you don't. For example, if you're shipping a bundle of voice conversations from one digital PBX to another across town in a T-1 format -- and both PBXs recognize the signal -- then you will probably not need a channel bank. You'll need a Channel Service Unit (CSU). If one, or both, of the PBXs is analog, then you will need a channel bank at the end of the transmission path whose PBX won't take a digital signal.

CHANNEL SERVICE UNIT
CSU. A device used to connect a digital phone line (T-1 or Switched 56 line) coming in from the phone company to either a multiplexer, channel bank or directly to another device producing a digital signal, e.g. a digital PBX, a PC, or data communications device. A CSU performs certain line-conditioning, and equalization functions, and responds to loopback commands sent from the central office. A CSU regenerates digital signals. It monitors them for problems. And it provides a way of testing your digital circuit. You can buy your own CSU or rent one from your local or long distance phone company.

CLASS
1. Custom Local Area Signaling Services. It is based on the availability of channel interoffice signaling. Class consists of number-translation services, such as call-forwarding and caller identification, available within a local exchange of Local Access and Transport Area (LATA). CLASS is a service mark of Bellcore. Some of the phone services which Bellcore promotes for CLASS are Automatic Callback, Automatic Recall, Calling Number Delivery, Customer Originated Trace, Distinctive Ringing/Call Waiting, Selective Call Forwarding and Selective Call Rejection.

2. In an object-oriented programming environment, a class defines the data content of a specific type of object, the code that manipulates it, and the public and private programming interfaces to that code.

CLIPPING
Clipping has two basic meanings. The first refers to the effect caused by a simplex (one way at a time) speakerphone. Here the conversation goes one way. When the other person wants to talk, the voice path has to reverse (to "flip"). While the flipping takes place, a few sounds are "clipped" from that person's conversation. This phenomenon happens on some long distance and many overseas channels. These channels are so expensive, they are simultaneously shared by many conversations. Gaps in your conversation are filled by other people's conversation. But when you start talking, the equipment has to recognize you're now talking, find some capacity for your conversation, and send it. In the process of doing this, your first word or part of your first word might be "clipped" and the conversation will sound "broken."

The second way your voice is clipped is what happens every day on the telephone. You're squeezing your own voice which typically spans 10,000 Hertz into a voice channel which is only 3,000 Hertz. This clips the extremes of your conversation -- the higher sounds. As a result, your voice sounds flatter over the phone. As you become more economical and try to squeeze your voice into smaller capacity channels, so it becomes increasingly clipped.

CMIP
Common Management Information Protocol. A protocol formally adopted by the International Standards Organization in Paris (ISO), used for exchanging network management information. Typically, this information is exchanged between two management stations. CMIP can, however, be used to exchange information between an application and a management station. CMIP has been designed for OSI networks, but it is transport independent. Theoretically, it could run across a variety of transports, including, for example, IBM's Systems Network Architecture.

CMTS
CMTS stands for the Cellular Mobile Telephone System. The original and still, most common CMTS is a low-powered, duplex, radio/telephone which operates between 800 and 900 MHZ, using multiple transceiver sites linked to a central computer for coordination. The sites, or "cells,", named for their honeycomb shape, cover a range of one to six, or more, miles in each direction. The cells overlap one another and operate at different transmitting and receiving frequencies in order to eliminate crosstalk when transmitting from cell to cell. Each cell can accommodate up to 45 different voice channel transceivers.

CO
Central Office. In North America, a CO is that location which houses a switch to serve local telephone subscribers. Sometimes the words "central office" are confused with the switch itself. In Europe and abroad, the words "central office"

are not known. The more common words are "public exchange." But those words tend to refer more to the switch itself, rather than the site, as in North America. CO was the name of a magazine published by Telecom Library Inc., the publisher of this dictionary.

CODE EXCITED LINEAR PREDICTION
CELP. An analog-to-digital voice coding scheme.

CLNP
Connectionless Network Protocol. An OSI network layer protocol that does not require a circuit to be established before data is transmitted.
The OSI protocol for providing the OSI Connectionless Network Service (data gram service). CLNP is the OSI equivalent to Internet IP, and is sometimes called ISO IP.

CPE
Customer Provided Equipment, or Customer Premise Equipment. Originally it referred to equipment on the customer's premises which had been bought from a vendor who was not the local phone company. Now it simply refers to telephone equipment -- key systems, PBXs, answering machines, etc. -- which reside on the customer's premises. "Premises" might be anything from an office to a factory to a home. GTE once used CPE to refer to "Company Provided Equipment." It doesn't any longer.

CPNI
Customer Proprietary Network Information. Information which is available to a telephone company by virtue of the telephone company's basic service customer relationship. This information may include the quantity, location, type and amount of use of local telephone service subscribed to, and information contained on telephone company bills. This is the definition of CPNI that the independent voice mail and live telephone answering industry uses.

CSTA
Computer Supported Telephony Application. A standard from the European Computer Manufacturers Association (ECMA) for linking computers to telephone systems. Basic CSTA is a set of API call agreed upon by the ECMA.

CSU
1. Channel Service Unit. Also called a Channel Service Unit/Data Service Unit or CSU/DSU because it contains a built-in DSU device. A device to terminate a digital channel on a customer's premises. It performs certain line coding, line-conditioning and equalization functions, and responds to loopback commands sent from the central office. A CSU sits between the digital line coming in from the central office and devices such as channel banks or data communications devices. A Channel Service Unit is found on every digital link and allows the transfer of data at a range greater than 56 Kbps. A 56 Kbps circuit would need a 56 Kbps DSU on both ends to transfer data from one end to the other. A CSU looks

like your basic "modem," except it can pass data at rates much greater and does not permit dial-up functions (unless it has an asynch dial-backup feature).

2. Channel Sharing Unit. Line bridging device that allows several inputs to share one output. CSUs exist to handle any input/output combination of sync or asynch terminals, computer ports, or modems and thus these units are variously called modem sharing units, digital bridges, port sharing units, digital sharing devices, modem contention units, multiple access units, control signal activated electronic switches or data-activated electronic switches.

CTI
Computer Telephone Integration. A term for connecting a computer (single workstation or file server on a local area network) to a telephone switch (e.g. a PBX or an ACD) and have the computer issue the switch commands to move calls around. The classic application for CTI is in call centers. Picture this: A call comes in. That call carries some form of caller ID -- either ANI or Class Caller ID. The switch "hears" the calling number, strips it off, sends it to the computer. The computer does a lookup, sends back the switch instructions on what to do with the call. The switch follows orders. It might send the call to a specialized agent or maybe just to the agent the caller dealt with last time. There are emerging standards for CTI.

CUSTOM ISDN
A version of ISDN BRI (Basic Rate Interface) provided off an AT&T 5ESS central office. It actually offers more features and is easier to install than a National ISDN-1 BRI line. We are all awaiting the specifications on National ISDN-2, which is meant to be "standard." Meantime, Custom ISDN is the most popular, most versatile and most understood ISDN service in North America.

CUT THROUGH
1. Cut-through is a voice processing term. It's what stops voice prompt playback when a key is pressed. Some of the speech recognition solutions also add cut-through that will stop voice prompt playback as soon as you start talking. Only voice cards that support continuous speech recognition are able to provide cut-through. Cut-though can be a problem in some cases. Imagine yourself at the airport trying to make a call using a speech recognition system. At the start of a new prompt, the airport public address system blares out a last boarding call for a flight. If cut-through is active, it would stop playing the prompt and wait on your response. Now what do you do?

2. The act of connecting one circuit to another, or a phone to a circuit. This is when a user dials the access code for the circuit and is immediately "cut through" to the tie line. The user controls the call. It is a tie line operation.

D CHANNEL
In an ISDN interface, the "D" channel (the Data channel) is used to carry control signals and customer call data in a packet switched mode. In the BRI (Basic Rate

Interface, i.e. the lowest ISDN service) the "D" channel operates at 16,000 bits per second, part of which will carry setup, teardown, ANI and other characteristics of the call. 9,600 bps will be free for a separate conversation by the user. In the PRI (Primary Rate Interface, i.e. ISDN equivalent of T-1), the "D" channel runs at 64,000 bits per second. The D channel provides the signaling information for each of the 23 voice channels (referred to as "B channels"). The actual data which travels on the D channel is much like that of a common serial port. Bytes are loaded from the network and shifted out to the customer site in a serial bit stream. The customer site of course responds with its serial bit stream, too. An example of a data packet sent from the network to indicate a new call has the following components:

-- Customer Site ID
-- Type of Channel Required (Usually a B channel)
-- Call Handle (Not unlike a file handle)
-- ANI and DNIS information
-- Channel Number Requested
-- A Request for a Response

This packet is responded to by the customer site with a format similar to:

-- Network ID
-- Channel Type is OK
-- Call Handle

The packets change as the state of the call changes, and finally ends with one side or the other sending a disconnect notice. The important concept here is the fact the information on the D channel could actually be anything -- any kind of serial data. It could just as well be sports scores! So with that in mind, consider the Channel Number Requested packet above. This is the networks' selected channel for the customer site to use. Normally, this number is between 1 and 23, but could be a higher number if needed. This is what NFAS is all about. NFAS (Non Facility Associated signaling, pronounced N-FAST without the T) allows a D channel to carry call information regarding channels which may not even exist in the same PBX or PC system.

D1, D1D, D2, D3, D4 and D5
T-1 framing formats developed for channel banks. All formats contain a framing bit in every 193rd bit position. The Superframe (introduced in D2 channel banks) is made up of 12 193-bit frames, with the 193rd bit sequence being repeated every 12 frames. D2 framing also introduced robbed bit signaling, where the eighth bit in frames 6 and 12 were "robbed" for signaling information (like dial pulses). D1D was introduced after D2 to allow backwards compatibility of Superframe concepts to D1 banks.

DATA OVER VOICE

A device that takes a voice grade line and multiplexes it so it can carry a voice and a data signal. Typically the data is carried in analog form. Thus, to put data on this type of circuit, you need a modem. It is called Data Over Voice, because the data streams (transmission and reception) travel at a higher frequency than the voice conversations using a technique called frequency division multiplexing (FDM).

DECT
Digital European Cordless Telecommunications. An interface specification under development and governing pan-European digital mobile telephony of the future. Based on advanced TDMA technology, DECT covers cordless PBXs, telepoint and residential cordless telephony. DECT frequency is 1800-1900 MHz.

DEDICATED CHANNEL OR CIRCUIT
A channel leased from a common carrier by an end user used exclusively by that end user. The channel is available for use 24 hours a day, seven days a week, 52 weeks of the year, assuming it works that efficiently.

DEFAULT CARRIER
Generic name given to the long distance carrier which will carry the traffic of customers who haven't pre-subscribed to a long distance carrier.

DIAL PULSING
A means of signaling consisting of regular momentary interruptions of a direct or alternating current at the sending end in which the number of interruptions corresponds to the value of the digit or character. In short, the old style of rotary dialing. Dial the number "five" and you'll hear five "clicks."

DIAL STRING
A Dial String is the sequence of characters sent to a device which can dial a phone number. Such a device might be a modem or a voice processing card. Here are some "digits" in a dial string: ! -- flashhook (TAPI standard); &
-- flashhook (Dialogic); T -- use tone dialing; , -- pause (typically of half a second to two seconds); W -- wait for dial tone.

DIAL TONE
The sound you hear when you pick up a telephone. Dial tone is a signal (350 + 440 Hz) from your local telephone company that it is alive and ready to receive the number you dial. If you have a PBX, dial tone will typically be provided by the PBX. Dial tone does not come from God or the telephone instrument on your desk. It comes from the switch to which your phone is connected to.

DID
Direct Inward Dialing. You can dial inside a company directly without going through the attendant. This feature used to be an exclusive feature of Centrex but it can now be provided by virtually all modern PBXs and some modern hybrids, but you must connect via specially configured DID lines from your local central office. A DID (Direct Inward Dial) trunk is a trunk from the Central office which

passes the last two to four digits of the Listed Directory Number to the PBX or hybrid phone system, and the digits may then be used verbatim or modified by phone system programming to be the equivalent of an internal extension. Therefore, an external caller may reach an internal extension by dialing a 7-digit central office number. Notice: DID is different from a DIL (Direct-In-Line) where a standard, both-way central office trunk is programmed to always ring a specific extension or hunt group. DID lines cannot be used for outdial operation, since there is no dialtone offered.

DIGITAL
1. In displays, the use of digits for direct readout.

2. In telecommunications, in recording or in computing, digital is the use of a binary code to represent information. See PCM (as in Pulse Code Modulation.) Analog signals -- like voice or music -- are encoded digitally by sampling the voice or music analog signal many times a second and assigning a number to each sample. Recording or transmitting information digitally has two major benefits. First, the signal can be reproduced precisely. In a long telecommunications transmission circuit, the signal will progressively lose its strength and progressively pick up distortions, static and other electrical interference "noises."

In analog transmission, the signal, along with all the garbage it picked up, is simply amplified. In digital transmission, the signal is first regenerated. It's put through a little "Yes-No" question. Is this signal a "one" or a "zero?" The signal is reconstructed (i.e. squared off) to what it was identically. Then it is amplified and sent along its way. So digital transmission is much "cleaner" than analog transmission. The second major benefit of digital is that the electronic circuitry to handle digital is getting cheaper and more powerful. It's the stuff of computers. Analog transmission equipment doesn't lend itself to the technical breakthroughs of recent years in digital.

DIGITAL SIGNAL PROCESSOR
A specialized digital microprocessor that performs calculations on digitized signals that were originally analog (e.g. voice) and then sends the results on. There are two main advantages of DSPs -- first, they have powerful mathematical computational abilities, much more than normal computer microprocessors. DSPs need to have heavy mathematical computation skills because manipulating analog signals requires it. For example, DSPs are often called upon to compress video signals. Each sample must be examined and processed. And all done in very little time. The second advantage of a DSP lies in the programmability of digital microprocessors. Just as digital microprocessors have operating systems, so DSPs are now acquiring their very own operating systems. DSPs are used extensively in telecommunications for tasks such as echo cancellation, call progress monitoring, voice processing and for the compression of voice and video signals. They are also used in devices from fetal monitors, to anti-skid brakes, seismic and vibration sensing gadgets, super-sensitive hearing aids, multimedia presentations and low cost desktop fax machines. DSPs are replacing the dedicated chipsets in modems

and fax machines with programmable modules -- which, from one minute to another, can become a fax machine, a modem, a teleconferencing device, an answering machine, a voice digitizer and device to store voice on a hard disk, to a proprietary electronic phone. DSPs will do (and are already doing) for the telecom industry what the general purpose microprocessor (e.g. Intel's 80286 or 80386) did for the personal computer industry. DSPs are made by Analog Devices, AT&T, Motorola, NEC and Texas Instruments, among others.

DIGITAL SUBSCRIBER LINE
A fancy name for an ISDN BRI channel. Here's AT&T's definition: "A three-channel digital line that links the ISDN customer's terminal to the telephone company switch with four ordinary copper telephone wires. Operated at the Basic Rate Interface (with two 64-kilobit per second circuit switched channels and one 16-kilobit packet switched channel), the DSL can carry both voice and data signals at the same time, in both directions, as well as the signaling date used for call information and customer data. With the introduction of the AT&T 5E5 generic, up to eight different users can be served by a single DSL."

DIL
Direct-In-Line. A standard, both-way central office trunk is programmed to always ring a specific extension or hunt group within the PBX. This contrasts with Direct Inward Dialing, which allows an external caller to reach an internal extension by dialing a 7-digit central office number. A DID (Direct Inward Dial) trunk is a trunk from the Central office which passes the last two to four digits of the Listed Directory Number into the PBX, thus allowing the PBX to switch the call to and thus ring the correct extension.

DIPHONES
Speech segment beginning in the middle of one phoneme and concluding in the middle of another.

E & M SIGNALING
In telephony, an arrangement that uses separate leads, called respectively the "E" lead and "M" lead, for signaling and supervisory purposes. The near end signals the far end by applying -48 volts dc (vdc) to the "M" lead, which results in a ground being applied to the far end's "E" lead. When -48 vdc is applied to the far end "M" lead, the near-end "E" lead is grounded. The "E" originally stood for "ear," i.e., when the near-end "E" lead was grounded, the far end was calling and "wanted your ear." The "M" originally stood for "mouth," because when the near-end wanted to call (i.e., speak to) the far end, -48 vdc was applied to that lead.

When a PBX wishes to connect to another PBX directly or to a remote PBX or extension telephone over a leased voice grade line, a channel on T-1, the PBX uses a special line interface which is quite different from that which it uses to interface to the phones it's attached directly to (i.e. with in-building wires). The basic reason for the difference between a normal extension interface and the long distance interface is that the signaling requirements differ -- even if the voice signal parameters such as level and two-wire, 4-wire remain the same. When dealing with tie lines or trunks it is costly, inefficient and too slow for a PBX to do what an extension telephone would do, i.e. go off hook, wait for dial tone, dial, wait for ringing to stop, etc. The E&M tie trunk interface device is the closest thing there is to a standard that exists in the PBX, T-1 multiplexer, voice digitizer telco world. But even then it comes in at least five different flavors. E&M signaling is the most common interface signaling method used to interconnect switching signaling systems with transmission signaling systems.

E-1
The European equivalent of the North American 1.544 Mbps T-1,except that E-1 carries information at the rate of 2.048 megabits per second. This is the rate used by European CEPT carriers to transmit 30 64 Kbps digital channels for voice or data calls, plus a 64 Kbps channel for signaling, and a 64 Kbps channel for framing (synchronization) and maintenance. CEPT stands for the Conference of European Postal and Telecommunication Administrations. Since robbed-bit signaling is not used (as it is for T-1 in North America) all 8 bits per channel are used to code the waveshape sample.

E-2
Interim data signal that carries four multiplexed E-1 signals. Effective data rate is 8.448 Mbps.

E-3
CEPT signal which carries 16 CEPT E-1s and overhead. Effective data rate is 34.368 Mbps.

ECHO CANCELLER
Device that allows for the isolation and filtering of unwanted signal caused by echoes from the main transmitted signal. In data communications networks, echo cancelers are used in the same way as PADS are in the network, but some brands of echo cancelers have the ability of being disabled by a 2100 Hertz tone transmitted by the data device prior to the exchange of the data device's handshaking protocol. If the echo canceler cannot be disabled by the data device, it will block the data call from completing.

ECM
Error Correction Mode. An enhancement to Group 3 fax machines. Encapsulated data within HDLC frames providing the received with an opportunity to check for, and request retransmission of garbled data. See FACSIMILE and V.17.

ECTF
Enterprise Computer Telephony Forum is a California non-profit mutual benefit corporation, formed to focus on the technical challenges of interoperability among Computer Telephony Integration (CTI) products. It was formed in the Spring of 1995. Information on the ECTF is available from The ECT Forum, Foster City, CA 415-578-6852. ectf@sbexpos.com. On July 31, 1995, the Versit companies

cooperating in computer telephony integration (Apple, AT&T, IBM, Novell and Siemens) announced that they will bring the results of their CTI development efforts (Versit CTI Encyclopedia including Versit TSAPI) to the ECTF, enabling review and use by the companies represented in this forum. In support of these efforts, the Versit CTI companies are also joining ECTF.

FAX SERVER

1. A relatively high-powered computer which sits on a LAN and has one or more PC fax boards in its slots. The fax server receives incoming faxes over phone lines, stores them on its hard disk and, if it knows for whom the faxes are meant, it will alert that person over the LAN. If it doesn't know for whom the faxes are meant, it may send the faxes to a printer or alert a supervisor to manually check the incoming faxes and distribute them

-- electronically or on paper. The fax server also accepts from workstations on the LAN, stores them and gets them ready for sending out over phone lines. It might send the faxes immediately or wait until later, when phone calls are cheaper.

2. A fax server is also a specialized interactive voice response system which you call. When you call it, it answers, reads you a menu of options

-- including various documents it can send you. You choose which documents you want by touchtoning in numbers. Then you designate to which fax machine you want the documents sent. The fax machine you designate might be the one you're calling from (i.e. you dialed using your fax machine's handset).

There are two types of interactive voice response fax servers. One is a one-call machine. The caller calls from his own fax machine. When he's chosen his faxes and he's ready to receive a fax, he simply hits the "Start" button on his fax machine and his machine receives the chosen faxes. There is also a two-call machine. The caller will call from a phone and touchtone in the phone number of a fax machine he wants the fax of his desired documents sent. One-call IVR fax servers are the newer breed, harder to build than the older two-call machines. There are obvious advantages to both. The one call machine -- in which the user pays the phone bill - - will, I suspect, become the more popular type.

FACSIMILE EQUIPMENT

FAX. Equipment which allows hard copy (written, typed or drawn material) to be sent through the switched telephone system and printed out elsewhere. Think of a fax machine as essentially two machines -- one for transmitting and one for receiving. The sending fax machines typically consists of a scanner for converting material to be faxed into digital bits, a digital signal processor (a single chip specialized microprocessor) for reducing those bits (encoding white space into a formula and not an endless series of bits representing white), and a modem for converting the bits into an analog signal for transmission over analog dial-up phone lines. The receiving fax consists of a modem and a printer which converts the incoming bits into black and white images on paper. More modern and more expensive machines also have memory -- such that if the machine runs out of paper, it will still continue to receive incoming faxes, storing those faxes into memory until someone fills the machine with paper and it prints the faxes out.

There are five internationally accepted specifications for facsimile equipment. Group 1, Group 2, Group 3, Group 3 Enhanced and Group 4. Only 1, 2, 3 and 3 Enhanced will work on "normal" analog dial-up phone lines. Group 4 is designed for digital lines running at 56/64 Kbps, e.g. ISDN lines. Among the analog line fax machines, Group 2 is faster than Group 1. Group 3 is faster than Group 2, etc. Virtually all machines sold today are Group 3, though an increasing percentage are Group 3 enhanced, which has speeded up Group 3's transmission speed from 9,600 bps to 14,400 bps and improved its error correction. Group 3 faxes send an 8-1/2 x 11 inch page over a normal phone line in about 20 seconds. How much time it actually takes depends on how much stuff is actually on the paper. Unlike older machines, Group 3 machines are "intelligent." They only transmit the information that's on the paper. They do not transmit white space, as earlier machines did. Group 3 fax machines are now available in "slimy" (i.e. chemically coated) paper and plain paper, the same paper your photocopy machines now uses.

When a fax machine calls a phone line, it emits a standard ITU-T-defined, "CNG tone" (calling tone) -- 1100 Hz tone every three seconds. When the receiving fax machine hears this tone, it knows it's an incoming fax call and it can automatically connect. With this tone it is possible to insert a "fax switch," which would "listen" for an incoming fax call and switch it to a fax machine if it heard the CNG tones or to something else -- like a phone or answering machine -- if it didn't. It is not possible to do this with a modem. A calling modem does not issue any tones whatsoever. A modem works backwards -- when the receiving modem answers the phone, it emits a tone.

Typically, a Group 3 machine can speak to a Group 2 and a Group 1 machine. A Group 2 can speak to a Group 1. Speaking down means slowing down. Fax machines are dropping in price. "Personal" fax machines are emerging. Most fax machines today at Group 3 or Group 3 enhanced.

A Group 4 standard has been promulgated by the ITU-T. Group 4 facsimile machines are 100% digital and directly attach to the B (bearer channel) of a digital ISDN line. They will transmit a sheet of 8 1/2 x 11 paper in under six seconds.

FDDI
Fiber Distributed Data Interface. FDDI is a 100 megabits per second fiber optic LAN. It is an ANSI standard. It uses a "counter-rotating" token ring topology. FDDI is ANSI X3T12 standard for a dual-ring LAN operating at 100 Mbps and using token passing; FDDI rings may use up to 200 km of optical fiber, or may employ twisted copper pairs for short hops. FDDI is compatible with the standards for the physical layer of the OSI model. An FDDI LAN is often known as a "backbone" LAN. It is used to join file servers together and to join LANs together. The theoretical limit of Ethernet, measured in 64 byte packets, is 14,800 packets per second (PPS). By

comparison, Token Ring is 30,000 and FDDI is 170,000 pps. See FDDI TERMS and FDDI-II.

FDDI-II
Fiber Distributed Data Interface-II is a recently standardized enhancement to FDDI. It still runs at 100 megabits per second on fiber or on twisted copper pairs, but in addition to transporting conventional packet data like other LANs, FDDI-II allows portions of the 100 Mbps bandwidth to carry low delay, constant bit rate, isochronous data like 64 Kbps telephone channels. This means the same LAN that carries computer packet data can carry live voice or live video calls. Some additional terms used with FDDI-II are: I-MAC which stands for Isochronous Media Access Control; P-MAC which stands for Packet Media Access Control; and WBC which stands for Wide Band Channel.

FEATURE GROUP A, B, C, D
FGA, FGB, FGC, FGD, are four separate switching arrangements available from local exchange carrier (LEC) end central offices to interexchange (long distance) carriers. These switching arrangements allow the LEC's end-users to make toll calls via their favorite LEC. Feature groups are described in a tariff filed by the National Exchange Carrier Association with the FCC. The feature group used by each IX (IntereXchange) carrier together with any special access surcharge determines the service they can provide their customers and the carrier common line access fee they will pay to the local exchange carrier involved. The most common Feature Group now is D. See the next four definitions. See FEATURE GROUP A, FEATURE GROUP B, FEATURE GROUP C, FEATURE GROUP D

FEATURE GROUP A
Offers access to the local exchange carrier's network through a subscriber-type line connection rather than a trunk. It is a continuation of the ENFIA arrangement used in the early days of OCCs, until equal access using an access tandem central office is available. Remember, without equal access the IX carrier had to require its customers to dial a local number to reach their long distance facilities, then dial an identification number, then dial long distance numbers of the called party desired. This service handicap, compared to AT&T's superior connections, qualifies the OCC for a discount off the FGA rate until access is equal. The IX carrier is billed by the LEC based upon actual monthly use rather than the ENFIA method of projected "minutes of use" rate.

FEATURE GROUP B
Is similar to FGA, but provides a higher quality trunk line connection from end CO to the IX carrier's facilities, instead of the subscriber-type line. The IX customer can originate a call from anywhere within the LATA, while FGA requires customers to initiate the call from within the local exchange of the exchange carrier connecting to the IXC. FGB billings to the IX are on a flat usage basis, and a discount is applicable.

FEATURE GROUP C

Is the traditional toll service arrangement offered by LECs to AT&T prior to breakup of the Bell System. Quality is superior, and the service includes automatic number identification of the calling party, answerback, and disconnection supervision, and the subscribers can use either a dial or touchtone pad. This FGC service is offered only to AT&T without a discount.

FEATURE GROUP D
Is the class of service associated with equal access arrangements. All IX carriers (i.e. long distance phone companies) enjoy identical connections to the local exchange carrier. All customers dial the same number of digits, and can reach the predetermined IX of their choice by dialing 1 plus the telephone number being called. Eventually, all other feature groups convert to FGD and the IX is billed for actual measured use, without discount. In some cases an IX carrier may desire to maintain FGA or FGB arrangements, but the FGD equal access rates will apply.

FOREIGN EXCHANGE TRUNK
A Foreign EXchange (FEX) trunk provides a direct connection between a PBX switch and a remote central office other than the central office that serves the location of the PBX.

FRACTIONAL T-1
Fractional T-1 refers to any data transmission rate between 56 kbps (DSO rate) and 1.544 megabits per second (Mbps), which is a full T-1. Fractional T-1 is simply a digital line that's not as fast as a T-1. Fractional T-1 is popular because it's typically provided by a phone company (local or long distance) at less money than a full T-1. FT-1 is typically used for LAN interconnection, video conferencing, high-speed mainframe connection and computer imaging. Fractional T-1 is typically provided on four-wire (two-pair) copper circuits.

FRACTIONAL T-3
A telephone company service in which portions of a T-3 (44.7364-megabits per second) transmission service are leased to provide a service similar to a T-1 (1.544-Mbps) or T-2 (3.152-Mbps) channel, but normally at a lower cost.

FRAME RELAY
Frame relay switching is a form of packet switching, but uses smaller packets and requires less error checking than traditional forms of packet switching. Frame Relay is very good at efficiently handling high-speed, bursty data over wide area networks. It offers lower costs and higher performance for those applications in contrast to the traditional point-to-point services. With frame relay, a pool of bandwidth is made instantly available to any of the concurrent data sessions sharing the circuit whenever a burst of data occurs. An addressed frame is sent into the network, which in turn interprets the address and sends the information to its destination at up to 2.048 Mbps.

Like traditional X.25 packet networks, frame relay networks use bandwidth only when there is traffic to send. Frame relay is often provided to the end user at three speeds -- 56/64 kilobits per second, 256 kbps and 1,024 Mbps. Frame relay does not support voice, because voice traffic is highly sensitive to variations in the transmission delay introduced by the packet networks, while such small variations are usually not as critical to data traffic. For voice to be supported satisfactorily in packet a network, each packet must have a time stamp which is monitored by the network. Frame relay lacks such a mechanism.

FULL DUPLEX
Transmission in two directions simultaneously, or, more technically, bidirectional, simultaneous two-way communications. The best two-way phone conversations take place on four-wire circuits, two for transmission in one direction and two for transmission in the other. All long distance circuits are four wire. Most local lines are two wire, which means they're a compromise. Most speakerphones are half-duplex, meaning they only transmit in one direction at one time. The speakerphone flips its direction based on who's talking, or, more precisely, who's talking the loudest. Full duplex speakerphones are the best, but they're expensive.

GO-MVIP
Global Organization for MVIP. GO-MVIP is a non-profit trade association established in 1993 to take over coordination of the Multi-Vendor Integration Protocol. Information is available by calling 1-800-NOW-MVIP (800-669-6847). See MVIP.

GRAPHICAL USER INTERFACE
GUI. A fancy name probably originated by Microsoft which lets users get into and out of programs and manipulate the commands in those programs by using a pointing device (often a mouse). Microsoft's own definition is more elaborate. Namely that GUI puts visual metaphor that uses icons representing actual desktop objects that the user can access and manipulate with a pointing device.

GROUND START
A way of signaling on subscriber trunks in which one side of the two wire trunk (typically the "Ring" conductor of the Tip and Ring) is momentarily grounded (often to a cold water pipe) to get dialtone. There are two types of switched trunks typically for lease by a local phone company -- ground start and loop start. PBXs work best on ground start trunks, though many will work -- albeit intermittently -- on both types. Normal single line phones and key systems typically work on loop start lines. You must be careful to order the correct type of trunk from your local phone company and correctly install your telephone system at your end -- so that they both match. In technical language, a ground start trunk initiates an outgoing trunk seizure by applying a maximum local resistance of 550 ohms to the tip conductor.

H-CHANNEL
The packet-switched channel on an ISDN BRI (Basic Rate Interface) which is designed to carry user information streams at varying rates, depending on type: H0 -- 384 Kbps; H11 -- 1,536 Kbps; and H12 -- 1,920 Kbps.

H.221
A framing recommendation which is part of the ITU-T's H.320 family of video interoperability Recommendations. The Recommendation specifies synchronous operation where the coder and decoder handshake and agree upon timing. Synchronization is arranged for individual B channels or bonded 384 Kbps (HO) connections.

H.230
A multiplexing recommendation which is part of the ITU-T's H.320 family of video interoperability recommendations. The recommendation specifies how individual frames of audiovisual information are to be multiplexed onto a digital channel.

H.231
A recommendation, formally added to the ITU-T's H.320 family of recommendations in March, 1993, which specifies the multipoint control unit used to bridge three or more H.320-compliant codecs together in a multipoint conference.

H.233
An Recommendation, part of the ITU-T's H.320 family, which specifies the encryption method to be used for protecting the confidentiality of video data in H.320-compliant exchanges. Also called H.KEY.

H.242
Part of the ITU-T's H.320 family of video interoperability Recommendations. This Recommendation specifies the protocol for establishing an audio session and taking it down after the communication has terminated.

H.261
The ITU-T's H.261 is the standards watershed in videoconferencing. Also known as p x 64, H.261 specifies the video coding algorithms, the picture format, and forward error correction techniques to make it possible for video codecs from different manufacturers to successfully communicate. Announced in November 1990, it relates to the decoding process used when decompressing video conferencing pictures, providing a uniform process for codecs to read the incoming signals. Other important standards are H.221: communications framing; H.230 control and indication signals and H.242d: call set-up and disconnect. Encryption, still-frame graphics coding and data transmission standards are still being developed.

H.320
The most common family of ITU-T videoconferencing standards. These standards allow dissimilar videoconferencing systems and videophones to communicate with each other. "H.320 compatible videoconferencing systems" are now the most common videoconferencing systems. They work on ISDN BRI circuits. I've personally had several H.320 compatible videoconferencing systems and

videophones on my desk and have received from and made videoconferencing calls to many different H320 compatible video phones. The quality is not brilliant. But you can recognize the person at the other end. And they can recognize you. Most H.320 systems allow you to bond together the two B channels of a 2B+D ISDN BRI channel and thus get better video.

H.324
Standard for analog telephone line based teleconferencing.

HANDSHAKING
The initial exchange between two data communications systems prior to and during data transmission to ensure proper data transmission. A handshake method is part of the complete transmission protocol. A serial (asynchronous) transmission protocol might include the handshake method (XON/XOFF), baud rate, parity setting, number of data bits and number of stop bits. Just as people shake hands, and go through a perfunctory "Hi, how are ya?", computers must go through a procedure of greeting the opposite party, verifying the identity of the other party, and other functions that can be described by this "humanizing term." As with human contacts, once the Handshaking is complete, the business of communications begins.

HAYES AT COMMAND SET
Before 1981, the modem was a dumb device. It had no memory or ability to recognize commands. It simply modulated and demodulated signals between the telephone line and the computer or terminal. In 1981, Hayes Microcomputer Products, Inc. in Norcross, GA produced the first "smart" modem, appropriately named the Smartmodem 300. It was "smart" because it understood commands, such as "ATD" which means "ATtention, Dial the phone." The Hayes Standard AT Command Set (its full name) -- a language for modems
-- has been accepted as a standard by the modem industry. And now many modems claim to be 100% Hayes compatible, which may mean they are and may mean they aren't. As in all cases of claimed compatibility, one should check. You'll find the complete Hayes AT Command Set spelled out in virtually every manual of every modem which purports to be "100% Hayes Compatible."

HIDDEN MARKOV METHOD
HMM. A common algorithm in voice recognition which uses probabilistic techniques for recognizing discrete and continuous speech.

HOME PAGE
The Internet is a gigantic network of connected computers. The World Wide Web is on the Internet. The Web is the universe of accessible information available on many computers attached to the Internet. The Web has a body of software, a set of protocols and a set of defined conventions for getting at the information on the Web. The Web uses hypertext and multimedia techniques to make the web easy for anyone to roam, browse and contribute to. The Web makes publishing information (i.e. making your information public) very easy. You'll need a

computer, a telecommunications connection to the Internet and software to make your information accessible to anyone browsing the Internet (also known as net'surfing). The first page that browsers see of the information you have posted on your computer attached to the World Wide Web is your "home page." It's a "welcome" page. It says "Welcome to my site, my home." It typically contains a table of contents to more information which a visitor (browser, surfer, etc.) will find at your site by clicking onto hypertext links you've created.

HOT SWAP
The process of replacing a failed component -- e.g. a RAID drive -- while the rest of the system (in this case, the disks) continue to provide function normally, i.e. providing data to the network users and providing a place for them to store their data.

HOTJAVA
HotJava is a Web browser that makes the Internet "come alive". HotJava builds on the Internet browsing techniques established by Mosaic and expands them by implementing the capability to add arbitrary behavior, which transforms static data into applications. The data viewed in other browsers is limited to text, illustrations, low-quality sounds and video. Using HotJava you can add applications to that range from interactive science experiments in educational material to games and specialized shopping applications. You can implement interactive advertising and customized newspapers. HotJava also provides a way for users to access these applications in a new way. Software transparently migrates across the network. There is no such thing as "installing" software. It just comes when you need it. Of course, some of it you must pay for. Content developers for the World Wide Web don't have to worry about whether or not some special software is installed in a user's system; it just gets there automatically. This transparent acquisition of applications frees developers from the boundaries of the fixed media types like images and text. It lets them do whatever they'd like. HotJava possesses these dynamic capabilities because it is written in a new software language called Java. Keeping the explanation simple, Java is a simplified, safe and portable version of C++. It has an architecture-neutral distribution format, meaning that compiled Java code runs on any CPU architecture.

HTML
HyperText Markup Language. This is the authoring software language used on the Internet's World Wide Web. HTML is used for creating World Wide Web pages. Imagine you're viewing a page of text. You run your mouse down the page to a heading or a phrase which will typically be in another color (i.e. green). You place your cursor on the word or phrase. Your cursor changes to a hand. You click on it. Suddenly a box opens up explaining the word. A simple example: You're reading this dictionary. You're reading this definition. You don't know what the World Wide Web is. You put your cursor on it, click and bingo, you have a definition. HTML is the software language that makes this possible. The linking of the phrases World Wide Web to another section of text is called Hypertext.

HTTP
HyperText Transfer Protocol. Invisible to the user, HTTP is the actual protocol used by the Web Server and the Client Browser to communicate over the "wire". In short, the protocol used for moving documents around the Internet.

IANA
Internet Assigned Numbers Authority. The entity responsible for assigning numbers in the Internet Suite of Protocols.

ICP
1. Intelligent Call Processing. The ability of the latest ACDs to intelligently route calls based on information provided by the caller, a database on callers and system parameters within the ACD such as call volumes within agent groups and number of agents available. 2. Instituto das Communicacoes de Portugal (The Portuguese Institute of Communications).
3. Intelligent Call Processor. The name of an AT&T service. ICP allows users to directly link their customer premise equipment to its network for individual call processing based on customer-specific information.

IN-BAND SIGNALING
Signaling made up of tones which pass within the voice frequency band and are carried along the same circuit as the talk path that is being established by the signals. Virtually all signaling -- request for service, dialing, disconnect, etc. -- is in-band signaling. Most of that signaling is MF -- multi-frequency dialing. The more modern form of signaling is out-of-band. Several local and long distance companies provide ANI (Automatic Number Identification) via in-band signaling. Some long distance companies provide it out-of-band, using the D-channel in a PRI ISDN loop.

INCOMING CALL IDENTIFICATION
ICI. Some way of telling the user who's calling. It might be the caller's extension number on an LCD screen or even the caller's name spelled out, e.g. "KATE BRODIE-DAVID CALLING." Today, most "incoming call identification" is done totally within one PBX. However, the days of ISDN and CCITT Signaling System Number 7 are arriving. They promise to deliver to us the phone number of everyone calling us -- from within the PBX or key system and from the outside world.

INFONET
A consortium of about 10 European telephone companies and MCI. InfoNet has built a worldwide packet switched network that is great for data communications. It offers one major advantage over packet switched networks owned by the individual phone companies -- You don't have to separately sign up to use InfoNet in all the countries it operates. You just dial Infonet up, punch in and communicate. Not having to sign up is a big savings in time, since it can take weeks to sign up for European packet switched networks and cost minimum

monthly fees. For MCI Mail subscribers, using Infonet means you only get one bill -- directly from MCI.

INTELLIGENT PERIPHERAL
IP. A network system in the Advanced Intelligent Network Release 1 architecture containing an Resource containing an Resource Control Execution Environment (RCEE) functional group that enables flexible information interactions between a user and the network.

INTEREXCHANGE CARRIER
IXC. A telephone company that is allowed to provide long-distance telephone service between LATAs but not within any one LATA.

INTERLATA
Telecommunications services that originate in one and terminate in another Local Access and Transport Area (LATA). Under provisions of Divestiture, the Bell operating companies cannot provide Inter-LATA service, but can provide Intra-LATA service. Some LATAs are very large. So some "local" phone companies provide the equivalent of long distance service. And some of these phone companies have different pricing packages. Some of these packages are cheap, but not highly-publicized.

INTERLATA CALL
A call that is placed within one LATA (Local Access Transport Area) and received in a different LATA. These calls are currently carried by a long distance company.

INTERNET
It is very hard to define the Internet in a way that is either meaningful or easy to grasp. To say the Internet is the world's largest computer network is to trivialize it. Once a cooperative research effort of the Federal Government (called ARPA), which tied universities to other universities to their military customers, the Internet has essentially become a new publishing and commerce medium. Several people (including myself) believe that its invention is as important to the dissemination of knowledge and to peoples' life styles as the invention of the Gutenberg Press was in 1455. At its heart, the Internet is several large computer networks joined together over high-speed data links ranging from ISDN to T1, T3, FDDI, SONET, SMDS, OT1, etc. The most prominent of these national nets are MILNET (Military Network), NSFNET (National Science Foundation NETwork), and CREN (Corporation for Research and Educational Networking). Commercial networks from AT&T, MCI, SPRINT, WILTEL, and many others are now beginning to carry the bulk of the traffic. Note: There "Internet" is now private. NSNET no longer does the funding it once did. But there are a myriad of regional and local campus networks joined in all over the world. And increasingly, businesses are joining their computers to the Internet. In 1995, the Government Accounting Office (GAO) said that the Internet linked 59,000 networks, 2.2 million computers and 15 million users in 92 countries. The Internet traces its origins to a network called Arpanet which was set up in 1969 by the Defense Department. The Internet's

backbones and university-government hookups are financed by federal government monies. But the Internet's networking technology is very smart. Every time, someone hooks a new computer to the Internet, the Internet adopts that hookup as its own and begins to route Internet traffic over that hookup and through that new computer. Thus as more computers are hooked to the Internet, its network grows exponentially. The invention of the Internet will go into the history books as one of the twentieth century's most important inventions.

INTERNET PROTOCOL (IP) ADDRESS
A unique, 32-bit number for a specific TCP/IP host on the Internet. IP addresses are normally printed in dotted decimal form, such as 128.127.50.224. Once your domain is assigned a group of numbers by the Internet's central registry, it can house one or several domains and/or hosts, i.e. computertelephony.com and teleconnect.com. People looking for those domains will be pointed to that server where they will find all information in the domain -- perhaps a home page, or a place to leave e-mail, etc. There are three classes of IP address A, B, and C -- the most common of which is a class "C" address block. A class "C" address block can address about 256 hosts (e.g., 128.10.10.*). a class "B" address block can contain about 256*256 (e.g., 128.10.*.*) hosts. Some ip addresses are reserved for broadcasts in respective domains.

INTERNET ASSIGNED NUMBERS AUTHORITY
IANA. The entity responsible for assigning numbers in the Internet Suite of Protocols.

INTERNET CONTROL MESSAGE PROTOCOL
ICMP. The protocol used to handle errors and control messages at the IP player. ICMP is actually part of the IP protocol.

INTERNET GATEWAY
Internet gateways are devices which typically sit on a local area network and handle all the translations between IPX traffic on your LAN (IPX is the NetWare protocol) and the TCP/IP traffic on the Internet. TCP/IP is the protocol used on the Internet.

INTERNET GROUP NAME
In Microsoft networking, a name registered by the domain controller that contains a list of the specific addresses of computers that have registered the name. The name has a 16th character ending in 0x1C.

INTERNET MIB SUBTREE
A tree-shaped data structure in which network devices on a local area network and their attributes can be identified within the confines of a network management scheme. The name of an object or attribute is derived from its location on this tree.

For example, an object in MIB-I might be named 1.2.1.1.1.0. the first 1 indicates the object is on the Internet. The 2 denotes that it falls within the Management category. The second 1 shows the object is part of the first fully defined MIB, known as MIB-I. The third 1 indicates which of the eight object groups is being referenced. And the fourth 1 is a textual description of the network component. The O indicates there is only one object instance. An object instance links a particular object to a specific node on the network. The numbering system is infinitely extendible to accommodate additions to this base identification scheme. This common naming structure permits equipment from a variety of vendors to be managed by a single management station that uses SNMP. The four main categories of the tree are Directory, Management, Experimental and Private/Enterprises.

INTERNET NUMBER
The dotted-quad address used to specify a certain system. The Internet number for cs.widener.edu is 147.31.130. A resolver is used to translate between hostnames and Internet addresses.

INTERNET PACKET EXCHANGE
IPX. Novell NetWare's native LAN communications protocol, used to move data between server and/or workstation programs running on different network nodes.

INTERNET PROTOCOL
IP. Part of the TCP/IP family of protocols describing software that tracks the Internet address of nodes, routes outgoing message, and recognizes incoming messages. Used in gateways to connect networks at OSI network Level 3 and above.

INTERNET PROTOCOL DATAGRAM
The fundamental unit of information passed across the Internet. Contains source and destination addresses along with data and a number of fields which define such things as the length of the datagram, the header checksum, and flags to say whether the datagram can be (or has been) fragmented, This is a self-contained packet, independent of other packets.

INTERNET PROTOCOL SUITE
A suite of network protocols which have been adopted as the main de facto protocols for LANs.

INTERNET RELAY CHAT
IRC. Sort of like CB radio, but run on the Internet, and far more confusing than CB radio.

INTERNET SERVER
An Internet server is a device which users on the Internet access to get services. Such services might be electronic mail, news, a Web page, etc. A company will have one or more Internet servers attached to the Internet when it wants to

deliver services to people on the Internet. Such Internet servers could be called e-mail servers, FTP servers, News servers and World Wide Web servers. Internet servers most commonly run on Unix. But Microsoft Windows NT is increasingly gaining popularity.

INTERNET SERVICE PROVIDER
ISP. A vendor who provides direct access to the Internet. The ISP also usually provides a core group of internet utilities and services like E-mail, News Group Readers and sometimes weather reports and local restaurant reviews. The user reaches his ISP by either dialing-up with their own computer, modem and phone line, or over a dedicated line installed by the telephone company. The ISP could also

INTERNETWORK ROUTER
In local area networking technology, an internetwork router is a device used for communications between networks. Messages for the connected network are addressed to the internetwork router, which chooses the best path to the selected destination via dynamic routing. Internetwork routers function at the network layer of the Open Systems Interconnection (OSI) model. Also known as a network router or simply as a router.

INTERRUPT LATENCY
The delay in servicing an interrupt request is known as interrupt latency. It is not a problem with devices that are not sensitive to timing inconsistencies (such as hard-disk controllers or video boards). But it is a problem with high-speed, asynchronous communications (9,600 bps and above), which are highly time-sensitive operations.

ISA
1. Interactive Services Association. http:/policy.net/isa

2. Industry Standard Architecture. The most common bus architecture on the motherboard of MS-DOS computers. The ISA bus was originally pioneered by IBM on its PC, then its XT and then its AT. ISA is also called classic bus. It comes in an 8-bit and 16-bit version. Most references to ISA mean the 16-bit version (which carries data at up to 5 megabytes per second). Many machines claiming ISA compatibility will have both 8- and 16-bit connectors on the motherboard. In 1987 IBM introduced a 32-bit bus which it called MCA for Micro Channel Architecture, which is the internal bus inside some of IBM's line of PS/2 MCA machines. But MCA isn't popular because it is incompatible with ISA, so the industry (excluding IBM) invented a 32-bit bus called EISA which stands for Extended Industry Standard Architecture, which is compatible with ISA. EISA, however, suffered from some of the same problems as the MCA bus, namely it was complicated to program to get the card's full benefit. As a result, other buses have been invented, including the VL bus from VESA (the Video Electronics Standards Association) and Intel's PCI bus. PCI stands for Peripheral Component Interconnect. Both the VL and PCI buses claim to be more than 20 times faster

than the ISA bus. The PCI bus, according to Intel, can transfer data at up to 132 megabytes per second. Until the advent of complex Windows graphics and imaging programs, the speed of the ISA was not a gating factor in a computer. As of writing, a VL bus could only handle three drop-in printed cards, while a PCI bus could handle 10.

ISDN

Integrated Services Digital Network. ISDN comes today in two basic flavors -- BRI, which is 144,000 bits per second and designed for the desktop, and PRI which is 1,544,000 bits per second and designed for telephone switches, computer telephony and voice processing systems. Neither ISDN BRI or ISDN PRI is a standard service, though there are several "standard" configurations. ISDN BRI is a wonderful service in your home or office because it can give you videoconferencing, and ultrafaster data communications. But it is not an easy service to get up and running. The best advice I can give you is: 1. Figure out what you want to do with your ISDN. 2. Find which equipment you're going to need that will do the best job for you. 3. Call the manufacturer of that equipment, tell him where you're located and ask him which ISDN service to order. 4. After he tells you, order your ISDN service from your local phone company. 5. Then buy the equipment. 6. Allow yourself at least a month to get up and running. 7. Any ISDN equipment you install in a PC will cause major interrupt problems. Make sure you know which interrupts your PC is using for what.

ISDN is essentially a totally new concept of what the world's telephone system should be. According to AT&T, today's public switched phone network has the following limitations: 1. Each voice line is only 4 KHz, which is very narrow, which limits also the speed you can send data across. 2. Most signaling is in-band signaling, which is very consuming of bandwidth (i.e. it's expensive and inefficient). 3. The little out-of-band signaling that exists today runs on lines separate to the network. This includes signaling for PBX attendants, hotel/motel, Centrex and PBX calling information. 4. Most users have separate voice and data networks, which is inefficient, expensive and limiting. 5. Premises telephone and data equipment must be separately administered from the network it runs on. 6. There is a wide and growing variety of voice, data and digital interface standards, many of which are incompatible.

ISDN's "vision" is to overcome these deficiencies in four ways: 1. By providing an internationally accepted standard for voice, data and signaling. That standard has pretty well achieved, though don't try and take North American ISDN equipment to Europe. 2. By making all transmission circuits end-to-end digital. 3. By adopting a standard out-of-band signaling system. 4. By bringing significantly more bandwidth to the desktop.

One of the best features of ISDN is the speed of dialing. Instead of 20 seconds for a call to go through on today's old analog network, with ISDN it takes less than a second. It's beautiful. Here are some sample ISDN services:

Call waiting: A line is busy. A call comes in. The user knows who is calling. He can then accept, reject, ignore, transfer the call.

Citywide Centrex: A myriad of services: Specialized numbering and dialing plans. Central management of all ISDN terminals, including PBXs, key systems, etc.

Credit card calling: Automatic billing of certain or all calls into accounts independent of the calling line/s.

Calling line identification presentation: Provides the calling party the ISDN "phone" number, possibly with additional address information, of the called party. Such information may flash across the screen of an ISDN phone or be announced by a synthesized voice. The called party can then accept, reject or transfer the call. If the called party is not there, then his/her phone will automatically record the incoming call's phone number and allow automatic callbacks when he/she returns or calls back in from elsewhere.

Calling line identification restriction: Restricts presentation of the calling party's ISDN "phone" number, possibly with additional address information, to the called party.

Closed user group: Restricts conversations to or among a select group of phone numbers, local, long distance or international.

Collaborative Computing. Work on the same document or drawing or design with someone 10,000 miles away. With ISDN, it doesn't really matter where members of the design team live.

Desktop videoconferencing. I have an ISDN desktop videoconferencing device on my desk. It's wonderful to see the person at the other end. It makes for a far more meaningful conversation.

E-Mail (a.k.a. Personal mailbox): ISDN can carry information to and from unattended phones as long as they're equipped with proper hardware and software.

Internet Access: It's much nicer to browse the Internet at 128 Kbps than at 28.8 Kbps which is the fastest I can do it today with the fastest analog modem I can buy.

Shared Screen -- Switched data services provided via ISDN lets two people in remote locations, both equipped with a computer terminal, view the same information on their screens and discuss its contents while making changes
-- all over one telephone line.

Simultaneous Data Calls: Two users can talk and exchange information over the D packet and/or the B circuit or packet switched channel.

There are two major problems to the widespread acceptance of ISDN: First, the cost of ISDN terminal equipment is too high. Second, the cost of upgrading central office hardware and software to ISDN is too high. Both costs are coming down. In early, 1995 Pacific Bell announced that it would install one million lines of ISDN BRI by the end of 1998. And, to make this happen, it dropped its ISDN monthly prices to an affordable $24.95 a month for residences and $26.50 a month for businesses.

There are three basic configurations you can get ISDN:

1. The 2B+D "S" interface (also called the "T" interface). The 2B+D is called the Basic Rate Interface (BRI). The "S" interface uses four unshielded normal telephone wires (two twisted wire pairs) to deliver two "Bearer" 64,000 bits per second channels and one "data" signaling channel of 16,000 bits per second. An S-interfaced phone can be located up to one kilometer from the central office switch driving it. Each of the two 64 Kbps "bearer" or B channels can be used to carry a voice conversation, or one high speed data or several data channels, which are multiplexed into zone 64 kbps high speed data line. The "D" channel of 16 kbps will carry control and signaling information to set up and break down the voice and data calls. The "D" channel can also carry data up to 9600 bits per second in addition to the control and signaling information. Signaling and control on the D channel conforms to a protocol (LAPD) and a messaging structure (Q.931). These two allow intelligent endpoints and switching nodes from different vendors to talk a common language and thus be able to transfer features across a network, from one switch to another, e.g. to transfer a Centrex call across town through several switches and to have it arrive at the end phone with the calling party's name.

2. The 2B+D "U" interface. This "U" interface delivers the same two 64 kbps bearer channels and one 16 kbps data channel, except that it uses 2-wires (one pair) and can work at 5-10 kilometers from the central office switch driving it. The "U" interface is the most common ISDN interface. It carries 160,000 bits per second from the central office to your home or office. Of those 160,000 bits, two are used for 64,000 bps Bearer (B) channels and one is used by the subscriber for 16,000 bps of data (the D channel). The other 16,000 bps is used by the network for signaling between the black box on the subscriber premises and the central office. The idea is to get the ISDN "U" interface working to 18,000 feet -- the average length of a North American subscriber local loop. You connect the two "U" wires (local loop pair) coming in from your local ISDN CO into a black box about the size of desk printing calculator, called an NT-1. Out the side of the black box comes four wires, which are called the "S Bus." Onto these four wires you can attach, in a loop configuration (also called single bus), as many as eight ISDN terminals -- telephones, fax machines, etc. See ISDN.

3. The 23B+D or 30B+D. This is called the Primary Rate Interface (PRI). At 23B+D, it is 1.544 megabits per second. At 30B+D, it is 2.048 megabits per second.

The first, 23B+D is the standard T-1 line in the U.S. which operates on two pairs. The second 30B+D is the standard T-1 line in Europe, which also operates on two pairs.

Integral to ISDN's ability to produce new customer services is CCITT Signaling System 7. This is a ITU-T recommendation which does two basic things: First, it removes all phone signaling from the present network onto a separate packet switched data network, thus providing enormous economies of bandwidth. Second, it broadens the information that is generated by a call, or call attempt. This information -- like the phone number of the person who's calling -- will significantly broaden the number of useful new services the ISDN telephone network of tomorrow will be able to deliver.

ISDN has "enjoyed" many "meanings," including I Still Don't Know to, its most recent, I Smell Dollars Now.

ISOCHRONOUS

Isochronous comes from the Greek "iso" (equal) and "chronous" (time). Isochronous transmission is used to move stuff which must get to its destination with absolutely no delays. Voice and video need isochronous transmission. Let me explain. In the beginning, the phone network switched a call from A to B. It kept the circuit open. Whatever you said at one end went to the other end at the speed of light, effectively instantly. Then they invented other methods of transmission, where the circuit isn't open 100% from end to end during the "conversation." One example is packet switching, used widely for sending data. If you're sending an electronic mail, it clearly doesn't matter if your electronic letter arrives half a second faster or slower. It does matter with voice and video.

In isochronous data transmission, timing is derived from the signal carrying the data. No timing or clock lead is provided at the customer interface. In isochronous data transmission, data has no embedded timing - send it slower and it is still valid, just late. Voice and video are intimately tied to timing. Send voice slower and it sounds very different. With TDM, there is a direct relationship between the signal rate used to digitize the voice samples and the bearer channel rate, allowing accurate reconstruction of the voice (or other signals) at the far end. With packet technologies no such relationship exists.

IVR

Interactive Voice Response. Think of IVR as a voice computer. Where a computer has a keyboard for entering information, an IVR uses remote touchtone telephones. Where a computer has a screen for showing the results, an IVR uses a prerecorded human voice that is stored (digitized) on a hard drive, in addition it can use a synthesized voice (computerized voice) for read back information that is constantly changing. (The synthesized voice is commonly referred to as Text-to-Speech.

Whatever a computer can do, an IVR can too -- from looking up train timetables to moving calls around an automatic call distributor (ACD). The only limitation on an IVR is that you can't present as many alternatives on a phone as you can on a screen. The caller's brain simply won't remember more than a few. With IVR, you have to present the menus in smaller chunks.

The benefits of Interactive Voice Response are obvious. By automating the retrieval and processing of information by phone, you can "give data a voice" and "add intelligence to the phone call." By doing that, you can:

Put information to work. The classic IVR "killer app" takes an existing database (e.g., a magazine's article archives, a freight company's package-tracking system) and makes it available by phone (or other media, such as fax, e-mail, or DSVD -- Digital Simultaneous Voice and Data). You can automate telephone-based tasks. From "bank by phone" to "find my package," to "sell me an airline ticket," to "validate my new credit card," IVR gives access to and takes in information; performs record-keeping, and makes sales, 24 hours a day -- supplementing or standing in for human personnel. IVR can add value to communications. Any call-handling operation can profit from IVR. Used as a front-end for an ACD, an IVR system can ask questions (e.g., "what's your product serial code?") that help routing and enable more intelligent and informed call processing (by people or automatic systems). IVR far supersedes more rudimentary technologies (such as Caller ID) in such applications. Used in place of traditional on-hold programming, IVR can add interactive value to what would otherwise be wait-time. The IVR can be used to distribute info, make callers aware of specials -- even provide entertainment. The result: fewer callers drop off queue; you make more sales.

JATE
The Japanese equivalent of the U.S. FCC part 68 certification for equipment to be attached to the Japanese telephone network. It stands for Japan Approvals Institute for Telecommunications Equipment. Getting JATE approval is expensive, complex, and immensely time consuming.

JAVA LANGUAGE
JAVA is a dynamic object-oriented, multithreaded high performance programming language which is architecture neutral and portable. It is robust and secure and works well in distributed environments.

JPEG
Worldwide standard for still image compression devised by the Joint Photographic Experts Group, experts sanctioned by the International Standards Organization (ISO) and the ITU-T. The JPEG compression standard works this way, according to the New York Times: A color image is converted into rows of dots, called pixels, each with a numerical value that represents brightness and color. The picture is then broken down into blocks, each 16 pixels x 16 pixels and then reduced to 8 pixels by 8 pixels by subtracting every other pixel. The software uses a formula that computes an average value for each block, permitting it to be

represented with less data. Further steps subtract even more information from the image. To retrieve the data, the process is simply reversed to decompress the image. A specialized chip decompresses the images hundreds of times faster than is possible on a standard desktop computer. PC Magazine defines JPEG as "a lossy image-compression algorithm that often reduces the size of bitmapped images by a factor of 10 or more with little or no discernible image degradation. JPEG compression works by filtering out an image's high-frequency information to reduce the volume of data and then compressing the resulting data with a lossless compression algorithm. Low-frequency information does more to define the characteristics of an image, so losing some high-frequency information doesn't necessarily affect the image quality."

JPEG++
Storm Technology's proprietary extension of the JPEG algorithm. It lets users determine the degree of compression that the foreground and background of an image receive; for example, in a portrait, you could compress the face in the foreground only slightly, while you could compress it in the background to a much higher degree.

LATA
Local Access and Transport Area. One of 161 local geographical areas in the US within which a local telephone company may offer telecommunications services -- local or long distance. AT&T is expressly prohibited from offering intraLATA calls by the terms of the Divestiture. Other competitors, such as MCI and Sprint, are not, though rules vary by state, according to state regulation.

LATENCY
A fancy term for waiting time. Real-time, interactive application such as desktop conferencing are sensitive to accumulated delay, which is referred to as latency. For example, telephone networks are engineered to provide less than 400 milliseconds (ms) round-trip latency. You can get latency in several ways:

1. From propagation delay -- the length of time it takes information to travel the distant of the line. This period is mostly determined by the speed of light; therefore, the propagation delay factor is not affected by the networking technology in use.

2. From transmission delay -- the length of time it takes to send the packet across thc given media. Transmission delay is determined by the speed of the media and the size of the packet.

3. From processing delay - the time required by a networking device for route lookup, changing the header, and other switching tasks. In some cases, the packet also must be manipulated; for instance, changing the encapsulation type, changing the hop count, and so on. Each of these steps can contribute to the processing delay.

4. Rotation delay. The delay in accessing data which comes from waiting for a disk to rotate to the currant location.

In a bridge or a router, latency is the amount of time elapsed between receiving and retransmitting the LAN packet. The length of time the packet is stuck in a bridge or router.

LEC
Local Exchange Carrier. The local phone companies, which can be either a Bell Operating Company (BOC) or an independent (e.g. GTE) which provide local transmission services. Prior to divestiture, the LECs were called telephone companies or telcos.

LEGACY
1. All the stuff you have on hand - equipment, software, files and paperwork. In short, everything of a data processing/telecommunications nature in your business today. The use of the word "legacy" suggests that you've inherited all this stuff from previous generations of obsolete management. And the idea is that you're forced to update it, without junking it altogether -- which is expensive and potentially problematical. All you have is the legacy of previous generations. Preserve it, because no one can afford to junk it. Or so the theory goes. 2. Microsoft defines legacy in its Windows 95 Resource Kit as hardware and devices cards that don't conform to its Plug and Play standard.

LINKON
Linkon Corporation, Fairfield, CT and New York, NY, manufactures DSP (digital signal processor) based communications processing hardware and software. The company introduced the first "universal port" design processing board for EISA/ISAbus and Sbus architecture in 1991. Their DSP "engine" is capable of processing voice, fax, modem, speaker independent speech recognition, speaker dependent speech recognition, speaker verification and text-to-speech simultaneously and concurrently during the same phone session without switching to an additional piece of hardware. Linkon sells through OEMs, SIs, VARs and distribution. Their customers add value by integrating this technology with application specific solutions for computer telephony.

LOCAL AREA SIGNALING SERVICES
LASS is a group of central office features provided now by virtually all central office switch makers that uses existing customer lines to provide some extra features to the end user (typically a business user). They are based on delivery of calling party number via the local signaling network. LASS can be implemented on a standalone single central office basis for intra office calls or on a multiple central office grouping in a LATA (what the local phone companies are allowed to serve) for interoffice calls. Local CCS7 (Common Channel Signaling Seven) is required for all configurations.

LOCAL CALL
Any call within the local service area of the calling phone. Individual local calls may or may not cost money. In many parts of the US, the phone company bills its local service as a "flat" monthly fee. This means you can make as many local calls per month as you wish and not pay extra. Increasingly this luxury is dying and local calls are costing money.

LOOP START
LS. You "start" (seize) a phone line or trunk by giving it a supervisory signal. That signal is typically taking your phone off hook. There are two ways you can do that -- ground start or loop start. With loop start, you seize a line by bridging through a resistance the tip and ring (both wires) of your telephone line. The Loop Start trunk is the most common type of trunk found in residential installations. The ring lead is connected to -48V and the tip lead is connected to OV (ground). To initiate a call, you form a "loop" ring through the telephone to the tip. Your central office rings a telephone by sending an AC voltage to the ringer within the telephone. When the telephone goes off-hook, the DC loop is formed. The central office detects the loop and the fact that it is drawing DC current and stops sending the ringing voltage. In ground start trunks, ground Starting is a handshaking routine that is performed by the central office and the PBX prior to making a phone call. The central office and the PBX agree to dedicate a path so incoming and outgoing calls cannot conflict, so "glare" cannot occur.

MAN
See METROPOLITAN AREA NETWORK.

MEDIA SERVER
A new term for a file server on a local area network which contains files containing voice, images, pictures, video, etc. In short, a media server is a repository for media of all types. Media servers are also called file servers. Here's a definition of Media Server, courtesy of Oracle Corporation, writing in early 1994: "The media server harnesses the power of the fastest computers ever built -- massively parallel computers. Media servers support diverse clients so that users don't have to discard existing systems. They provide storage, network interfaces and memory -- plus support for all forms of multimedia information. Because they use thousands of low-cost microprocessors, media servers offer astonishing performance at low cost. Example, an nCUBE Model 20 or IBM SP/2."

METROPOLITAN AREA NETWORK
MAN. A high-speed data intra-city network that links multiple locations within a campus, city, or LATA. Typically extends as far as 50-kilometers, operates at speeds from 1 Mbit/s to 200 Mbps and provides an integrated set of services for real-time data, voice and image transmission. The IEEE 802.6 standard defines MAN standards and SMDS is the MAN service offered by local phone companies. Private and public MANs may use the ANSI FDDI standard.

METROPOLITAN FIBER RING

A metropolitan fiber ring is an advanced, high-speed local network that can also be used to connect businesses and residences directly to a long distance carrier's network, and provide alternatives to the local telecommunications services they have today. This definition, courtesy MCI.

MIRRORING
A fault tolerance method in which a backup data storage device maintains data identical to that on the primary device and can replace the primary if it fails. Mirroring will typically cost you a 50% performance degradation when your write to disk and 0% performance degradation when you are reading.

MIXED MODE
An imprecise term which suggests that one digital bit stream can carry voice, data, facsimile and video signals.

MODIFIED HUFFMAN
A one-dimensional data-compression scheme that compresses data in a horizontal direction only. Allows no transmission of redundant data.

MODULAR
Equipment is said to be modular when it is made of "plug-in units" which can be added together to make the system larger, improve its capabilities or expand its size. There are very few phone systems that are truly modular.

MODULATION PROTOCOLS
Modem stands for MOdulator/DEModulator. A modem converts digital signals generated by the computer into analog signals which can be transmitted over an analog telephone line. It also transforms incoming analog signals into their digital signals for inputting into a computer. The specific techniques used to encode the digital bits into analog signals are called modulation protocols. The various modulation protocols define the exact methods of encoding and the data transfer speed. In fact, you cannot have a modem without modulation protocols. A modem typically supports more than one modulation protocol.

MPEG
MPEG is commonly known as a series of hardware and software standards designed to reduce the storage requirements of digital video, i.e. video recorded digitally or converted into digital bits. MPEG is most commonly known as an compression scheme for full motion video. The word MPEG is actually the acronym for the Motion Pictures Experts Group, a joint committee of the International Standards Organization (ISO) and the International Electrotechnical Commission (EG). The first MPEG specification, known as MPEG-1, was introduced by this committee in 1991. The common goal of all MPEG compression is to convert the equivalent of about 7.7 meg down to under 150 Kb, which represents a compression ratio of about 52 to one. The two requirements of MPEG-1 are 30 frames per second of Standard Image Format (SIG) of 352 pixels x 240 pixels and CD-quality sound at 44.1 Khz, 16 bit stereo. MPEG image scheme

offers more compression than the other poplar JPEG image compression scheme, which is largely for still images. MPEG takes advantage of the fact that full motion video is made up of many successive frames consisting of large areas that are not changed -- like blue sky background. While JPEG compresses each still frame in a video sequence as much as possible, MPEG performs "differencing," noting differences between consecutive frames. If two consecutive frames are identical, the second can be stored in remarkably few bits. MPEG condenses moving images about three times more tightly than JPEG.

MPS
Multi Page Signal. A frame sent in fax transmission if the sender has more pages to transmit.

MTSO
Mobile Telephone Switching Office. This central office houses the field monitoring and relay stations for switching calls between the cellular and wire-based (land-line) central office. The MTSO controls the entire operation of a cellular system. It is a sophisticated computer that monitors all cellular calls, keeps track of the location of all cellular-equipped vehicles traveling in the system, arranges handoffs, keeps track of billing information, etc.

MU-LAW
The PCM voice coding and companding standard used in Japan and North America. A PCM encoding algorithm where each digitized sample is represented by eight bits, thus a 64k transmission rate -- the standard rate for encoding voice -- is actually 8K samples/second. A sample consists of a sign bit, a three bit segment specifying a National logarithmic range, and a four bit step offset into the range. All bits of the sample are inverted before transmission.

MULTI-RATE ISDN
"Nx64" or switched-fractional T1 lets users combine multiple ISDN PRI "B" channels on a call-by-call basis. Since each "B" channel operates at 64 kbps, users could get from 128 kbps to 1.544 Mbps of bandwidth in 64 kbps increments.

MULTI-TASKING
The concurrent management of two or more distinct tasks by a computer. Although a computer with a single processing unit (as virtually all PCs are) can only execute one application's code at a given moment, a multi tasking operating system can load and manage the execution of multiple applications, allocating processing cycles to each in sequence. Because of the processing speed of computers, the apparent result is the simultaneous processing of multiple tasks. Standard mode Windows performs multi tasking only in the form of context switching. 386 enhanced mode allows multi tasking in the form of time slicing.

MULTI-THREADING

Concurrent processing of more than one message by an application program. One of OS/2's advantages over Windows is that IBM designed it as a multi-threaded operating system. Each program in OS/2 can start two or more threads, which carry out various interrelated tasks with less overhead than two separate programs would require. For example, a communications program could have three threads running: one that waits for characters to be received, another that monitors the keyboard, and a third that displays information. This is more efficient than running multiple tasks because it doesn't require the overhead of an operating system context switch.

In short, a new and complex method of programming. Here is a definition from Sun Microsystems. A traditional UNIX process has a single thread of control. A thread of control, or more simply a thread, is a sequence of instructions being executed in a program. A thread has a program counter (PC) and a stack to keep track of local variables and return addresses. A multi-threaded UNIX process is no longer a thread of control of itself; instead, it is associated with one or more threads. Threads execute independently. There is in general no way to predict how the instructions of different threads are interleaved, though they have execution priorities that can influence the relative speed of execution. In general, the number or identities of threads that an application process chooses to apply to a problem are invisible from outside the process. Threads can be viewed as execution resources that may be applied to solving the problem at hand.

Threads share the process instructions and most of its data. A change in shared data by one thread can be seen by the other threads in the process. Threads also share the operating system. Each sees the same open files. For example, if one thread opens a file, another thread can read it. Because threads share so much of the process state, threads can affect each other in sometimes surprising ways. Programming with threads requires more care and discipline than ordinary programming because there is no system enforced protection between threads.

MULTIPROCESSING
A type of computing characterized by systems that use more than one CPU to execute applications. Multi processing is not multi tasking, which is the ability to have more one application running on a system at the same time. The technique is not associated with multi processing, nor does it require multi processing to take place. Multi tasking typically uses a computer with one CPU (e.g. your desktop or laptop). Multi processing uses a computer with several CPUs, often a server.

MULTIPURPOSE INTERNET MAIL EXTENSION
MIME. An extension to electronic mail (Internet and other e-mail systems) which provides the ability to transfer non-ASCII data, such as graphics, software, audio, binary and fax. MIME was developed and adopted by the Internet Engineering Task Force. It was designed for transmitting mixed-media files across TCP/IP networks. The MIME protocol, which is actually an extension to SMTP, covers binary, audio and video data. Essentially what happens with Mime is that you pick

a binary file (any file that isn't ASCII) to send along with your e-mail. Your e-mail software then converts your binary file to ASCII text which can easily be transmitted across one or more e-mail systems. All e-mail networks will transmit ASCII (i.e. ASCII 128 and below). The technique that makes MIME encoding possible is called UUencoding. Here is an example. Notice that every line begins with a M. Every line is the same length. At one end you UUencode the file (and it looks as below). When you receive the file, you must UUdecode it. Sometimes the coding and encoding is done automatically by your e-mail program. Sometimes you must decode it manually. Don't try and UUdecode the following file. It won't work. I chopped the middle of the file out. It's 64K bytes long. I'm running it below purely as an example. See UUENCODE.

```
----------------------------------------------------------
          Uuencoded File Attachment: ANTARESN.DOC
---------------------------------------------------------- begin 660 ANTARESN.DOC
MVZ4M````"00`````````````````````@`$```$%``!["P````````````````
M```````````($#`````````````````````````````````````````````````*
M``!_```*``!_`\*`````\*`````\*`````\*`````\*```.`(T*````
M______________________________________________________________
M______________________________________________________________
9_________________
________________T16
`

end
```

MVIP

Multi-Vendor Integration Protocol. MVIP is a family of standards to let telephony products from different vendors inter-operate within a computer or group of computers. MVIP started in 1990 with a telephony bus for use inside a single computer. Picture a printed circuit card that fits into an empty slot in a personal computer. The slot carries information to and from the computer. This is called the data bus. Printed circuit cards that do voice processing typically have a second "bus" -- the voice bus. That "bus" is actually a ribbon cable which connects one voice processing card to another. The ribbon cable is typically connected to the top of the printed circuit card, while the data bus is at the bottom. As of this writing (late 1993), there are at least four such buses defined. Several (AEB, PEB, SCSA) have been defined by Dialogic Corporation, Parsippany, NJ for connecting devices with Dialogic products and Dialogic compatible devices. The MVIP Bus was defined by Natural MicroSystems, Natick, MA with assistance from Mitel, Promptus Communications and Rhetorex as a vendor-independent means of connecting telephony devices within a computer chassis. MVIP was introduced by a seven company group in 1990 and has been distributed by Natural MicroSystems, Mitel and NTT International (part of the Japanese telephone company). By 1993, MVIP had attracted several hundred participating companies including companies manufacturing a variety of board-level MVIP products including telephone line interfaces, voice boards, FAX boards, video codecs, data multiplexers and LAN/WAN interfaces. MVIP now has its own trade association --

the Global Organization for MVIP (see GO-MVIP), which develops extensions such as higher-level APIs and multi-chassis switching. Here is a write-up on the original MVIP Bus from Mitel's Communicating Objects Division:

"The MVIP Bus consists of communications hardware and software that allows printed circuit cards from multiple vendors to exchange information in a standardized digital format. The MVIP bus consists of eight 2 megabyte serial highways and clock signals that are routed from one card to another over a ribbon cable. Each of these highways is partitioned into 32 channels for a total capacity of 256 full-duplex voice channels on the MVIP bus. These serial link from one card to another. They are electronically compatible with Mitel's ST_BUS specification for inter-chip communications. By letting expansion cards exchange data directly, the MVIP bus opens the PC architecture to voice/data applications that would otherwise overburden the PC processor with data transfers. The MVIP bus is equivalent to an extra backplane that is capable of routing circuit switched data.

"MVIP systems generally have two types of cards ; network cards and resource cards. They differ by the switching they provide and in the way they are wired to the bus. Network cards almost always provide more flexible switching and can drive either the input or the output side of the bus, although they usually drive the output side of the bus. Resource cards usually provide very little switching and are only able to drive the input side of the bus. Resource cards usually rely on the network cards to do most or all of the switching on the MVIP bus."

According to Brough Turner, chairman of the MVIP Technical Committee, MVIP Evangelist and vice president of Natural MicroSystems, "The original MVIP bus distributes switching elements but only requires switches on telephone network interface cards. This has simplified access to MVIP for many communications service providers and has permitted easy inter-connection of pre-existing and proprietary technology to the common bus. More recently, the Flexible MVIP Interface Circuit has reduced the complete MVIP interface, including clocks and switching, to one chip, further simplifying MVIP connections.

"MVIP's distributed switching architecture provides software controlled digital switching within the PC chassis. MVIP software driver standards simplify access for application developers and support the integration of components from multiple vendors. As a result of work by more than a dozen companies active in the MVIP Technical Committee, MVIP now addresses multi-chassis connections and higher-level APIs. Multi-Chassis MVIP (MC-MVIP) provides telephony connections and distributed digital switching between MVIP-based PCs and supports interconnection of MVIP PCs with proprietary telephony equipment including PBXs, ACDs, and voice processing systems. MC-MVIP includes redundant clocks and specifically permits plugging and unplugging connections while a multi-chassis system is operating. This allows application developers to construct fault-tolerant systems."

NETRA

The Netra Internet Server is a simple, integrated secure solution that provides Internet and WWW access from PC and Macintosh LANs. It consists of a Sun Netra server configured with the hardware and software necessary to interface the LAN to the Internet through a connection from an Internet service provider. The Netra Internet Server allows businesses and users to access other WWW sites and to set up their own WWW Home Page to publish information. The Netra Internet Server enables an easy and transparent access to the Internet from LAN desktop computers through such key client applications as Mosaic, Netscape or Gopher. These client applications are not bundled with Netra and must be either downloaded from the Internet, if publicly available, or purchased from third parties. The Netra Internet Server includes WWW server software; anonymous file transfer protocol (FTP) server software; and e-mail, FTP and Telnet capabilities. Base-level server security is included in the product. The optional FireWall software package provides security for the entire LAN. The Netra Internet Server comes equipped to function as a primary or secondary DNS server. The Netra Internet Server also comes with ease-of-use administration tools. Its software includes a graphical HTML (Hyper Text Markup Language) setup.

NETWORK INTERFACE MODULE
Electronic circuitry connecting a system (typically a PC) to the telephone network. Network interface modules come in as many versions as there are ways of connecting to the telephone network. They range from a simple loop start telephone line to complex ISDN PRI circuits. Usually the network interface modules slides into one of the expansion slots inside a PC. The card transmits and receives messages from the resource modules and provides access to the telephone network.

NETWORK MANAGEMENT
A set of procedures, software, equipment and operations designed to keep a network operating near maximum efficiency. Network management generally falls into five areas:

1. CONFIGURATION MANAGEMENT deals with installing, initializing, "boot" loading, modifying and tracking the configuration parameters of network hardware and software.

2. FAULT LOCATION and REPAIR MANAGEMENT tools let you find out what's going wrong with what equipment or lines and give you the ability to fix those resources -- by re-routing traffic on different lines or reporting problems to the carrier, or suggesting to whoever that certain equipments should be replaced. Fault location and management tools have strong error and alarm characteristics.

3. SECURITY MANAGEMENT tools allow the network manager to restrict access to various resources in the network. There are devices such as password protection schemes, giving users different levels of access to different network resources.

4. PERFORMANCE MANAGEMENT tools provide real-time and historical statistical information about the network's operation. Such tools show, for example, how many packets are being transmitted at any given moment, the number of users logged into a specific server and use of network lines.

5. ACCOUNTING MANAGEMENT applications help users allocate the costs of the various network resources -- from lines to PBXs, from access to a mainframe to time used on printers.

NETWORK TERMINATION TYPE 1
NT-1 or NT1. The NT-1 is the first customer premise device on a two-wire ISDN circuit coming in from the ISDN central office. It does several things. It converts the two-wire ISDN circuit (called "U" interface) to four-wire so you can hook up several terminals -- like a voice phone and a videophone. The NT-1 typically has several lights on it which indicate if it's working. An ISDN central office can usually "talk" to the NT-1 and do testing and maintenance by instructing the NT-1 to loop signals back to the central office. An NT-1 will support up to eight terminal devices, though I've never seen it work with that many. Sometimes, NT1s are built into ISDN terminal devices, like phones or PC cards. The basic NT1 functions are:

* Line transmission termination.

* Layer 1 line maintenance functions and performance monitoring.

* Layer 1 multiplexing, and

* Interface termination, including multi drop termination employing layer 1 contention resolution.

NFS
Network File System. One of many distributed-file-system protocols that allow a computer on a network to use the files and peripherals of another networked computer as if they were local. This protocol was developed by Sun Micro-Systems and adopted by other vendors.

NNI
1. Network Node Interface. An Asynchronous Transfer Mode (ATM) term. The interface between two public network pieces of equipment (contrast that to UNI, which stands for User Network Interface).

2. Network to Network Interface. A protocol defined by the Frame Relay Forum and the ATM forum to govern how ATM switches establish connection and how ATM signaling requests are routed through an ATM network. Equivalent to a routing protocol in a router environment.

3. Network to Network Interface. There are two types of network interfaces specified by frame relay standards. The first is a user-to-network (UNI) interface

and the second is a network-to-network interface (NNI). The NNI describes the connection between two public frame relay services, and includes elements such as bidirectional polling, to assist the network services providers with gaining information on the status of the public networks being interconnected.

NNTP
Network News Transport Protocol. An extension of the TCP/IP local area network protocol that provides a network news transport service. NNTP is the standard for Internet exchange of Usenet messages, published in RFC-977, Network News Transfer Protocol: A Proposed Standard for the Stream-Based Transmission of News.

NO ANSWER TRANSFER
A service provided by a cellular carrier that automatically transfers an incoming cellular call to another phone number if the cellular subscriber is unable to answer. Most no-answer transfer systems will automatically transfer an incoming cellular call to another phone number if the cellular subscriber is unable to answer (it's not turned on) or if it's not answered after the third or fourth ring if the phone is turned on.

NODAL ARCHITECTURES
Also called Hub architectures. Nodal network architectures means that network traffic from many locations are connected into a single network site for consolidation and aggregation to higher speed circuits. The traffic is then transmitted from this hub site to a central network site, or to other hub sites. Nodal architectures save money, but also increase the risk of a single network failure affecting multiple network locations.

NODAL CLOCK
The principal clock or alternate clock located at a particular node that provides the timing reference for all major functions at that node.

NON-BLOCKING
A PBX is Non-Blocking if all callers can be switched through the PBX even during the busiest time of day. However, the caller may find no trunks are available and thus may not be able to complete the call. The trunk availability is a different problem than the ability of the switch to handle all the traffic it gets. But if your call does not go through do you care? A more technical explanation from InteCom of non-blocking: "Indicates that the internal network of the switching system is such that the total number of available transmission paths is equal to the number of ports. Therefore, all ports can have simultaneous access through the network."

NON-ISDN LINE
Any connection from a CPE to a central office switch that is not served by D-channel signaling.

NON-ISDN TRUNK

In ISDN language, a non-ISDN trunk is any trunk not served by either SS7 or D-channel signaling.

NON-WIRELINE
Also called the block "A" carrier. The "A" originally stood for "alternate." The FCC, in setting up the licensing and systems in each market. It reserved one for the local wireline telephone company, and opened the second system -- the Block A system -- to other interested applicants. The distinction between Block A and Block B is meaningful only during the licensing phase at the FCC. Once a system is constructed, it can be sold to anyone. Thus in some markets today, both the A and B systems are owned by a telephone company. One happens to be the local phone company, and the other is a phone company that decided to buy a cellular system outside its home territory. Non-Wireline, or Block A systems, operate on the radio frequencies from 824 to 849 Megahertz.

NORTH AMERICAN NUMBERING PLAN
NANP. The method of identifying telephone trunks in the public network of North America, called World Numbering Zone 1 by the ITU-T. The Plan has three ways of identifying phone numbers in North America -- a three digit area code, a three digit exchange or central office code and four digit subscriber code. Other countries have much more complicated numbering schemes. There are some countries, for example, where the length of the area code actually exceeds the subscriber code. There are many countries where there is no consistency in the length of phone numbers. Some are nine-digit. Some are 12-digit, etc. All these varying number lengths may be within 100 miles of each other.

Under the NAMP format prior to January 1, 1995, the second digit of the area code was always a one or a zero. With this system, approximately one billion telephone numbers, 152 area code combinations and 640 prefixes were available. Numbers started to run out as a result of increased use of fax, modem and cellular lines. With the new Numbering Plan introduced on January 1, 1995, the area code can be any combination of three digits. This will allow more than 6 billion telephone numbers and 792 area code combinations.

NT-1 or NT1
The NT-1 is the first customer premise device on a two-wire ISDN circuit coming in from the ISDN central office. It does several things. It converts the two-wire ISDN circuit (called "U" interface) to four-wire so you can hook up several terminals -- like a voice phone and a videophone. The NT-1 typically has several lights on it which indicate if it's working. An ISDN central office can usually "talk" to the NT-1 and do testing and maintenance by instructing the NT-1 to loop signals back to the central office. An NT-1 will support up to eight terminal devices, though I've never seen it work with that many. Sometimes, NT1s are built into ISDN terminal devices, like phones or PC cards. The basic NT1 functions are:

* Line transmission termination.

* Layer 1 line maintenance functions and performance monitoring.

* Layer 1 multiplexing, and

* Interface termination, including multi drop termination employing layer 1 contention resolution.

NT-2
In ISDN, the Network Termination type 2 is an intelligent customer premise device terminates the 4-wire T interface. The NT-2 can perform switching and concentration, such as a digital PBX. It typically terminates primary rate access lines from the local ISDN CO switch. It may include the Layer 1 functions of a NT-1 and the protocol handling functions of Layers 2 and 3.

NT1
Network Termination Type 1. A small box that attaches to an ISDN BRI line. It is the first piece of ISDN equipment at your office. It terminates the ISDN loop from the central office and provides the interface into which you, the user, can plug a telephone, a data, video device, a PC, etc. It contains a power supply and changes the signaling on the ISDN line from the two wires that come in from the central office to the four wires that most ISDN terminal equipment needs. These days some ISDN terminal devices contain their own built-in NT1.

NT2
Network Termination 2. A unit that provides switching concentration of subscriber lines at the S-interface. This unit performs the functions of a customer-premises switch or multiplexer to multiplex B channel(s) and D channel onto one physical path and to route calls to the appropriate B or D channel.

NUI
Network User Identifier. A unique alphanumeric number provided to dial-up users to identify them to packet switched networks around the world. The number is used to get onto the network and for billing.

NUMBER ADMINISTRATION AND SERVICE CENTER
NASC. Provides centralized administration of the Service Management System (SMS) database of 800 numbers. The NASC keeps track of the 800 numbers that are in use, or available for use, by new 800 customers.

NUMBER PORTABILITY
The technology that will allow a telephone number to travel with a customer from place to place and, for 800 numbers, from one long-distance company to another. Number portability for 800 numbers is scheduled to begin March, 1993.

OAI
Open Application Interface. Basically one or many openings in a telephone system that lets you link a computer to that phone system and lets the computer

command the phone system to answer, delay, switch, hold etc. calls. The term is also called PHI -- as in PBX-Host-Interface. The term OAI was first used by PBX makers, NEC and InteCom. And now the term has become somewhat generic, like all good things. Essentially every manufacturer of phone systems is evolving towards open application interfaces of their own. According to Probe Research, there are really two separate "markets" for OAI or PHI:

First, there's Horizontal/Office Automation applications. These are applications that support business functions across organizational groups or industry verticals in inter- or intra-department business settings. Examples include voice mail, electronic mail, message centers, corporate telephone directories, automated screen-based dialing, personal productivity tools, conferencing, PBX feature enhancements, ANI interfaces, time clocks, 911 emergency service and compound image (data, text, image and voice) processing. I see these applications as all those useful, productivity-enhancing things I always wished my telephone systems would do if only they would let me program the thing. (Until OAI, all phone systems were totally closed architecture.)

Second, there's transaction applications. These are applications that support an actual business transaction -- customer service, inbound telephone order taking, outbound telemarketing, market research, data gathering, inventory inquiry, account time billing, credit collections, locator services with or without transfer to the local dealer. These always require access to a computer database. These applications are generally complicated, time-sensitive and customized for each installation.

A sample OAI arrangement: In early 1993, Novell and AT&T inked a deal to put telephony onto Novell LANs. The Telephony Server NetWare Loadable Module (NLM) will be the first product. It is an AT&T PC-card sitting in the Novell File server. The card connects to the ASAI (Adjunct Switch Applications Interface) port on the AT&T Definity PBX. Anyone with a PC on the network and an AT&T phone on their desk can use telephone features, zzsuch as auto-dialing, conference calling and message management (a new term for integrating voice, fax and e-mail on your desktop PC via your LAN). The Novell/AT&T deal intends to create open Application Programming Interfaces (APIs) that third party developers can work with. A Novell/AT&T example of what could be developed: A user could select names from a directory on his PC. He could tell the Definity PBX through the PC over the LAN to place a conference call to those names. At the same time, a program running under NetWare would automatically send an e-mail to the people, alerting them to the conference call and giving them the agenda. All participants would have access to both the document and the conference call simultaneously.

OFF/ON HOOK
A modem term. Modem operations which are the equivalent of manually lifting a phone receiver (taking it off hook) and replacing it (going on hook).

OFF-HOOK
When the handset is lifted from its cradle it's Off-Hook. When you lift the handset, the hookswitch is moved by a spring (except in Rolm phones) and alerts the central office that the user wants the phone to do something like receive an incoming call or dial an outgoing call. A dial tone is a sign saying "Give me an order." The term "off-hook" originated when the early handsets were actually suspended from a metal hook on the phone. When the handset is removed from its hook or its cradle (in modern phones), it completes the electrical loop, thus signaling the central office that it wishes dial tone. Some leased line channels work by lifting the handset, signaling the central office at the other end which rings the phone at the other end. Some phones have autodialers in them. Lifting the phone signals the phone to dial that one number. An example is a phone without a dial at an airport, which automatically dials the local taxi company. All this by simply lifting the handset at one end -- going "off-hook."

OFF-LINE
Any equipment not actively connected to a phone line but which can be activated to work with that system is Off-Line. This concept also applies to computer systems. For example, a modem attached to or built into a microcomputer can be plugged permanently into a phone line. But the microcomputer can be used for word processing most of the time. While it's doing word processing, it is "off-line." When the user loads the communications software and turns on the modem, the microcomputer is now said to be "on line." Off-line computer storage is a place to put stuff which a computer cannot access "on-line," like a hard-disk. Off-line storage might be microfilm or microfiche.

OFF-PREMISES EXTENSION
OPX. Now also called OPS for Off-Premises Station. A telephone located in a different office or building from the main phone system. The OPX is connected by a phone line dedicated to it. It acts as if it were in the same place as the main phone system and can use its full capabilities. Here's another explanation a reader sent in. OPX is the appearance of an actual telephone line (such as 212-691-8215) in two physically separate locations. For example, this line (212-691-8215) could appear and ring in my office and at my home without my home phone being a part of the office telephone system. An OPX is commonly used for answering services or for small businesses. Bell operating phone companies are sharply increasing the charge for dedicated OPS lines -- often by several hundred percent per year.

ON-NET
Telephone calls which stay on a customer's private network, traveling by private line from beginning to end are said to be on-net. Here's MCI's definition: Billable calls to MCI-tariffed cities, including those cities reached via leased lines. Can be MCI On-Net or WATS On-Net. Classified as Tier 1 for tariff purposes.

ON-RAMPS

A way of getting onto The Information Superhighway. Such an on-ramp could be anything from a phone line to a two-way cable TV channel. Many companies are trying to create on-ramps, however they define them.

ONE CALL
Fax-back system that requires you to call from your fax machine to send documents on the same line after picking from menu of verbal prompts. the other fax-back system is called a two-call system. You dial from one line and tell the fax-back machine on the other end that you want it to send your requested fax to second number, i.e. one where your fax machine will receive your requested message.

ONE NUMBER SYSTEMS
Also called Follow Me Systems. Follow me systems and services are based on the premise that people are mobile (e.g., they move around a lot in and out of the office), and have many phone numbers or places they might be. A person could have an office number, a cellular number, a voice mail number, a home number, and a pager number. Which phone number will a caller be at? Follow me systems will "track-down" the user being called no matter where they are and connect the caller to the user. The caller need only dial a single phone number. Usually, network or local switch provided call data (or data gathered by a voice response unit) is used to identify each call as being intended for a specific user. Based on options the user has selected, the caller will hear an answering prompt customized to that user, and the one number system will then automatically attempt to locate the user at one of several locations. The tracking-down process varies considerably between follow me systems. Some systems try multiple locations at once, others will try the possible destination locations sequentially. Almost all follow me services provide the caller an exit to voice mail at various points of the call.

OPERATING SYSTEM
A software program which manages the basic operations of a computer system. It figures how the computer main memory will be apportioned, how and in what order it will handle tasks assigned to it, how it will manage the flow of information into and out of the main processor, how it will get material to the printer for printing, to the screen for viewing, how it will receive information from the keyboard, etc. In short, the operating system handles the computer's basic housekeeping. MS-DOS, UNIX, PICK, etc. are operating systems.

OSI STANDARDS
The International Standards Organization (ISO) has established the Open Systems Interconnection (OSI). The idea of OSI is to provide a network design framework to allow equipment from different vendors to be able to communicate. Codex of Mansfield, MA. has published an excellent booklet called "The Basics Booklet of Local Area Networking". Here is a shortened excerpt of what Codex says about standards. "Standards allow us to buy items such as batteries and bulbs. Many of us have learned "the hard way" that the lack of computer standards can make it

impossible for computers from different vendors to talk to each other. Because a major goal of a LAN (Local Area Network) is to connect varied systems, standards are being developed to specify the set of rules networks will follow. The OSI Model is a design in which groups of protocols, or rules for communicating, are arranged in layers. Each layer performs a specific data communications function. The concept of layered protocols is analogous to the steps we follow in making a phone call:

Step 1 -- Listen for dial tone.

Step 2 -- Dial a phone number.

Step 3 -- Wait for a ring.

Step 4 -- Exchange greetings.

Step 5 -- Communicate message.

Step 6 -- Say Good-bye.

Step 7 -- Hang up.

Each of these steps, or OSI "layers," builds upon the one below it. Although each step must be performed in preset order, within each layer there are several options. Within the OSI model, there are seven layers. The first three are the PHYSICAL, DATA LINK, and NETWORK layers, all of which are concerned with data transmission and routing. The last three -- SESSION, PRESENTATION and APPLICATIONS -- focus on user applications. The fourth layer TRANSMISSION provides an interface between the first and last three layers. The X.25 PROTOCOL which created a standard for data transmission and routing is equivalent to the last three layers of the OSI Model. The OSI model is quickly becoming the standard for how LAN products should be built."

OUTPULSE DIAL
A pushbutton dial which allows rotary dial users the convenience of "Touch-Tone" dialing. Pushing the buttons makes the phone pretend to be a rotary dial phone. This is necessary because touch-tones are not recognized everywhere.

PACKETIZED VOICE
First read the definition of "Packet Switching" just above. The idea is to digitize voice and then slice it up into packets and sent those packets from the sender by various routes and assemble them as they get to the receiver. Packet switching for data makes sense. Packet switching for voice has not made sense because the voice is too sensitive to small delays. Yet the logic and the economics are there. And experiments continue.

PART 68 REQUIREMENTS

Specifications established by the FCC as the minimum acceptable protection communications equipment must provide the telephone network. Meeting these requirements does not certify that equipment performs any task. Part 68 is the section of Title 47 of the Code of Federal Regulations governing the direct connection of telecommunications equipment and premises wiring with the public switched telephone network and certain private line services, e.g., foreign exchange lines (customer premises end), the station end of off-premises stations associated with PBX and Centrex services, trunk-to-station tie lines (trunk end only), and switched service network station lines (common control switching arrangements); and the direct connection of all PBX (or similar) systems to private line services for tie trunk type interfaces, off-premises station lines, automatic identified outward dialing and message registration. These rules provide the technical, procedural and labeling standards under which direct electrical connection of customer-provided telephone equipment, systems, and protective apparatus may be made to the nationwide network without causing harm and without a requirement for protective circuit arrangements in the service provider's network. Form 730 Application Guide is a collection of literature you'll need to register your telephone/telecom equipment under Part 68 of Title 47 at the Federal Communications Commissions. To get this material (it's free) drop a line or call the Federal Communications Commission, Washington DC 20554. As I write this edition, the person at the FCC in charge is William H. Von Alven, (202-634-1833) who also puts out a very useful newsletter for Part 68 applicants. The newsletter is called "The Billboard." You can file Form 730 yourself, but the Form 730 Application Guide also contains a list of Part 68 Certification Laboratories, a list of technical references and a list of reference sources.

PBX
Private Branch eXchange. A private (i.e. you, as against the phone company owns it), branch (meaning it is a small phone company central office), exchange (a central office was originally called a public exchange, or simply an exchange). In other words, a PBX is a small version of the phone company's larger central switching office. A PBX is also called a Private Automatic Branch Exchange, though that has now become an obsolete term. In the very old days, you called the operator to make an external call, except in Europe. Then later someone made a phone system that you simply dialed nine (or another digit -- in Europe it's often zero), got a second dial tone and dialed some more digits to dial out, locally or long distance. So, the early name of Private Branch Exchange (which needed an operator) became Private AUTOMATIC Branch Exchange (which didn't need an operator). Now, all PBXs are automatic. And now they're all called PBXs, except overseas where they still have PBXs that are not automatic.

At the time of the Carterfone decision in the summer of 1968, PBXs were electro-mechanical step-by-step monsters. They were 100% the monopoly of the local phone company. AT&T was the major manufacturer with over 90% of all the PBXs in the U.S. GTE was next. But the Carterfone decision allowed anyone to make and sell a PBX. And the resulting inflow of manufacturers and outflow of innovation caused PBXs to go through five, six or seven generations -- depending on which guru you listen to. (See my definition for GENERATIONS in this dictionary). Anyway, by the fall of 1991 when I wrote this fourth edition of this dictionary, PBXs were thoroughly digital, very reliable, and very full featured. There wasn't much you couldn't do with them. They had oodles of features. You could combine them and make your company a mini-network. And you could buy electronic phones that made getting to all the features that much easier. Sadly, by the late 1980s the manufacturers seemed to have finished innovating and were into price cutting. As a result, the secondary market in telephone systems was booming. Fortunately, that isn't the end of the story. For some of the manufacturers in the late 1980s figured that if they opened their PBXs' architecture to outside computers, their customers could realize some significant benefits. (You must remember that up until this time, PBXs were one of the last remaining special purpose computers that had totally closed architecture. No one else could program them other than their makers.) Some of the benefits customers could realize from open architecture included:

* Simultaneous voice call and data screen transfer.

* Automated dial-outs from computer databases of phone numbers and automatic transfers to idle operators.

* Transfers to experts based on responses to questions, not on phone numbers.

And a million more benefits. We discuss them at our annual trade show called TELECOM DEVELOPERS held in May each year. Call 212-691-8215 for more information. For more on open architecture, see OAI.

PBX DRIVER PROFILES
Telephony Services (also called TSAPI) is software which AT&T invented to run on NetWare servers and allow those servers to communicate with PBXs. According to a presentation made by Oliver Tavakoli of Novell in late August, 1994, "The purpose of TSAPI is to make it easier for developers to create telephony applications. Given the broad scope of TSAPI and the number of ways in which PBX vendors can package and deliver solutions, applications being developed against TSAPI are unlikely to work with a large cross-section of PBXs. This is particularly problematic for application developers wishing to add simple telephony features to a product that is not telephony centric. In order to spur the growth of CTI application development, Novell has created 'profiles' that are placed on top of TSAPI. The profiles listed in this document (i.e. the speech) are based on documents created by other organizations:

"- A NOTA document entitled 'PABX Driver NLM Conference Meeting and Draft Recommendations dated July 20th, 1994

"- An ECMA document entitled 'Technical report on CSTA Scenarios/Second Draft' dated July 1994.

"Each profile is made up of the following components: Functions (and matching confirmation events) included in the profile; Unsolicited events that may be encountered in the event flow scenarios in the profile; Event flow scenarios that attempt to describe what should occur in the common scenarios encountered when using the functions in the profile.

"Group A Profile: Using the functions and event flows in the Group A profile, an application should be able to: provide "screen pops" to the end user; make outbound calls originating at an end user's device; provide hands-free operation for answering the phone and making calls (assuming the PBX and end user's device support hands-free operation)

"Group B Profile: Using the functions and event flows in the Group B profile, an application should be able to: perform operations involving two calls on a single device (the connection to one of the calls is in the held state); perform operations involving more than two parties on the same call; obtain information and unsolicited events from a call-centric view (as opposed to a device-centric view).

"Group C Profile: Using the functions and event flows in the Group C profile, an application should be able to: perform the rest of the telephony functions usually done from a phone -- pickup, group pickup, call completion and perform more sophisticated monitoring functions.

(There is apparently no Group D profile.)

"Group E Profile: Using the functions in the Group E ('E' for Environmental) profile, an application should be able to: activate and deactivate features of a device, query the state of features on a device, query the generic information about a device."

Once the profiles are finalized, they will be incorporated into Novell's PBX Driver certification process; groups supported by the driver will be included in literature describing all certified drivers.

PCM
Pulse Code Modulation. The most common method of encoding an analog voice signal into a digital bit stream. First, the amplitude of the voice conversation is sampled. This is called PAM, Pulse Amplitude Modulation. This PAM sample is then coded (quantized) into a binary (digital) number. This digital number consists of zeros and ones. The voice signal can then be switched, transmitted and stored digitally. There are three basic advantages to PCM voice. They are the three basic advantages of digital switching and transmission. First, it is less expensive to switch and transmit a digital signal. Second, by making an analog voice signal into a digital signal, you can interleave it with other digital signals -- such as those from computers or facsimile machines. Third, a voice signal which is switched and transmitted end-to-end in a digital format will usually come through "cleaner," i.e. have less noise, than one transmitted and switched in

analog. The reason is simple: An electrical signal loses strength over a distance. It must then be amplified. In analog transmission, everything is amplified, including the noise and static the signal has collected along the way. In digital transmission, the signal is "regenerated," i.e. put back together again, by comparing the incoming signal to a logical question: Is it a one or a zero? Then, the signal is regenerated, amplified and sent along its way.

PCM refers to a technique of digitization. It does not refer to a universally accepted standard of digitizing voice. The most common PCM method is to sample a voice conversation at 8000 times a seconds. The theory is that if the sampling is at least twice the highest frequency on the channel, then the result sounds OK. Thus, the highest frequency on a voice phone line is 4,000 Hertz. So one must sample it at 8,000 times a second. Many PCM digital voice conversations are typically put on one communications channel. In North America, the most typical channel is called the T-1 (also spelled T1). It places 24 voice conversations on two pairs of copper wires (one for receiving and one for transmitting). It contains 8000 frames each of 8 bits of 24 voice channels plus one framing (synchronizing bit) bit which equals 1.544 Mbps, i.e. 8000 x (8 x 24 + 1) equals 1.544 megabits.

Countries outside of the United States and North America use a different scheme for multiplexing voice conversations. It is based not on 24 voice channels, but on 32. This scheme keeps two of the 32 channels for control, actually transmitting 30 voice conversations at a data rate of 2.048 Mbps. The European system is calculated as 8 bits x 32 channels x 8000 frames per second. European PCM multiplexing is not compatible with North American multiplexing. The two systems cannot be directly connected. Some PBXs in the U.S. conform to the U.S. standard only. Some (very few) conform to both. Both the European and North American T-1 "standards" have now been accepted as ISDN "standards." In addition to PCM, there are many other ways of digitally encoding voice. PCM remains the most common. See T-1 and VOICE COMPRESSION.

PERSONAL COMMUNICATIONS NETWORKS
PCN. A new type of wireless telephone system that would use light, inexpensive handheld handsets and communicate via low-power antennas. PCN is primarily seen as a city communications system, with far less range than cellular. Subscribers would be able to make and perhaps receive calls while they are traveling, as they can do today with cellular radio systems, but at a low price. There's talk that they'll put PCN antennas in communities and issue everyone in the community with a PCN phone that would hub off the PCN antenna. This would save the phone company wiring up each house. In this way, the PCN phone would resemble the common household cordless phone, but with a larger range. Dr. Sorin Cohn of Northern Telecom calls the new low-power, wireless, personal communications systems an "enabler of unplanned growth." Nice definition.

In the fall of 1992 MCI broadened the definition of PCN. It proposed a national consortium of local PCNs joined together by a long distance carrier (namely it). In its release, MCI said that "PCNs are the next generation of digital wireless communications technology. PCNs use less power and are less expensive than the current cellular technology and permit the use of inexpensive pocket telephones with much longer battery life than cellular portables. PCN phones will have many more features than today's cellular or conventional telephones, and unlike cellular phones, will be usable in most areas of the world. PCNs will operate in the same frequency band in most countries, (1850-1990 MHz) while cellular is operating in several different frequency bands in various countries, and thus is not portable from country to country."

PHASE A
Phase A is the first part of a fax machine's call process. It is the call establishment. It occurs when transmitting and receiving units connect over the phone line, recognizing one another as fax machines. This is the start of the handshaking procedure. See PHASE B.

PHASE B
Phase B is the second part of a fax machine's call process. It is the premessage procedure, where the answering machine identifies itself, describing its capabilities in a burst of digital information packed in frames conforming to the HDLC standard. See PHASE C.

PHASE C
Phase C is the third part of a fax machine's call process. It is the fax transmission portion of the operation. This step consists of two parts C1 and C2 which take place simultaneously. Phase C1 deals with synchronization, line monitoring and problem detection. Phase C2 includes data transmission. See PHASE D.

PHASE D
Phase D is the fourth part of a fax machine's call process. This phase begins once a page has been transmitted. Both the sender and receiver revert to using HDLC packets as during Phase B. If the sender has further pages to transmit, it sends an MPS and Phase C recommences for the following page. See PHASE E.

PHONEME
A voice recognition term. The minimal significant structural unit in the sound system of any language that can be used to distinguish one word from another. For example, the p of pit and the b of bit are considered two separate phonemes, while the p of spin is not. These minimal sound units comprise words.

PLMN
Public Land Mobile Network. A mobile telephone communications network established by a provider to facilitate mobile telecommunications services. This includes equipment, operations, and staff. A single provider may have more than one PLMN.

POP

1. Point Of Presence. The IXC equivalent of a local phone company's central office. The POP is a long distance carrier's office in your local community (defined as your LATA). A POP is the place your long distance carrier, called an IntereXchange Carrier (IXC), terminates your long distance lines just before those lines are connected to your local phone company's lines or to your own direct hookup. Each IXC can have multiple POPs within one LATA. All long distance phone connections go through the POPs.

2. Short for "population." One "pop" equals one person. In the cellular industry, systems are valued financially based on the population of the market served.

3. Post Office Protocol. An Application protocol offering mailbox retrieval service. Used by many Internet personal computer users. POP is increasingly being replaced by IMAP.

PORT

1. An entrance to or an exit from a network.

2. The physical or electrical interface through which one gains access. A point in the computer or telephone system where data may be accessed. Peripherals -- like call accounting devices -- are connected to ports. The two most common ports are the parallel and serial ports.

3. The interface between a process or program and a communications or transmission facility.

4. To move a process, program or subroutine from one processor to controller to another ("port it over").

5. Network access point for data entry or exit. In Internet terms, it is the identifier (16-bit unsigned integer) used by Internet transport protocols to distinguish among multiple simultaneous connections to a single destination host.

POTS

Plain Old Telephone Service. Pronounced POTS, like in pots and pans. The basic service supplying standard single line telephones, telephone lines and access to the public switched network. Nothing fancy. No added features. Just receive and place calls. Nothing like Call Waiting or Call Forwarding. They are not POTS services. All POTS lines work on loop start signaling.

PREDICTIVE DIALING

An automated method of making many outbound calls without people and then passing answered calls to a person as the calls are answered. Here's the story: Imagine a bunch of operators having to call a bunch of people. Those calls may be for collections. They may be for employee callups. They may be for alumnae fund raising. When it's done manual, here's how it works: Before each call operators spend time reviewing paper records or computer terminal screens, selecting the person to be called, finding the phone number, dialing the numbers, listening to rings, listening to phone company intercepts, busy signals and answering machines. Operators also spend time updating the records after each call. Predictive dialing automates this process, with the computer choosing the person to be called and dialing the number and only passing it to an operator when a real live human being answers. There are enormous productivity gains made by screening out answering machines, busy signals, network busy signals, non-completed calls, operator intercepts etc. The result is productivity increases of 200% to 300%. According to generally accepted industry lore, a well-run manual dialing center can get its people talking on the phone for 25 minutes an hour. With a predictive dialer you can get them on the phone making sales, collecting money, etc. for 55 minutes an hour. It's a major productivity gain.

True predictive dialing should not be confused with automated dialing. True predictive dialing has complex mathematical algorithms that consider, in real time, the number of available telephone lines, the number of available operators, the probability of getting no answer, a busy signal, a disconnected number, operator intercept or an answering machine, the time between calls required for maximum operator efficiency, the length of an average conversation and the average length of time the operators need to enter the relevant data. Some predictive dialing systems constantly adjust the dialing rate by monitoring changes in all these factors.

Some people don't like the term "predictive dialing," since they think it's getting "a bad rap" in Washington, DC by being associated with junk phone calls. As a result some people would prefer to call it Computer Aided Dialing.

PROMPT
An audible or visible signal to the system user that some process is complete or some user action is required. Also used to signify a need for further input and/or location of needed input.

PROSODY
Intonation. In text to speech, prosody refers to how natural it sounds --
the ups and downs of the sentence.

PROTOCOL
A procedure for adding order to the exchange of data. A protocol is a specific set of rules, procedures or conventions relating to format and timing of data transmission between two devices. A standard procedure that two data devices must accept and use to be able to understand each other. Sort of like us both speaking English so we can communicate. The protocols for data communications cover such things as framing, error handling, transparency and line control. There are three basic types of protocol: character-oriented, byte-oriented and bit-oriented.

Protocols break a file into equal parts called blocks or packets. These packets are sent and the receiving computer checks the arriving packet and sends an acknowledgement (ACK) back to the sending computer. Because modems use phone lines to transfer data, noise or interference on the line will often mess up the block. When a block is damaged in transit, an error occurs. The purpose of a protocol is to set up a mathematical way of measuring if the block came through accurately. And if it didn't, ask the distant end to re-transmit the block until it gets it right.

PSTN
Public Switched Telephone Network. An abbreviation used by the ITU-T, PSTN simply refers to the local phone company.

PTT
Post Telephone & Telegraph administration. The PTTs, usually controlled by their governments, provide telephone and telecommunications services in most foreign countries. In ITU-T documents, these are the Administrations referred to as Operating Administrations. The term Operating Administrations also refers to "Private Recognized Operating Agencies" which are the private companies that provide communications services in those very few countries that allow private ownership of telecommunications equipment.

RBOC
Regional Bell Operating Company. There are seven RBOCs each of which own two or more BOCs (Bell Operating Companies). The RBOCs were carved out of the old AT&T/Bell System by Judge Harold Greene when he signed off on the divestiture of the Bell operating companies from AT&T at the end of 1983. There is nothing magical about seven -- nor the grouping of BOCs into RBOCs
-- except the Judge wanted to keep them all roughly the same size. The seven RBOCs are Ameritech, Bell Atlantic, BellSouth, NYNEX, Pacific Telesis, Southwestern Bell and US West. In early October, 1994, Southwestern Bell changed its name to SBC Communications Inc. Its telephone companies, it said, would still operate under the Southwestern name.

REJECTION
A word used in voice recognition to mean a type of recognition classification where the input utterance did not meet the criteria necessary to be classified as a word in the active vocabulary. Usually the speaker is asked to repeat the utterance.

RELAY
An electrically activated switch used to operate a circuit. It connects one set of wires to another. Usually, the relay is operated by low voltage electric current and is used to open or close another circuit, which is of much higher voltage. Older telephone switches used many relays to switch (i.e. complete) their calls. Relays come in many forms. There are hermetically-sealed relays, in which thin metal

contacts are sealed in an airtight glass or metal enclosure. There are also mercury relays in which a small tube of mercury tilts and completes or breaks a circuit.

RESPONSE TIME
The time it takes a system to react to a given input. In voice recognition, response time typically refers to the amount of time required for a word (or utterance) to be recognized once the end of the word is detected. True response time is longer because silence often must occur before the end of the word can be declared. When operating a terminal connected to a computer, response time would be the time between the operator pressing the last key of a series of keys and the appearance of a response on the operator's display. In a data communications system, response time includes the transmission time, the processing time, the searching for records time and the transmission time back to the originator. Response time is very critical in applications like airline reservation systems. Here the customer is on the phone awaiting a reply. That time is critical in whether the customer perceives he's getting good or bad service. Response times of more than three seconds are not acceptable in situations where the customer is waiting on the phone to buy something. Response time is a function, inter alia, of the number of phone lines you lease or use. You can save a lot of phone line costs by cutting back on lines. But you'll extend response time. Life, as always, is a trade-off.

RING
1. As in Tip and Ring. One of the two wires (the two are Tip and Ring) needed to set up a telephone connection.

2. Also a reference to the ringing of the telephone set.

3. The design of a Local Area Network (LAN) in which the wiring loops from one workstation to another, forming a circle (thus, the term "ring"). In a ring LAN, data is sent from workstation to workstation around the loop in the same direction. Each workstation (which is usually a PC) acts as a repeater by re-sending messages to the next PC in the ring. The more PC's, the slower the LAN. Network control is distributed in a ring network. Since the message passes through each PC, loss of one PC may disable the entire network. However, most ring LANs recover very quickly should one PC die or be turned off. If it dies, you can remove it physically from the network. If it's off, the network senses that and the token ignores that machine. In some token LANs, the LAN will close around a dead workstation and join the two workstations on either side together. If you lose the PC doing the control functions, another PC will jump in and take over. This is how the IBM Token-Passing Ring works.

RING BACK TONE
The sound you hear when you're calling someone else's phone. The tone you hear is generated by a device at your central office and may bear no relationship to the sound the phone at the other end is emitting -- or not emitting. If your call didn't go through the first time, always call back at least once.

RISC
Reduced Instruction Set Computer. A microprocessor architecture that favors the speed at which individual instructions execute over the robustness of the instruction set. Computers based on RISC use an unusual high speed processing technology that uses a far simpler set of operating commands. These commands greatly speed a computer's performance, especially for calculation-intensive operations such as those performed by scientists and computer-aided design (CAD) and computer-aided manufacture (CAM) engineers. RISC is a design that achieves high performance by doing the most common computer operations very quickly. In contrast, the microprocessors used in most PCs are based on a design called CISC (Complex Instruction Set Computing). CISC does not execute instructions as quickly as RISC but it has more commands and accomplishes more with each command. Programs written for RISC are typically not compatible with those written for CISC processors. RISC is the prevailing technology for workstations today. The RISC semiconductor was an IBM baby, born in its Yorktown Heights, NY lab in 1974. But internal arguments over, and even whether the chip should be used kept IBM fiddling while Sun and other companies decisively powered ahead. IBM got its first good RISC product, the RS/6000, to market in 1990. The PowerPC architecture is one example of a RISC microprocessor design.

RJ
Registered Jacks. They're telephone and data plugs registered with the FCC. RJ-XX (where X is a number) are probably the most common plugs in the world. The most common are RJ-11 and RJ-45. See below.

RJ-11
RJ-11 is a six conductor modular jack that is typically wired for four conductors (i.e. four wires). The RJ-11 jack (also called plug) is the most common telephone jack in the world. The RJ-11 is typically used for connecting telephone instruments, modems and fax machines to a female RJ-11 jack on the wall or in the floor. That jack in turn is connected to twisted wire coming in from "the network" -- which might be a PBX or the local telephone company central office. In a home installation, the red and green pair would be used for carrying the phone conversation and the black and white might be used for carrying low voltage from a plugged-in power transformer to light buttons on the phone. In many offices, the tip and ring were used for the voice conversation and the black and white were used for signaling. Increasingly, these days more and more office phone systems use only one pair, i.e. the red and green conductors.

RJ-45
The RJ-45 is the 8-pin connector used for data transmission over standard telephone wire. That wire could be flat or twisted. And it's very important that you know what you're working with. You can easily use flat wire for serial data communications up to 19.2 Kbps. Up to that speed you're connecting with your wire to a dataPBX, a modem, a printer or a printer buffer. If you wish to connect to a 10BaseT local area network, which you also do with a RJ-45, you must use

twisted wire. You can typically tell the difference by looking at the cable. If it's flat grey satin (like a typical phone wire, only bigger) than it's probably untwisted. If it's circular, then it's probably twisted and therefore good for LANs. RJ-45 connectors come into two varieties -- keyed and non-keyed. Keyed means that the male RJ-45 plug has a small, square bump on its end and the female RJ-45 plug is shaped to accommodate the plug. A keyed RJ-45 plug will not fit into a female, non-keyed (i.e. normal) RJ-45.

ROBBED-BIT SIGNALING
This explanation from Gary Maier of Dianatel: ISDN is the key to future sophisticated telephone network services with its dynamic, highly configurable T-1 connection (also called PRI connection). Since T-1 is a common method of carrying 24 telephone circuits, many wonder about the uses for ISDN, especially when they learn ISDN signaling requires an entire voice channel, reducing today's T-1 from 24 voice channels to 23. But the popular signaling mechanism of "robbed bit" signaling in T-1 has serious limitations. Robbed bit signaling typically uses bits known as the A and B bits. These bits are sent by each side of a T-1 termination and are buried in the voice data of each voice channel in the T-1 circuit. Hence the term "robbed bit" as the bits are stolen from the voice data. Since the bits are stolen so infrequently, the voice quality is not compromised by much. But the available signaling combinations are limited to ringing, hang wink, and pulse digit dialing. In fact, the limitations are obvious when one recognizes DNIS and ANI information are sent as DTMF tones.

This introduces a problem: time. Each DTMF tone requires at least 100 milliseconds to send, which in a DNIS and ANI situation with 20 DTMFs will take at least two full seconds. There is also a margin for error in transmission or detection, resulting in DNIS or ANI failures. With the explosion of telephone related services, the telephone companies are turning to ISDN PRI to provide the more complicated and exact signaling required for new services. ISDN employs a more robust method of signaling. ISDN uses a T-1 circuit as 23 voice channels and one signaling channel. The term 23B plus D refers to 23 bearer (voice) channels and 1 Data (signaling) channel. The data channel carries the signaling information at a rate of 64 kilobits per second. This speed is many times greater than some of the most powerful modems available. Because of this high speed, telephone calls can be placed more quickly, and because of the protocol used, DNIS or ANI transmission failures are impossible.

Additionally, since no bits are "robbed" from the voice channels, the voice quality is better than that of Robbed Bit signaling on today's T-1 circuits. Also, computer modems and high speed faxes can use the voice channel for sending digital data instead of the traditional analog bit "noise." Therefore, ISDN PRI offers the end user countless new service capabilities. One channel could be used for faxing, another for modem data, several for video, another for a LAN and the remainder for voice. Suddenly, the average T-1 circuit becomes a pipeline for all communications! Increasingly long distance carriers are using ISDN PRI to provide inbound 800 calls with ANI and DNIS and re-routing skills.

ROTARY DIAL
The circular telephone dial. As it returns to its normal position (after being turned) it opens and closes the electrical loop sent by the central office. Rotary dial telephones momentarily break the DC circuit (stop current flow) to represent the digits dialed. The circuit is broken three times for the digit 3. The CO counts these evenly-spaced breaks and determines which digit has been dialed. You can hear the "clicks". The number "seven," for example consists of seven "opens and closes," or seven clicks. You can dial on a rotary phone without using the rotary dial. Simply depress the switch hook quickly, allowing pauses in between to signify that you're about to send a new digit. It's a good party trick.

ROUTING TABLE
1. Incoming Phone Calls: A routing table is a user definable list of steps which are treatment instructions for an incoming call. Ideally these steps should be addressed and the call treatment begun before the call is answered. A routing table should consist of a minimum of steps that include agent groups, voice response devices, announcements (delay and informational) music on hold, intraflow and interflow steps, route dialing (machine based call forwarding). A significant issue in the structure of routing tables is "look-back" capability, where no one previously interrogated resource is abandoned by the system (i.e. an agent group is now ignored, even though an agent is now available, because the ACD does not consider previous steps in the routing table).

2. Outgoing Phone Calls: For a specific calling site, this table lists the long distance routing choices for each location to be dialed. There may be only one choice (route) listed for some or all destinations or there may be several choices for some destinations. (It depends how many outgoing lines and how many outgoing trunk groups you have.) If there are several choices then they will be ranked by some criteria (least cost, best quality, etc.).

3. In data communications, a routing table is a table in a router or some other internetworking device that keeps track of routes (and, in some cases, metrics associated with those routes) to particular network destinations.

SCAI
Switch to Computer Applications Interface. A protocol that defines how switches talk to outboard computers, i.e. computers which are external to the switch and contain such a database of customer buying information. Using SCAI, calls and data screens about a calling customer can be presented to the agent simultaneously.

SCbus
An SCSA definition. The standard bus for communication within an SCSA node. The SCbus features a hybrid bus architecture consisting of a serial Message Bus for control and signaling, and a 16-wire TDM data bus. The SCbus is a serial time division multiplexed bus for carrying information between hardware devices in a

signal processing node. The SCbus can support up to 1024 bidirectional timeslots in a PC implementation or up to 2048 timeslots in a backplane implementation. The SCbus uses 16 synchronous data lines for carrying data and a dedicated messaging channel (SCbus Message Channel) for carrying signaling information and messages between devices.

SCCP
Signaling Connection Control Part. Part of the ITU-T #7 signaling protocol. and of the SS7 protocol. It provides additional routing and management functions for transfer of messages other than call set-up between signaling points.

SCE
Service Creation Environments. A term used in the jargon of intelligent networks (IN) to allow outside developers to define and create new value-added (i.e. intelligent) services.

SCP
1. Service Control Point. Also called Signal Control Point. A remote database within the System Signaling 7 network. The SCP supplies the translation and routing data needed to deliver advanced network services. The SCP translates an 800-IN-WATS number to the required routing number. It is separated from the actual switch, making it easier to introduce new services on the network.

SCSA
Signal Computing System Architecture. SCSA is a comprehensive architecture that describes how both hardware and software building blocks work together. It focuses on "Signal Computing" devices, which refer to any devices that are required to transmit information over the telephone network. Information can be transmitted via data modems, fax, voice or even video. SCSA defines how all these devices work together. Signal computing systems combine three major elements for call processing. Network interfaces provide for the input and output of signals transmitted and switched in telecommunications networks. Digital signal processors and software algorithms transform the signals through low-level manipulation. Application programs provide computer control of the processed signals to bring value to the end user.

SCSA is the common set of standards that telecommunication system manufacturers and computing system manufacturers can use to create computer telephony systems. The theory is no single company today can create the total solution for all customers. SCSA represents the common ground between the two fields so that manufacturers from each area can safely develop products that will work with other manufacturers. SCSA's coverage extends from low-level bus and hardware interfaces, like the inter-board switching bus that enables boards from different suppliers to work together, to high-level application programming and software interfaces, so that software designed to work with one set of hardware products, will work with different hardware. Dialogic Corporation of Parsippany, NJ announced SCSA in the Spring of 1993. Dialogic said that SCSA was defined and created with input from a number of leading computer and switch manufacturers, call processing suppliers, and technology developers. In many cases, SCSA has drawn on existing standards, like the T.611 fax standard endorsed by the European Computer Manufacturers Association, and in other cases SCSA has extended standards to make them more useful for call processing suppliers and users.

SCSA describes all elements of the system architecture from the electrical characteristics of the SCbus and SCxbus to the high level application programming interfaces (APIs). According to TELECONNECT Magazine, this SCSA standard is remarkable for several things:

1. On the day of its announcement over 60 telecom and voice processing companies publicly endorsed SCSA. In early 1994, over 150 companies public endorsed it.

2. With SCSA -- a standard for PC/LANs and VME-backplaned computers -- you can build much larger telecom switches and much larger call and voice processing boxes. Previous standards, like AEB, PEB and MVIP, were basically limited to what you could do with one PC. Now PCs can be joined together. With SCSA, you can put 16 T-1 lines, or 512 voice lines in one PC and join together 16 PCs, for a total of 16 x 16 x 24 = 6,144 lines! That's a central office built out of networked PCs. A mainframe built out of a LAN. The SCSA joining is not via LAN or LAN-emulation. That would be too slow and the transmission too bursty (great for data, lousy for voice). It's via an SCbus -- something that looks and works like a PBX backplane.

3. SCSA incorporates virtually every other standard in PC-based switching -- including the most popular ones, Mitel's ST-Bus, MVIP, Siemens PCM Highway, AEB and PEB.

4. It's a lot faster and more reliable. All signaling is out of band. There's clock fall back and time slot bundling. It's more modular, meaning you can start with one PC and grow one at a time. That makes it more "modular" (scaleable is the new word). It's also hot pluggable. You don't have to turn off to upgrade.

5. It has applications portability. Tandem, the highly-successful fault tolerant minicomputer maker, has an SCSA application in a call center. They call it the Tandem Non-Step Call Center. It uses the Tandem 2400 VRU and the 4800 VRU.

SCSA is open, truly open. All its specs and all levels of its specs are available. To that extent, SCSA represents a remarkable gamble by its creator, Dialogic, a telecom/voice processing hardware company. It is encouraging competing manufacturers to build hardware to its specs and gambling that it won't be left in the dust, as IBM was with its PC. (Compaq, not IBM, built the first '386 PC.)

SCSA, as an idea, is revolutionary (for telecom). No one in telecom has ever promulgated an open standard everyone can adopt -- hardware and software vendors. Dialogic has done it to create opportunities by providing great economies of scale for the developers. Write one application, create one applications generator, design one piece of hardware. Erector set telecom/voice processing! Build small. Build large. Just join the bits and pieces together.

SEARCH ENGINE (ON THE WEB)
An Internet World Wide Web term. A search engine is a program that returns a list of Web Sites (URLs) that match some user-selected criteria such as "contains the words cotton and blouse."

SECOND DIALTONE
1. Dialtone given to the caller, on a phone system (e.g. PBX, Centrex or hybrid), after dialing an access code (e.g. 8 or 9) to make a call out of the system (e.g. a local or long distance). Sometimes referred to as outside dialtone. 2. Dialtone returned to a caller after they've dialed a local or long distance number and reached some type of switching device. That switching device might allow you to dial into a fax machine, a modem, a phone or an answering machine. It might allow you to dial into one of several cash registers or soda machines (to check if they need being refilled). It might even allow you to dial long distance through a voice mail system or through a long distance phone company. As you dial through networks you might encounter not only second, but also third and fourth dialtones.

SERVICE ACCESS CODE
SACs are 3-digit codes in the NPA (N00) format that are used as the first three digits of a 10-digit address, and that are assigned for special network uses. Whereas NPA codes are normally used for identifying specific geographical areas, certain SACs have been allocated in the NANP (North American Numbering Plan) to identify generic services or provide access capability. Currently only four SACs have been assigned and are in use: 600, 700, 800, and 900.

SERVICE CONTROL POINT
SCP. The local versions of the national SMS/800 number database. SCPs contain the intelligence to screen the full ten digits of an 800 number and route calls to the appropriate, customer-designated long distance carrier. Bellcore defines SCP as the network system in the Advanced Intelligent Network Release 1 architecture that contains SLEE (Service Logic Execution Environment) functionality and communicates with AIN Release 1 Switching Systems in processing AIN Release 1 calls.

SHORT TONES
First, we invented touchtone, also called DTMF, Dual Tone Multi Frequency tones. You'd punch your number with tones, instead of dialing them. Then someone thought you could control telephone response gadgets, like voice mail, interactive voice response, etc. with touch tones. For these gadgets to work, they had to

"hear" the tones you sent. No one really set standards as to the minimum length tone they would hear. But it was generally conceded that they were to be 120 milliseconds. So some manufacturers of telephone equipment started to make phone equipment that, if you pushed a touchtone button, the machine would only sent a touchtone of 120 millisecond duration. That was called a short tone. It wasn't very useful because the manufacturers quickly discovered that many pieces of equipment couldn't respond that quickly. And the manufacturers got complains that their customers couldn't call their voice mail, their bank, etc. As a result, some manufacturers of equipment brought out new hardware (replacing the old) to allow you to send "long tones," which are now defined as touchtones that last for as long as you hold down the button -- just as it is (and has always been) on a normal single line, non-electronic, non-digital telephone.

SIGNAL CONTROL POINT
SCP. Computers that hold databases in which customer-specific information used by the "advanced intelligent network" (AIN) to route calls is stored. The AIN refers to a specific architecture -- typically promulgated and created by a local or long distance phone company -- that provides core capabilities in which customer-specific information held in databases within the network is used to intelligently process calls. An example of an AIN service is an 800 service that routes calls based on where the calls are coming from. The routing information is stored in the SCP, which is typically tied into the Signaling System 7 network.

SIGNALING CONNECTION CONTROL PART
SCCP. Part of the SS7 protocol that provides communication between signaling nodes by adding circuit and routing information to the signaling message. The ISDN-UP (Integrated Services Digital Network User Part) and TCAP (Transaction Capabilities Application Part) use the SCCP (Signaling Connection Control Part) and the MTP (Message Transfer Part) to transport information. Definition from Bellcore in reference to its concept of the Advanced Intelligent Network.

SIGNALING LINK SELECTION CODE
SLS. The part of a routing label that identifies the SS7 signaling link on which the message should be sent.

SIGNALING POINT
A node in a SS7 signaling network that either originates and receives signaling messages, or transfers signaling messages from one signaling link to another, or both. SPs are located at each switch in a Signaling System 7 network. They interface the switch with the Signal Transfer Points (STPs).

SIGNALING POINT CODE
A binary code uniquely identifying a SS7 signaling point in a signaling network. This code is used, according to its position in the label, either as destination point code or as originating point code.

SIGNALING POINT INTERFACE

SPOI. The demarcation point on the SS7 signaling link between a LEC network and a Wireless Services Provider (WSP) network. The point established the technical interface and can designate the test point and operational division of responsibility for the signaling.

SIGNALING SYSTEM 7
SS7. All phone systems need signaling. According to James Henry Green, author of the Dow Jones-Irwin Handbook of Telecommunications, signals have three basic functions:

1. SUPERVISING. Monitoring the status of a line or circuit to see if it is busy, idle or requesting service. Supervision is a term derived from the job telephone operators perform in manually monitoring circuits on a switchboard. On switchboards, supervisory signals are shown by a lit lamp indicating a request for service on an incoming line or an on-hook condition of a switchboard cord circuit. In the network (i.e. the automated part of the network), supervisory signals are indicated by the voltage level on signaling leads, or the on-hook/off-hook status of signaling tones or bits.

2. ALERTING. Indicates the arrival of an incoming call. Alerting signals are bells, buzzers, whoofers, tones, strobes and lights.

3. ADDRESSING. Transmitting routing and destination signals over the network. Addressing signals are in the form of dial pulses, tone pulses or data pulses over loops, trunks and signaling networks.

Most signaling today is MF (multi-frequency) and SF (single frequency) and is inband. This means that it goes along and occupies the same circuits as those which carry voice conversations. There are two problems with this. First, about 35% of all toll calls are not completed because the phone doesn't answer or is busy, or there are equipment problems along the way. The circuit time used in signaling is substantial, expensive and wasteful. Second, inband signaling is vulnerable to fraud. So the idea of out-of-band signaling came about. It got the name of Common Channel Interoffice Signaling (CCIS) because it used a communications network totally separate from the switched voice network. In North America, CCIS started out as an AT&T packet switched network operating at 4800 bits per second. Each of the packet switches in this network (they are no longer exclusively AT&T's) are called Signal Transfer Points -- STPs. CCIS has the following advantages over SF/MF signaling:

Fraud is reduced. "Talk-off" is reduced. (Talk-off occurs when your voice contains enough 2600 Hz energy to activate the tone-detecting circuits in the central office.) Signaling is faster allowing circuits and conversations to be set up and torn down (i.e. disconnected) faster. Signals can be sent in both directions simultaneously and during voice conversation if necessary. Network management information is routed over the CCIS network. For example, when trunks fail, switching systems can be told with CCIS data messages to reroute traffic around problem areas.

The older CCIS signaling is being replaced with a newer out-of-band signaling system called CCITT Signaling System 7. According to an AT&T technical paper delivered at the International Switching Symposium in Spring, 1987, CCITT Signaling System 7 is being required by telecommunications administrations worldwide (i.e. all the local country-owned telephone companies) for their networks. AT&T continued with the introduction of digital switches and transmission equipment with 56 Kbps and 64 Kbps transmission rates, the International Telegraph and Telephone Consultative Committee (CCITT in French) in 1980 approved the CCITT 7 recommendations optimized for digital networks. This new protocol uses destination routing, octet oriented fields, variable length messages and a maximum message length allowing for 256 bytes of data. Addition of flow control, connectionless services and Integrated Services Digital Network (ISDN) capabilities were approved by CCITT in 1984. A major characteristic of CCITT #7 is its layered functional structure. Its transport functions are divided into four levels, three of which constitute the Message Transfer Part (MTP). The fourth consists of a common Signaling Connection Control Part (SCCP).

The SS7 protocol consists of four basic sub-protocols:

* Message Transfer Part (MTP), which provides functions for basic routing of signaling messages between signaling points.

* Signaling Connection Control Part (SCCP), which provides additional routing and management functions for transfer of messages other than call setup between signaling points.

* Integrated Services Digital Network User Part (ISUP), which provides for transfer of call setup signaling information between signaling points.

* Transaction Capabilities Application Part (TCAP), which provides for transfer of non-circuit related information between signaling points.

Signaling System 7 provides two major capabilities:

1. Fast call setup, via high-speed circuit-switched connections.

2. Transaction capabilities which deal with remote data base interactions. What this means in its simplest terms and in one simple application is that Signaling System 7 information can tell the called party who's calling and, more important, tell the called party's computer. A scenario: when you call a direct mail order business, Signaling System 7 will send a signal as to which phone is calling. The agent's CRT screen will pop the caller's name and perhaps the caller's most recent buying information. The agent may answer the phone "Good morning, Mr. Newton. Did you enjoy the three khaki pants we sent you last week?..." Signaling System 7 will be an integral part of ISDN. It will enable us to extend full PBX and

Centrex-based services like call forwarding, call waiting, call screening, call transfer, etc. outside the switch to the full international network. In effect, with Signaling System 7, the entire network will acquire the "smarts" of today's smartest electronic digital PBX.

SINGLE NUMBER SERVICE
An optional feature for 800 IN-WATS Services which allows a subscriber who has or wants to have both intrastate and interstate 800 service to use the same 1-800 number for both services. If you'd like to buy another copy of this dictionary, call 1-800-LIBRARY. That phone number will be answered at our office in New York City. It will work for calls from inside and from outside New York State.

SIR
1. Speaker Independent Recognition.

2. An SMDS term meaning Sustained Information Rate. Determined at time of subscription, the SIR defines the long term average throughout an SMDS access line can carry. SIR is enforced through a Credit Manager resident of the SMDS network switches, and the Access Class-4 Mbps, 10 Mbps, 16 Mbps, 25 Mbps or 34 Mbps- the subscriber selects.

SIT TONES
1. Standard Information Tones. These are tones sent out by a central office to a pay phone to indicate that the dialed call has been answered by the distant phone, etc. 2. Special Information Tones. These are tones for identifying network provided announcements. Here's Bellcore's explanation: Automated detection devices cannot distinguish recorded voice from live voice answer unless a machine-detectable signal is included with the recorded announcement. The CCITT, which specifies signals that may be applied to international circuits, has defined Special Information Tones for identifying network provided announcement. The SIT used to precede machine-generated announcements also alerts the calling customer that a machine-generated announcement follows. Since SIT consists of a sequence of three precisely defined tones, SIT can be machine-detected, and therefore machine-generated announcements preceded by a SIT can be classified. At least four SIT encodings have been defined: Vacant Code (VC), Intercept (IC), Reorder (RO) and No Circuit (NC). With the exception of some small stored Program Control Systems (SPCSs) and some customer negotiated announcements, Bell operating companies in North America now precede appropriate announcements with encoded SITs to detect and classify announcements.

SMDI
Station Message Desk Interface or Simplified Message Desk Interface. The SMIS is the data link from the central office if you have ESSX, Centrex or Centron (etc.) that gives you your stutter dial tone or message waiting light. In essence, SMDI is a data line from the central office containing information and instructions to your on-premises voice mail box. With SMDI, the calling person is not required to re-

enter the called phone number (or in any other way identify the called party) once the call terminates to the messaging system.

SOHO
Small Office Home Office. An acronym for a new market opening up, which is part work at home, part commute from home. In New York City, there's an area called SOHO. It stands for South of Houston Street.

SPAN LINE
A T-1 link.

SPC
1. Stored Program Control. All phone systems these days are SPCs. There's stored software, which is the program, which controls the computer or microprocessor which in turn controls the operation of the switch. Thus switches are stored program control.

2. Signal Processing Component.

SPEAKER DEPENDENT VOICE RECOGNITION
Technology capable of recognizing speech from a given user or others who sound like this user after completion of an enrollment procedure. It is not voice verification although it is sometimes confused with this technology.

SPEAKER IDENTIFICATION
Speaker identification is used to determine the identity of a known speaker. It is accomplished by taking spoken input and searching a database of all known system users for a match. Due to its speaker dependent recognition characteristics, you must first be enrolled as a user prior to using the system. To enroll as a user, an individual is required to speak one or more password phrases which are recorded. These phrases create a reference templates which are stored in the system user database for later use during identification sessions. When in operation, the individual using the system is prompted for a specific password or password phrase. When speaking the prompted password as input it creates a new template. This template is then compared to all reference templates in the system for that particular password. The reference template with the closest match is selected. The uniqueness of each user's voice and the finite number of users of the system makes the identification accuracy quite high. With speaker identification the speaker does not claim to be a particular individual. He or she is identified from a group of common users. For the most part, this technology is used for hands free operation of a system where messages and other information specific to that identified individual are pulled-up for use at that time.

SPEAKER INDEPENDENT VOICE RECOGNITION
SIR or SIVR. Technology capable of recognizing any user without prior training or knowledge of the user. SIR converts speech to accurate and meaningful textual information (typically ASCII). SIR is used to accept input from callers to voice

processors where the callers are using rotary dial phones instead of touchtone phones. SIR can substitute for the numbers on the DTMF keypad and can add the benefit of a few basic voice commands, e.g., Yes, No, Help, etc.

Because computer processing demands are formidable with speaker independent recognition, accurate speaker independent products are created with limited vocabularies. In contrast, trainable or speaker dependent recognizers can feature larger vocabularies at lower prices. SIR has been slowly gaining acceptance in telephone applications. SIR is increasingly used in automated operator assistance applications. SIR will see increased use as system builders respond to pressures to provide voice processing functions to the enormous rotary phone installed base domestically and abroad.

SPEECH CONCATENATION
A term used in voice processing for economical digitized speech playback that uses independently recorded files of phrases or file segments linked together under application program control to produce a customized response in natural sounding language. For example, order status, bank balances, bus schedules or lottery results, etc. Concatenation is done for speed and economy. It lends itself to limited and structured vocabularies that are best stored in RAM (Random Access Memory) or speedily accessible from disk. Concatenation does not replace Text-To-Speech (TTS) as a method of getting the voice processor to deliver its responses. Concatenation, however, can be an excellent complement to TTS when a voice application demands broad, real time vocabulary production.

SSP
Service Switching Point. Also called Signal Switching Point. A switch that can recognize IN (Intelligent Network) calls and route and connect them under the direction of an SCP (Service Control Point. A computer database that holds information on IN (Intelligent Network) services and subscribers. The SCP is separated from the actual SCP switch, making it easier to introduce new services on the network.

ST-BUS
Serial Telecom Bus. Mitel Semiconductor, which makes telecom componentry, has structured its digital component product line around the ST-BUS. The ST-BUS is a high speed, synchronous serial bus for transporting information in a digital format. Whether the digital information is voice, data, or video -- or a mixture of each -- the ST-BUS is designed to accommodate it. The ST-BUS consists of one or several serial data streams with a framing signal and clock signals. The framing signal always has a period of 125 us, resulting in 8,000 frames per second, with the original ST-BUS clock rate of 2.048 Mbit/s (thirty two 64 kbit/s channels). The ST-BUS standard now includes higher speed modes of 4.096 or 8.192 Mbit/s ST-BUS, resulting in 64 or 128 channels of 64kbit/s, respectively. This provides the bandwidth necessary for newer multimedia applications. According to Mitel, the advantages of using the ST-BUS are:

1. Printed circuit board area devoted to information transfer between functional modules is minimized.

2. Fewer tracks, backplane connections, and intra-shelf cables are needed compared to systems that use parallel paths.

3. The ST-BUS is designed to be divided down into individual channels of 64 kbit/s, resulting in improved efficiency and lower cost when several information paths are able to share the same ST-BUS.

4. Additional glue logic is not required when using ST-BUS compatible components.

5. From an IC perspective, the ST-BUS results in lower pin counts, improved reliability, and less power consumption.

SWITCHED 56
A switched data service which lets you dial someone else and transmit at 56 kilobits per second. It is a circuit switched service, letting the user transmit data at full duplex, digital synchronous 56 Kbps for usually the price of a phone call. Switched 56 can be delivered on one pair or two pairs. Switched 56 is used for videoconferencing, high speed data transfer, digital audio broadcasting, Group IV fax and remote LAN access for telecommuters. Switched 56 data service is probably the most widely used switched digital service available in North America. It has been deployed by telephone companies in strange, expensive ways -- like, overlay networks, etc. -- that many phone companies are not pleased with. (I'm not quite sure why, except for a general telephone company aversion to overlay networks.) Most phone companies are phasing Switched 56 service out in favor of ISDN, which has the major advantage of potentially delivering more bandwidth, i.e. 144 Kbps versus 56 Kbps. The phone companies are doing this by raising the price of their Switched 56 service and dropping the price of their ISDN service. Over time, Switched 56 will disappear for good in North America and ISDN will become the pre-eminent service. Over time.

T CARRIER
Generic name for any of several digitally multiplexed carrier systems. The designators for T carrier in the North American digital hierarchy correspond to the designators for the digital signal (DS) level hierarchy. T carrier systems were originally designed to transmit digitized voice signals. Current applications also include digital data transmission. The table below lists the designators and rates for current T carrier systems. If an "F" precedes the "T", it's an optical fiber cable system, but the same rates.

NORTH AMERICAN DESIGNATOR (DS LEVEL)

T1 (DS 1)	1.544 Mbps	24 voice channels
T1C	3.152 Mbps	48 voice channels

T2 (DS 2)	6.312 Mbps	96 voice channels
T3 (DS 3)	44.736 Mbps	672 voice channels
T4 (DS 4)	274.176 Mbps	4032 voice channels

JAPANESE HIERARCHY

DS 1	1.544 Mbps	24 voice channels
DS 2	6.312 Mbps	96 voice channels
DS 3	32.064 Mbps	480 voice channels
DS 4	97.728 Mbps	1440 voice channels
DS 5	400.352 Mbps	5760 voice channels

EUROPEAN HIERARCHY (CEPT)

DS 1	2.048 Mbps	30 voice channels
DS 2	8.448 Mbps	120 voice channels
DS 3	34.368 Mbps	480 voice channels
DS 4	139.268 Mbps	1920 voice channels
DS 5	565.148 Mbps	7680 voice channels

T-1

Also spelled T1. A digital transmission link with a capacity of 1.544 Mbps (1,544,000 bits per second). T-1 uses two pairs of normal twisted wires, the same as you'd find in your house. T-1 normally can handle 24 voice conversations, each one digitized at 64 Kbps. But, with more advanced digital voice encoding techniques, it can handle more voice channels. T-1 is a standard for digital transmission in the United States, Canada, Hong Kong and Japan. (For the complete T Carrier hierarchy see the definition for T Carrier above.)

T-1 lines are used for connecting networks across remote distances. Bridges and routers are used to connect LANs over T-1 networks. There are faster services available. T-1 links can often be connected directly to new PBXs and many new forms of short haul transmission, such as short haul microwave systems. It is not compatible with T-1 outside the United States and Canada. In Europe T-1 is called E-1 or E1.

Outside of the United States and Canada, the "T-1" line bit rate is usually 2,048,000 bits per second. France and West Germany impose slight variations that make their formats unique. Only one element remains constant -- the DS-0. The 64 kilobit per channel is universal. Most often it represents a PCM voice signal sampled at 8,000 times per second. However, the form of PCM encoding differs between T-1 (mu-law) and E-1 (A-law companding). According to Bill Flanagan's book, the differences are not so great that a multiplexer cannot convert between them. Conversion of E-1 to T-1 involves both the compression law and the signaling format.

At the higher rate of 2,048,000, 32 time slots are defined at the CEPT interface, but two are used for signaling and other housekeeping chores. Typically 30 channels are left for user information -- voice, video, data, etc. CEPT is the Conference of European Postal and Telecommunications administrations. Standards-setting body whose membership includes European Post, Telephone, and Telegraphy Authorities (PTTs).

T-1 FRAMING
Digitization and coding of analog voice signals requires 8,000 samples per second (two times the highest voice frequency of 4,000 Hz) and its coding in 8-bit words yields the fundamental T-1 building block of 64 Kbps for voice. This is termed a Level 0 Signal and is represented by DS-0 (Digital Signal at Level 0). Combining 24 such voice channels into a serial bit stream using Time Division Multiplexing (TDM) is performed on a frame-by-frame basis. A frame is a sample of all 24 channels (24 x 8 = 192) plus a synchronization bit called a framing bit, which yields a block of 193 bits. Frames are transmitted at a rate of 8,000 per second (corresponding to the required sampling rate), thus creating a 1.544 Mbps (8,000 x 193 = 1,544 Mbps) transmission rate, the standard North American T-1 rate. This rate is termed DS-1.

TALK OFF
Talk off (also called Talk-Off) is one hazard of inband signaling. The classic definition of talk-off is that it occurs when your voice has enough 2600 Hz energy to activate the 2600 Hz tone-detecting circuits in the central office. The 2600 Hz tone is used for inband signaling. In voice processing, talk-off happens when a person is recording an audio file (say, leaving a voice mail message) and the frequencies in his voice happen to match those of a touch-tone digit. The voice board reacts as if a touchtone digit were press. Which means it may terminate the recording immediately and move to another menu selection.

TARIFF
Documents filed by a regulated telephone company with a state public utility commission or the Federal Communications Commission. The tariff, a public document, details services, equipment and pricing offered by the telephone company (a common carrier) to all potential customers. Being a "common carrier" means it (the phone company) must offer its services to everybody at the prices and at the conditions outlined in its public tariffs. Tariffs do not carry the weight of law behind them. If you or the telephone company violate them, no one will go to jail. The worst that can happen to you, as a subscriber, is that your service will be cut off, or threatened to be cut off. Regulatory authorities do not normally approve tariffs. They accept them -- until they are successfully challenged before a hearing of the regulatory body or in court (usually Federal Court). Many tariffs were accepted by regulatory commissions only to be struck down in court as unlawful, discriminatory, not cost-justified, etc. Monies collected under the tariff have been refunded and unnecessary equipment removed. In these new, competitive days, many telephone companies are violating their own tariffs by charging less money than their tariffs say they should, or bundling

services together at a discount. They are also providing service and equipment on terms less onerous than outlined in their tariffs. Many users now regard tariffs as starting bargaining points, rather than ending bargaining points.

TCP/IP

Transmission Control Protocol/Internet Protocol. A set of protocols developed by the Department of Defense to link dissimilar computers across many kinds of networks, including unreliable ones and ones connected to dissimilar LANs. TCP/IP is the protocol used on the Internet. It is, in essence, the glue that binds the Internet. Developed in the 1970s by the U.S. Department of Defense's Advanced Research Projects Agency (DARPA) as a military standard protocol, its assurance of multi vendor connectivity has made it popular among commercial users as well, who have adopted TCP/IP as an interim step while awaiting the availability of OSI products. Consequently, TCP/IP now is supported by many manufacturers of minicomputers, personal computers, mainframes, technical workstations and data communications equipment. It is also the protocol commonly used over Ethernet (as well as X.25) networks. It has been implemented on everything from PC LANs to minis and mainframes. Although committed to an eventual migration to an OSI architecture, TCP/IP currently divides networking functionality into only four layers:

A Network Interface Layer that corresponds to the OSI Physical and Data Link Layers. This layer manages the exchange of data between a device and the network to which it is attached and routes data between devices on the same network.

An Internet Layer which corresponds to the OSI network layer. The Internet Protocol (IP) subset of the TCP/IP suite runs at this layer. IP provides the addressing needed to allow routers to forward packets across a multiple LAN inter network. In IEEE terms, it provides connectionless datagram service, which means it attempts to deliver every packet, but has no provision for retransmitting lost or damaged packets. IP leaves such error correction, if required, to higher level protocols, such as TCP.

IP addresses are 32 bits in length and have two parts: the Network Identifier (Net ID) and the Host Identifier (Host ID). Assigned by a central authority, the Net ID specifies the address, unique across the Internet, for each network or related group of networks. Assigned by the local network administrator, the Host ID specifies a particular host, station or node within a given network and need only be unique within that network.

A Transport Layer, which corresponds to the OSI Transport Layer. The Transmission Control Protocol (TCP) subset runs at this layer. TCP provides end-to-end connectivity between data source and destination with detection of, and recovery from, lost, duplicated, or corrupted packets -- thus offering the error control lacking in lower level IP routing. In TCP, message blocks from applications are divided into smaller segments, each with a sequence number that indicates the order of the segment within the block. The destination device examines the message segments and, when a complete sequence of segments is received, sends an acknowledgement (ACK) to the source, containing the number of the next byte expected at the destination.

An Application Layer, which corresponds to the session, presentation and application layers of the OSI model. This layer manages the function required by the user programs and includes protocols for remote log-in (Telnet), file transfer (FTP), and electronic mail (SMTP).

TELEPHONE LINE SIMULATOR
Also called ring-down box. Ring down boxes, also known as CO simulators, are simple devices used for generating calls from a POTS line to a computer telephony system (or vice versa). When one side goes offhook, the ring down box will "ring" the other side. When both sides are offhook, both sides are coupled together and the line is powered. Ring down boxes are available with various options and configurations. These include the ability to provide dialtone to the caller side (required to test applications with modems, faxes, or other automated outdialing devices), caller ID, and disconnect supervision. They are generally available in one to four line sizes, although special configurations may support more. Ring-down boxes are used for giving demonstrations and testing. We use them in our test labs to testdrive new computer telephony systems.

TELEPHONY SERVER
A telephony server is a computer whose major function is to control, add intelligence, store, forward and manipulate the various voice, data, fax and e-mail calls flowing into and out of a computer telephony system. The traditional function of a telephony server is to move call control commands from client workstations on a LAN to an attached PBX or ACD. (This is what it does under the paradigm called "Telephony Services.") A telephony server can also be a voice response system. It can also be a fax on demand system. It can also be a conferencing device. It can also be switch. And it can be all these capabilities, which traditionally run on physically separate servers, all rolled into one machine, called generically a "telephony server."

TIE LINE
A dedicated circuit linking two points without having to dial the normal phone number. A tie line may be accessed by lifting a telephone handset or by pushing one, two or three buttons.

TIME DIVISION MULTIPLEX
TDM. A technique for transmitting a number of separate data, voice and/or video signals simultaneously over one communications medium by quickly interleaving a piece of each signal one after another. Here's our problem. We have to transport the freight of five manufacturers from Chicago to New York. Each manufacturer's freight will fit into 20 rail boxcars. We have three basic solutions. First, build five separate railway lines from Chicago to New York. Second, rent five

engines and schlepp five complete trains to New York on one railway track. Or, third, join all the boxcars together into one train of 100 boxcars and run them on one track. The train might look like this: Engine, Boxcar from Producer A, Box Car from Producer B, Producer C, Producer D, Producer E, and then the order begins again...Boxcar from Producer A, Producer B...Moving one large train of 100 boxcars is likely to be cheaper and more efficient than moving five smaller trains each of 20 boxcars on five separate railway tracks. Time Division Multiplexing, thus, represents substantial savings over have five separate networks (five separate tracks) and sending five separate transmissions (five separate trains).

This is what Time Division Multiplexing is all about. And the analogy is perfect. Take one large train (fast communications channel) and interleave pieces (boxcars) from each conversation one after another. If you do this fast enough, you'll never notice you've broken the conversations apart, moved them separately, and then put them back together at the distant end. In TDM, you "sample" each voice conversation, interleave the samples, send them on their way, then reconstruct the several conversations at the other end. There are several ways to do the sampling. You can sample eight bits (one byte) of each conversation, or you can sample one bit. The former is called word interleaving; the latter bit interleaving. The basic goal of multiplexing -- whether it be time division multiplexing, or any other form
-- is to save money, to cram more conversations (voice, data, video or facsimile) onto fewer phone lines. To substitute electronics for copper.

TIME SLOT
1. In time division multiplexing or switching, the slot belonging to a voice, data or video conversation. It can be occupied with conversation or left blank. But the slot is always present. You can tell the capacity of the switch or the transmission channel by figuring how many slots are present.

2. An SCSA term. The smallest switchable data unit on the SCbus or SCxbus Data Bus. A time slot consists of eight consecutive bits of data. One time slot is equivalent to a data path with a bandwidth of 64 Kbps.

TIP & RING
An old fashioned way of saying "plus" and "minus," or ground and positive in electrical circuits. Tip and Ring are telephony terms. They derive their names from the operator's cordboard plug. The tip wire was connected to the tip of the plug, and the ring wire was connected to the slip ring around the jack. A third conductor on some jacks was called the sleeve. That's it. Nothing more sinister. Nothing more interesting.

TOUCHTONE
Touchtone is not a trademark of AT&T, despite what editions one through six of Newton's Telecom Dictionary said. It is a generic term for pushbutton telephones and pushbutton telecommunications services and the term "touchtone" may be used by anyone. At one stage it was a trademark of AT&T. At divestiture in 1984,

AT&T gave it to the public. And that's who owns it now. The public. If you don't believe this, call Frank L. Politano, AT&T Trademark and Copyright Counsel, in Basking Ridge, New Jersey. For a full explanation of touchtone, see DTMF, which stands for Dual Tone Multi Frequency signaling, i.e. touchtone.

TRELLIS CODING
A method of forward error correction used in certain high-speed modems where each signal element is assigned a coded binary value representing that element's phase and amplitude. It allows the receiving modem to determine, based on the value of the preceding signal, whether or not a given signal element is received in error.

TRELLIS CODING MODULATION
TCM. A modem modulation technique in which sophisticated mathematics are used to predict the best fit between the incoming signal and a large set of possible combinations of amplitude and phase changes. TCM provides for transmission speeds of 14,400 bps and above on single voice grade phone lines. See V.32 and V.32 bis.

TRUNK GROUP
A group of essentially like trunks that go between the same two geographical points. They have similar electrical characteristics. A trunk group performs the same function as a single trunk, except that on a trunk group you can carry multiple conversations. You use a trunk group when your traffic demands it. Typically, the trunks in a trunk group are accessed the same way. You dial your Band 5 WATS trunk group by dialing 62, for example. If the first trunk of that group is busy, you choose the second, then the third, etc.

TRUNK GROUP MULTIPLEXER
TGM. A time division multiplexer whose function is to combine individual digital trunk groups into a higher rate bit stream for transmission over wideband digital communication links.

U-LAW
A voice amplitude compression/expansion quasi-logarithmic curve, based on the approximation with 15 linear segments. Used for PCM encoding/decoding in North America.

V FAST
A new higher speed over-normal-phone-line modem called V.Fast Class (V.FC) for 28,800 bits per second speed. See V.34.

V SERIES RECOMMENDATIONS
ITU-T standards dealing with data communications operation over the telephone network. The idea of standards is simple. If you have them and if every manufacturer conforms, then every modem can talk to every other one. That's the idea. But it's not always that simple. Sometimes you have to conform to several

standards. For example, in the higher speed modems, for example those at 9,600 bps, you have to conform to speed (that's one standard). You have to conform to error control. That's another standard. And you also have to conform to data compression -- if you are using data compression. ISDN terminal adapters are V series recommendations, too. ITU-T uses the term "bis" to designate the second in a family of related standards and "ter" designates the third in a family.

V.110
Terminal rate adaptation protocols for the ISDN B channel with a V-type interface. Includes V.120.

V.120
Terminal rate adaptation protocols for the ISDN B channel with a V-type interface. Includes V.110.

V.13
ITU-T standard for simulated carrier control. Allows a full-duplex modem to be used to emulate a half-duplex modem with interchange circuits changing at appropriate times.

V.14
ITU-T standard for asynchronous-to-synchronous conversion without error control. Allows a modem that is actually synchronous to be used to carry start/stop (async) characters. If a V.42 modem connects with another modem that doesn't have error-control, it falls back to V.14 operation to work without error-control.

V.17
New ITU-T standard for simplex (one-way transmission) modulation technique for use in extended Group 3 Facsimile applications only. Provides 7200, 9600, 12000, and 14400 bps trellis-coded modulation (the modulation scheme is similar to V.33), MMR (Modified Modified Read) compression and error-correction mode (ECM).

V.21
ITU-T standard for 300 bit per second duplex modems for use on the switched telephone network. V.21 modulation is used in a half-duplex mode for Group 3 fax negotiation and control procedures (ITU-T T.30). Modems made in the U.S. or Canada follow the Bell 103 standard. However, the modem can be set to answer V.21 calls from overseas.

V.21 CH 2
ITU-T standard for 300 bps modem, describing the operation of modems at 300 bps, and used for critical control and handshaking functions. This low speed is highly tolerant of noise and impairments on the phone line. Fax machines use only Channel 2 of the V.21 recommendations (half duplex channel).

V.21 FAX

An ITU-T standard for facsimile operations at 300 bps.

V.22

ITU-T standard for 1,200 bit per second duplex modems for use on the switched telephone network and on leased circuits. V.22 is compatible with the Bell 212A standard observed in the U.S. and Canada.

V.22 bis

ITU-T standard for 2,400 bit per second duplex modems for use on the switched telephone network. V.22 bis also provides for 1200bps operation for V.22 compatibility. Bis is used by the ITU-T to designate the second in a family of related standards. "ter" designates the third in a family. The standard includes an automatic link negotiation fallback to 1200 bps and compatibility with Bell 212A/V.22 modems.

V.23

V.23 is the standard for a modem with a 600 bps or 1200 bps "forward channel" and a 75 bps "reverse" channel for use on the switched telephone network.

V.24

ITU-T definitions for interchange circuits between data terminal equipment (DTE) and data communications equipment (DCE) equipment. In data communications, V.24 is a set of standards specifying the characteristics for interfaces. Those standards include descriptions of the various functions provided by each of the pins. This standard is similar (but not identical) to the RS-232-C as established by the American TIA/EIA --- Telecommunications Industry Association / Electronics Industries Association.

V.25

Automatic calling and/or answering equipment on the general switched telephone network, including disabling of echo suppressors on manually established calls. Among other things, V.25 specifies an answer tone different from the Bell answer tone. Many modems, including U.S. Robotics modems, can be set with the BO command so that they use the V.25 2100 Hz tone when answering overseas calls.

V.25 bis

An ITU-T standard for synchronous communications between the mainframe or host and the modem using the HDLC or character-oriented protocol. Modulation depends on the serial port rate and setting of the transmitting clock source.

V.26

V.26 is the ITU-T standard for 2400 bps modem for use on 4-wire leased lines.

V.26 bis

ITU-T standard for 1.2/2.4 Kbps modem. It is important to note that V.26 bis is a half-duplex modem (1200 or 2400 bps in only one direction at a time); it provides an optional 75 bps reverse channel as well.

V.26 ter
V.26 ter is a FULL DUPLEX 2400 bps modem, like V.22 bis. The difference is that V.26 ter uses echo cancellation (like V.32) instead of frequency division (like V.22 bis), making it more expensive than V.22 bis. It was intended to serve as a fallback mode from V.32, but most manufacturers ignored it and provide V.22 bis as a fallback instead (V.26 ter is used only in a few installations in France, as far as we know).

V.27
ITU-T standard for 4,800 bits per second modem with manual equalizer for use on leased telephone-type circuits. May be full-duplex on four wire leased lines, or half-duplex on two wire lines.

V.27 bis
ITU-T standard for 2.4/4.8-kbit/s modem with automatic equalizer for use on leased telephone-type circuits. 2.4 Kbps modem for 4-wire leased circuits. Either speed (2400 is a fallback) can be used on either 4-wire leased lines (full duplex) or 2-wire leased lines (half-duplex). It also provides an optional 75 bps reverse channel.

V.27 ter
ITU-T standard for 2.4/4.8-kbit/s modem for use on the switched telephone network. Half-Duplex only. V.27 ter is the modulation scheme used in Group 3 Facsimile for image transfer at 2400 and 4800 bps. 4800 bps is a common "fallback" speed.

V.28
V.28, entitled "Electrical Characteristics for Unbalanced Double-Current Interchange Circuits" provides the ITU-T equivalent of the electrical characteristics defined in EIA-232.

V.29
ITU-T standard for 9,600 bits per second modem for use on point-to-point leased circuits. Virtually all 9,600 bps leased line modems adhere to this standard. V.29 uses a carrier frequency of 1700 Hz which is varied in both phase and amplitude. V.29 also provides fallback rates of 4800 and 7200 bps. V.29 can be full-duplex on 4-wire leased circuits, or half-duplex on two wire and dial up circuits. V.29 is the modulation technique used in Group 3 fax for image transfer at 7200 and 9600bps.

V.3
ITU-T specification that describes communications control procedures implemented in 7-bit ASCII code.

V.32

ITU-T standard for 9,600 bit per second two wire full duplex modem operating on regular dial up lines or 2-wire leased lines. If you're buying a 9,600 bps modem for use on the normal dial up switched phone lines, make sure it conforms to V.32. If your modem also conforms to V.42 bis, you should be able to transmit and receive at up to 38,400 bps with other modems that conform to these two specifications. I personally use a number of V.32/V.42 bis modem and they work wonderfully fast. V.32 also provides fallback operation at 4,800 bps.

V.32 bis

New higher speed ITU-T standard for full-duplex transmission on two wire leased and dial up lines at 4,800, 7,200, 9,600, 12,000, and 14,400 bps. Provides backward compatibility with V.32. Modems running at V.32 bis at its highest speed of 14,400 bps are actually transmitting that many bits per seconds. They do not rely on compression to achieve that high speed. However, with data compression -- such as V.42 and V.42 bis -- they can achieve higher speeds. The V.32 bis standard also includes "rapid rate renegotiation" feature to allow quick and smooth rate changes when line conditions change. See MODULATION PROTOCOLS, V.42 and V.42 bis.

V.32 terbo

Modulation scheme that extends the V.32 connection range: 4800, 7200, 9600, 12K and 14.4K bps. V.32 bis terbo modems fall back to the next lower speed when line quality is impaired, and fall back further as necessary. They fall forward to the next higher speed when line quality improves.

V.33

ITU-T standard for 14,400 and 12,000 bps modem for use on four wire leased lines.

V.34

V.34 is a new international standard for dial up modems of up to 28,800 bits per second. Since the standard suggests speeds twice as fast as the top standard they replace, they carry the nickname "V.Fast." New V.34 modems have a feature called line probing that will allows them to identify the capacities and quality of the specific phone line and adjust themselves to allow, for each individual connection, for maximum throughput. The standard also supports a half-duplex mode of operation for fax applications. The new V.34 technology includes an optional auxiliary channel with a synchronous data signaling rate of 200 bits/second. Data conveyed on this channel consists of modem control data. V.34 modems contain multidimensional trellis coding, which is used to gain higher immunity to noise and other phone line impairments. V.34 modems are the first modems to identify themselves to telephone network equipment (handshaking). V.34 technology has been long in coming and has had to overcome many obstacles. At one point, members of the modem manufacturing industry became so impatient, that some of them began shipping their own proprietary versions of what they thought V.34/V.Fast/28,800 bps modems would be. Many of these modems are only compatible, at higher than 14,400 bps speeds, with themselves. See V.34bis.

V.34 bis
A faster version of the data communications standard, V.34, which supports up to 28,800 bps. V.34bis adds two optional higher data rates to V.34. These speeds are 31,200 bit/s and 33,600 bit/s.

V.35
ITU-T standard for trunk interface between a network access device and a packet network that defines signaling for data rates greater than 19.2 Kbps. V.35 was a definition of a GROUP band modem (meaning, one that used the bandwidth of several telephone circuits). V.36 and V.37 are other group band modems. V.35 just happened to describe, in an appendix, electrical characteristics for a high speed interface. IBM decided to use this for other things, but it was never standardized for those purposes. The ITU-T, in the 1988 Blue Book, says "It is the opinion of the ITU-T that the information contained in Recommendation V.35 is out of date. Therefore it is not recommended to use the techniques described in this Recommendation for new designs. Alternative techniques are described in Recommendations V.36 and V.37. It should be noted that other Recommendations make reference to the electrical characteristics described in Appendix II to this Recommendation [V.35]. As these characteristics are expected to allow interworking with V.11 characteristics, use of V.11 circuits is recommended in those cases." V.35 is no longer published by the ITU-T.

V.42 ERROR CORRECTION
ITU-T error-correction standard specifying both MNP4 and LAP-M. The ITU-T title says "Error-correcting procedures for DCEs using Asynchronous-to-Synchronous Conversion". It also notes in the text that it applies only to full-duplex devices. The ITU-T modulation schemes with which V.42 may be used are V.22, V.22 bis, V.26 ter, and V.32, and V.32 bis. LAPM, based on HDLC, is the "primary" protocol, on which all future extensions will be based. The Alternative Protocol specified in Annex A of the Recommendation is for backward compatibility with the "installed base" of error-correcting modems. See V.42 bis.

V.42 bis DATA COMPRESSION
ITU-T data compression standard. It compresses files "on the fly" at an average ratio of 3.5:1 and can yield file transfer speeds of up to 9,600 bps on a 2,400 bps modem, 38,400 bits per second with a 9,600 bps modem, 57,600 bps with a 14,400 bps V.32 bis modem, or 115,600 bit/s on a 28,800 bps modem. On-the-fly data compression only has value if you use it to transfer and receive material that is not already compressed. Compressing stuff a second time yields no significant improvement in speed (assuming your compression technique the first time around worked). So the decision to buy a V.42 bis modem depends on the material you're working with and your pocketbook. V.42 bis modems are more expensive.

V.42 bis was approved by the ITU-T because of its technical merits. Existing data compression methods (MNP 5 for example) only provided up to two-to-one compression. Also, V.42 bis provides for built-in "feedback" mechanisms, so that the modem can monitor its own compression performance. If the DTE starts send pre-compressed or otherwise uncompressible data, V.42 bis can automatically suspend its operation to avoid expansion of the data. It continues to monitor performance even when sending data "in the clear," and when a performance improvement can be gained by reactivating compression, it will do so automatically.

V.42 bis was selected because it would work with a wide variety of different implementations -- different amounts of memory, different processor speeds, etc. Because of this, there WILL be differences between various manufacturer's products in terms of THROUGHPUT performance (although they will all properly compress and decompress, some will do it faster than others). If maximum throughput is important, you should check published benchmark tests to find the modem that provides the best performance.

V.54
ITU-T standard for loop test devices in modems, DCEs (Data Communications Equipments) and DTEs (Data Terminal Equipment). Defines local and remote loopbacks. There are four basic tests -- a local digital loopback test that is used to test the DTE's send and receive circuits; a local analog loopback test that is used to test the local modem's operation; a remote analog loopback test that is used to test the communication link to the remote modem; and a remote digital loopback test that is used to test the remote modem's operation. If a modem has V.54 capability (most V.32 and V.32 bis modems do), its manual should include documentation on performing the various tests. Version 7 of the Norton Utilities (from Symantec) also includes a local digital loopback test for your PC's COM ports, for which you will need the optional jumper plug offered with the software. Where a modem supports local digital loopback testing, it simulates the jumper plug and does not, therefore, need to be disconnected.

V.FAST
V.FC. An interim modem standard to support speeds to 28,800 bits per second for uncompressed data transmission rates over regular dial up, voice-grade lines. V.FAST stands for Very Fast. V.Fast was a "standard" that only a few manufacturers of modems adopted. These manufacturers adopted V.Fast because they were impatient with the ITU's slowness. Eventually, however the ITU did adopt a new standard, called V.34. See V.34 and V.34bis.

V.FC
Version Fast Class. It is an interim standard that was developed for use until the ITU-T ratified V.Fast, i.e. V.34, which is the speed that a V.34 modem communicates at -- namely at 28,800 bits per second. V.FC was eventually obsoleted by V.34, which the ITU-T eventually adopted. See V.34.

VIDEO DIAL TONE
Video dial tone in telco-speak means the phone company, in competition with the cable TV business, provides video to houses and offices. It does not affect the content of that video signal in any way. Thus the term video dial tone, which is like voice dial tone, whose content the phone company also does not affect or change in any way, shape or form.

VIDEOCONFERENCING
Video and audio communication between two or more people via a videocodec (coder/decoder) at either end and linked by digital circuits. Formerly needing at least T-1 speeds (1.54 megabits per second), systems are now available offering acceptable quality for general use at 128 Kbit/s and reasonable 7 KHz audio. Factors influencing the growth of videoconferencing are improved compression technology, reduced cost through VLSI chip technology, lower-cost switched digital networks -- particularly T-1, fractional T-1, and ISDN -- and the emergence of standards.

VIDEOCONFERENCING STANDARDS
ITU-T H.261 was the standards watershed. Announced in November 1990, it relates to the decoding process used when decompressing videoconferencing pictures, providing a uniform process for codecs to read the incoming signals. Originally defined by Compression Labs Inc. Other important standards are H.221: communications framing; H.230 control and indication signals and H.242d: call setup and disconnect. Encryption, still-frame graphics coding and data transmission standards are still being developed.

VIRTUAL PRIVATE NETWORK
VPN. A carrier-provided service in which the public switched network provides capabilities similar to those of private lines, such as conditioning, error testing, and higher speed, full-duplex, four wire transmission with a line quality adequate for data. A virtual private network eliminates or partly eliminates the need for fixed point-to-point private line because it provides on demand dial up circuits or bandwidths that can be dynamically allocated. AT&T, a major provider of virtually private networks, defines them as the equivalent of a private network designed logically within a public network,thus achieving the economy of scale of a public network while offering the user control of the simulated private network. Virtual private network resources are occupied only while information is transiting the network.

VIRTUAL NETWORK
A network that is programmed, not hard-wired, to meet a customer's specifications. Created on as-needed basis. Also called Software Defined Network by AT&T.

VIRTUAL PRIVATE NETWORK
VPN. A carrier-provided service in which the public switched network provides capabilities similar to those of private lines, such as conditioning, error testing, and

higher speed, full-duplex, four wire transmission with a line quality adequate for data. A virtual private network eliminates or partly eliminates the need for fixed point-to-point private line because it provides on demand dial up circuits or bandwidths that can be dynamically allocated. AT&T, a major provider of virtually private networks, defines them as the equivalent of a private network designed logically within a public network,thus achieving the economy of scale of a public network while offering the user control of the simulated private network. Virtual private network resources are occupied only while information is transiting the network.

VISUAL VOICE MESSAGING
A term created by Microsoft as part of its At Work announcement in June of 1993. There'll be At Work-based visual voice messaging servers sitting on a LAN. Messages for PC users on the LAN will be able to be displayed in a list, much like electronic mail, including the caller's name or number, the time he or she called and the length of the call. This information would let the user browse all messages and select the order for listening to the messages. Administrative options, such as creating a new greeting, will be accessed with a single button. Operations that are difficult today, such as forwarding a voice message to multiple people, will be dramatically simplified, according to Microsoft. One will simply select the recipients from the phone book and broadcast the message. Using visual voice messaging, users will be able to bypass today's inconsistent,
time consuming and confusing audio menus and access their voice messages with the push of a button or the click of a mouse on a Windows type icon. Messages will be able to be retrieved in any order and even delivered to a single mailbox along with other messages such as e-mail and faxes. These visual voice messaging servers will, according to Microsoft, provide applications beyond basic voice messaging, such as supporting voice annotation of PC documents or reading electronic mail over the phone to a traveler.

VITA
VFEA International Trade Association. A widely supported industry trade group in Scottsdale, AZ. VITA is chartered to promote the growth and technical excellence of the VME bus and Futurebus based microcomputer board market. VITA is chartered to submit standards for ANSI registration.

VME
Acronym for "VersaModule-Europe". A one through 21 slot, mechanical and electrical bus standard originally developed by the Munich, Germany division of Motorola in the late 70s. VME uses most of the bus structure from then current Motorola's VersaBus board standard along with the newly developed DIN 41612 standard pin-in-socket connector for enhanced reliability. After years of work, VME was finally adopted by the ANSI/IEEE in 1987 (as ANSI/IEEE-1014). VME is known in Europe as the IEC 821 bus. This makes it an open standard. The VME backplane runs at 80 Mbytes per second. It is the most common bus on big open computers (i.e. ones larger than the PC). As of writing, there were over 300 vendors offering more than 3,000 off-the-shelf VME products. The IEEE standard is

soon to lapse and be replaced by an extended VME64 specification, now in ANSI ballot being conducted by VITA.

VME64
An enhanced VME bus standard which includes multiplexed address and data cycles with 40 and 64 bit address modes and 64 bit data transfer modes allowing up to 80 MB/s transfer speed. This standard is under the ANSI ballot process conducted by VITA.

VME64 EXTENSIONS
A VITA draft standard that provides extra functionality to VME64 including 5 row J1/P1 and J2/P2 connectors that support live insertion on both 3U and 6U VME boards. Other features: 3.3V power, more grounds, ETL (slew rate) drivers, geographic addressing (slot ID) as well as support for parity, a serial diagnostic bus, JTAG test support and lots of user I/O. Some mechanical features: locking extractors, RFI gasketing, and ESD chassis discharge strips.

VOCABULARY DEVELOPMENT
Development of specific word sets to be used for speaker independent recognition applications.

VOICE ACTIVATED DIALING
A feature that permits you to dial a number by calling that number out to your cellular phone, instead of punching it in yourself.

VOICE APPLICATIONS PROGRAM
System software providing the necessary logic to carry out the functions requested by telephone system users. It is responsible for actual call processing, making the various voice connections and providing user features, such as Call Forwarding, Speed Dialing, Conference, etc.

VOICE BOARD
Also called a voice card or speech card. A Voice Board is an IBM PC- or AT-compatible expansion card which can perform voice processing functions. A voice board has several important characteristics: It has a computer bus connection. It has a telephone line interface. It typically has a voice bus connection. And it supports one of several operating systems, e.g. MS-DOS, UNIX. At a minimum, a voice board will usually include support for going on and off-hook (answering, initiating and terminating a call); notification of call termination (hang-up detection); sending flash hook; and dialing digits (touchtone and rotary).

VOICE BUS
Picture an open PC. Peer down into it. At the bottom of the PC, you'll see a printed circuit board containing chips and empty connectors. That board is called the motherboard. Fatherboards are inserted into the connectors on the motherboard. These fatherboards do things on the PC -- like pump out video to your screen or material to your printer or your local area network. The motherboard controls

which device does what WHEN by sending signals along the motherboard's data bus -- basically a circuit that connects all the various fatherboards through their connectors. That data bus was not designed for voice. For voice you need another bus. Several voice processing manufacturers have addressed that need by creating a voice bus at the top of their PC-based voice processing cards. They have tiny pins sticking out of their cards. You attach a ribbon cable from one set of pins on one voice processing card to the next set on the adjacent card and then the next. There are several voice bus "standards." Two come from Dialogic. One is called AEB, Analog Expansion Bus. And one is called PEB, PC Expansion Bus (a digital version). One comes from a consortium of companies and is called MVIP. There are many advantages to having a voice bus. It gives you enormous flexibility to mix and match voice processing boards, like voice recognition, voice synthesis, switching, voice storage, etc. You can build really powerful voice processing systems inside today's fast '386 and '486 PCs with the great variety of voice processing now available. For more information on this exciting field, read TELECONNECT Magazine. 212-691-8215.

VOICE COMPRESSION
Refers to the process of electronically modifying a 64 Kbps PCM voice channel to obtain a channel of 32 Kbps or less for the purpose of increased efficiency in transmission.

VOICE DIALING
The ability to tell your phone to dial by talking to it. Say, "Call Police" and it will automatically dial the police. This feature has enormous benefits for handicapped people. It will have greater benefits for normal people when the technology of voice recognition improves.

VOICE DIGITIZATION
The conversion of an analog voice signal into binary (digital) bits for storage or transmission.

VOICE DTMF FORMS APPLICATIONS
This Voice DTMF (DUAL TONE MULTIPLE FREQUENCY) application allows a use of a voice mail system to take specific information from its customers 24 hours a day. By prompting callers to respond by speaking or pressing the keys of their touchtone phones, a city department, for example, could plan service calls, building inspections or send out appropriate forms.

VOICE MAIL
Voice Mail allows you to receive, edit and forward messages to one or more voice mailboxes in your company or in your universe of friends. With voice mail, employees can have their own private mailboxes. Here's an explanation of how it works: You call a number. A machine answers. "Sorry. I'm not in. Leave me a message and I'll call you back." It could be a $50 answering machine. Or it could be a $200,000 voice mail "system." The primary purpose is the same -- to leave someone a message. After that, the differences become profound. a voice mail

system lets you handle a voice message as you would a paper message. You can copy it, store it, send it to one or many people, with or without your own comments. When voice mail helps business, it has enormous benefits. When it's abused -- such as when people "hide" behind it and never return their messages -- it's useless. Some people hate voice mail. Some people love it. It's clearly here to stay.

VOICE MAIL SYSTEM
A device to record, store and retrieve voice messages. There are two types of voice mail devices -- those which are "stand alone" and those which profess some integration with the user's phone system. A stand alone voice mail is not dissimilar to a collection of single person answering machines, with several added features. You can instruct the machines (voice mail boxes) to forward messages among themselves. You can organize to allocate your friends and business acquaintances their own mail boxes so they can dial, leave messages, pick up messages from you, pass messages to you, etc. You can also edit messages, add comments and deliver messages to a mailbox at a pre-arranged time. Messages can be tagged "urgent" or "non-urgent" or stored for future listening. The range of voice mail options varies among manufacturers.

An integrated voice mail system includes two additional features. First, it will tell you if you have any messages. It does this by lighting a light on your phone and/or putting a message on your phone's alphanumeric display. Second, if your phone rings for a certain number of rings (you set the number), the phone will transfer your caller automatically to your voice mail box, which will answer the phone, deliver a little "I am away" message and then receive and record the caller's message.

There are other levels of integration. You might have a phone which has "soft" buttons and an alphanumeric display. That display might label your phone's soft buttons like those on a cassette recorder -- forward, reverse, slow, fast, stop, etc. so you can go through your messages any way you like. Telenova has such a phone. It's very impressive.

There are pros and cons to voice mail systems. Some employees will hide behind them, forwarding calls from their customers into voice mail boxes and never returning them. Some employees will make good use of them. They dial in for their messages, research what the customer wants and return the voice mail calls quickly. Many voice mail systems are being combined with automated attendants. Many are being combined with interactive voice processing systems, including sophisticated tie-ins to mainframe databases. Some people hate voice mail systems. Others love them. It all depends on how the system is used, managed and sold.

VOICE MESSAGE SERVICE

A leased service typically over dial up phone lines which provides the ability for a phone user to access a voice mail system and leave a message for a particular phone user.

VOICE PRINT
A voice recognition term. A voice print is a speech template used to "train" systems, in particular voice patterns. When a system is operating, the user's speech is compared to the stored voice prints. If they match, the system recognizes the word and executes the command.

VOICE PROCESSING
Think of voice processing as voice computer. Where a computer has a keyboard for entering information, a voice processing system recognizes touchtones from remote telephones. It may also recognize spoken words. Where a computer has a screen for showing results, a voice processing system uses a digitized synthesized voice to "read" the screen to the distant caller. Whatever a computer can do, a voice processing system can too, from looking up train timetables to moving calls around a business (auto attendant) to taking messages (voice mail). The only limitation on a voice processing system is that you can't present as many alternatives on a phone as you can on a screen. The caller's brain simply can't remember more than a few. With voice processing, you have to present the menus in smaller chunks.

Voice processing is the broad term made up of two narrower terms -- call processing and content processing. Call processing consists of physically moving the call around. Think of call processing as switching. Content consists of actually doing something to the call's content, like digitizing it and storing it on a hard disk, or editing it, or recognizing it (voice recognition) or some purpose (e.g. using it as input into a computer program.)

VOICE RECOGNITION
The ability of a machine to recognize your particular voice. This contrasts with speech recognition, which is different. Speech recognition is the ability of a machine to understand human speech -- yours and most everyone else's. Voice recognition needs training. Speech recognition doesn't.

VOICE RESPONSE UNIT
VRU. Think of a Voice Response Unit (also called Interactive Voice Response Unit) as a voice computer. Where a computer has a keyboard for entering information, an IVR uses remote touchtone telephones. Where a computer has a screen for showing the results, an IVR uses a digitized synthesized voice to "read" the screen to the distant caller. An IVR can do whatever a computer can, from looking up train timetables to moving calls around an automatic call distributor (ACD). The only limitation on an IVR is that you can't present as many alternatives on a phone as you can on a screen. The caller's brain simply won't remember more than a few. With IVR, you have to present the menus in smaller chunks.

VOICE SERVER
A PC sitting on a LAN (Local Area Network) and containing voice files which are accessible by the PCs on the LAN. Such voice files may be transmitted on the LAN or over phone lines under the control of the PCs on the LAN. A voice server might contain voice mail. It might contain voice annotated electronic mail. Its primary function is to store voice in such a way that it's accessible easily. Voice servers are typically faster, have more disk capacity and more backup provisions than normal PCs. According to a letter I received in early May, 1993 from the lawyers for a company called Digital Sound Corporation, that company owns federal trademark registration number 1,324,258 for the mark Voiceserver, spelled as one word, not two.

VSELP
Vector Sum Exited Linear Prediction. A speech coding technique used in U.S. and proposed Japanese DMR standards. Second generation European DMR will probably use some version of VSELP.

VOICE STORE AND FORWARD
Voice mail. A PBX service that allows voice messages to be stored digitally in secondary storage and retrieved remotely by dialing access and identification codes.

VOICE VERIFICATION
The process of verifying one's claimed identity through analyzing voice patterns.

VPDN
Virtual Private Data Network. A private data communications network built on public switching and transport facilities rather than dedicated leased facilities such as T1s.

VPN
Virtual Private Network. Virtual Private Network is a software-defined network offering the appearance, functionality and usefulness of a dedicated private network, at a price savings. Here's how it works: Your company buys a bunch of leased lines from your offices to the nearest local offices of your chosen long distance carrier. You're in your New York offices. You want to dial your offices in Chicago. You pick up the phone, dial perhaps seven digits. The phone rings in Chicago. What's happened is that your local PBX has recognized that call as belonging to your VPN. So it shunts the calls over the dedicated local loop to your long distance carrier. Your carrier then checks your dialed number, perhaps changing it with the aid of a database look up table, and completes the call over the carrier's own switched telephone facilities (fiber optic, microwave, copper, etc.). These are the same facilities which you and I use when we dial 1 and the long distance number (assuming we're equal accessed to that carrier).

WAIS

Wide-Area Information Servers. A very powerful system for looking up information in databases (or libraries) across the Internet. WAIS allows you to perform a keyword search. WAIS is like an index, whereas Gopher, which is sometimes used as a complement to WAIS, is like a table of contents.

WAIT STATE
A period of time when the processor does nothing; it simply waits. A wait state is used to synchronize circuitry or devices operating at different speeds. Wait states are introduced into computers to compensate for the fact that the central microprocessor might be faster than the memory chips next to it. For example, wait states used in memory access slow down the CPU so that all components seem to be running at the same speed. A wait state is a "missed beat" in the cycle of information to and from the CPU that is necessary for a memory transaction to be completed.

WARBLE TONE
A tone changing in frequency at a slow enough rate to give the effect of warbling. A warble tone is the sound of an electronic ringer, according to many people.

WATS
Wide Area Telecommunications Service. Basically, a discounted toll service provided by all long distance and local phone companies. AT&T started WATS but forgot to trademark the name, so now every supplier uses it as a generic name. There are two types of WATS services -- in and out WATS, i.e. those WATS lines that allow you to dial out and those on which you receive incoming calls (the typical 800 line service). You subscribe to in- and out-WATS services separately. In the old days you needed separate in and out lines to handle the in and out WATS services. But these days you can choose to have in- and out-WATS on the same line. This is not particularly brilliant traffic engineering, since you can't receive an incoming 800 call if you're making an outgoing call. But I do know someone who has an 800 line on his cellular phone! Many users inside companies think their company's WATS lines (and thus their WATS calls) are free, so they speak longer. This can kill the idea of buying WATS lines to save money. In the old days, interstate WATS was charged at effectively a flat rate and thus, there was some reason to believe that marginal WATS calls were 'free.' These days EVERY WATS call costs money. EVERY one! Without exception.

WEB SERVICE PROVIDER
A vendor who provides customers with Web Pages on the vendor's computer/s. Frequently, a Web Service Provider will provide additional services such as design help and usage statistics. Often they will just provide the computer space and leave the rest to you. A Web Service Provider may or may not also be an Internet Service Provider.

WEB SITE
Any machine on the Internet that is running a Web Server to respond to requests from remote Web Browsers is a Web Site. In more common usage it refers to

individual sets of Web Pages that can be visited with Web Browsers. It is also spelled as one word, namely website.

WEBMASTER
An Internet term. The Webmaster is the administrator responsible for the management and often design of a company's World Wide Web site.

WHISPER TECHNOLOGY
A call comes into a call center. The voice response unit prompts the caller to the enter their account number. When the call is transferred to the agent, the VRU "whispers" the account number to the agent, who then manually types it into his computer. This technology is now obsolete, since VRUs can now transfer their account number directly into the agent's database and have the look up done automatically. And the call is transferred simultaneously.

WHITE FACSIMILE TRANSMISSION
In an amplitude-modulated facsimile system, that form of transmission in which the maximum transmitted power corresponds to the minimum density of the subject copy. In a frequency-modulated system, that form of transmission in which the lowest transmitted frequency corresponds to the minimum density of the subject copy.

WHITEBOARD
a device which lets you share images, text and data simultaneously as you speak on the phone with someone else. That someone might be in the next office. Or that someone might be 3,000 miles away. The transport mechanism might be a local area network or an analog phone line running a special modem designed for whiteboarding or it might be an ISDN digital line running special PC software and hardware. The concept of whiteboarding is new; there are no standards. As a result to do whiteboarding successfully, you typically need the same equipment (hardware and software) on either end. Whiteboarding has the potential to be one of the most successful "multimedia" applications around.

WIDE AREA NETWORK
WAN. An data network typically extending a LAN (local area network) outside the building, over telephone common carrier lines to link to other LANs in remote buildings in possibly remote cities. A WAN typically uses common-carrier lines. A LAN doesn't. WANs typically run over leased phone lines -- from one analog phone line to T1 (1.544 Mbps). The jump between a local area network and a WAN is made through a device called a bridge or a router. Bridges operate independently of the protocol employed. They will work, according to Jeff Weiss, of Cryptall Communications, with all present and expected future communications packages. Routers are specific to the protocol being employed. New routing software is needed for each new protocol or protocol deviation.

WINDOWS TELEPHONY

Introduced in the spring of 1993 jointly by Microsoft and Intel, Windows Telephony is a piece of software called a Windows Telephony DLL AND two standards. The first standard is the Service Provider Interface (SPI). If a hardware manufacturer's product honors that SPI, that product can happily talk to the Windows Telephony DLL. The second standard is called the Application Programming Interface and it is directed at software developers who write applications programs. If those developers' programs adhere to the API, they can take advantage of the Windows Telephony DLL to drive whatever telephony devices or services adhere to the SPI. The Windows Telephony API is affectionately called TAPI. DLL stands for Dynamic Link Library. It is a Windows feature that allows executable code modules to be loaded on demand and linked at run time.

Windows Telephony should bring about an explosion of shrink-wrapped Windows based telephone software applications -- from simple personal rolodexes to power dialers, to customized phone systems for banks and for bakers. It should also bring about an explosion of new telephony hardware devices -- from telephones that look more like PCs than phones, to PCs that are phones, to blackbox telephony devices that hook to laptops and transform hotel phones.

Windows Telephony effectively removes earlier overwhelming barriers to creating PC-driven telephony applications, namely the wide enormity of telephony "network" services -- from the many telephone company interfaces (POTS to T-1), to the many more proprietary interfaces behind dozens of proprietary PBXs, key systems and hybrid phone systems.

The goal is to bundle the Windows Telephony DLL in the next major release of Windows, sometime in 1994. It will also be included in Windows NT and Windows NT Server. Although the Windows NT code may be different, the API and SPI interfaces will be the same, thus causing no re-write of software code or necessitating redesign of telephony hardware.

The original work on the Windows Telephony DLL was done by Herman D'Hooge, a senior software architect with Intel's Architecture Development Lab in Hillsboro, Oregon. The final effort is a result of joint development effort with Microsoft, where the team was headed by Charles Fitzgerald. It also includes input from 40-odd companies -- including virtually all major telecom switch vendors and several major telephony developers.

The goal of joint Microsoft/Intel Windows Telephony is to get rid of the bottleneck to bringing the power of the PC to telephony. Intel and Microsoft believe that the bottleneck exists because of two factors:

First, it has been incredibly difficult to interface to the variety of telecom switches in existence today. For example, no manufacturer's switch will talk to another's manufacturer's proprietary phone.

Second, it has been incredibly redundant and time consuming for software to talk to the various switches. The big analogy is word processing in the old days. In those days, each word processing software company could easily spend 99% of his R&D budget writing drivers to get his program to work with yet another new printer. That is no longer necessary under Windows. Windows takes care of interfacing the printers. All you have to do, as a developer is to make sure you conform to Windows specs.

According to Microsoft, Windows Telephony products will include ones, such as those including:

* Visual interface to telephone features.
* Personal communication management. With a graphical user interface, people can have their PCs handle incoming telephone calls, automatically controlling which calls reach them. For example, people will be able to ensure that they receive important calls by requesting that certain calls be forwarded automatically to locations at which they expect to be working. * Telephone network access. A personal information management can be used not only to look up phone numbers, but also to actually place calls to those numbers.
* Integrated messaging. People will be able to check their messages -- electronic mail, fax mail and voice mail -- from a single place, namely their telephony empowered PC. Also voice mail messages can be accessed randomly, which is far more efficient than the serial access provided on most telephone based voice mail-mail systems.
* Integrated meetings. Here's Microsoft's explanation: "One of the most attractive capabilities of the computer is that it can store, communicate and present information that spans the entire spectrum of media -- text, data, graphics, voice and video in any combination. By itself, the telephone can communicate and present voice information only. By combining the functions of the computer and the telephone, people in geographically separate locations can participate in interactive meetings and share visual as well as audio information. That means they can hold meetings over the telephone network that are nearly as rich in information content as in-person meetings."

WINK
A signal sent between two telecommunications devices as part of a hand-shaking protocol. It is a momentary interruption in SF (Single Frequency) tone, indicating that the distant central office is ready to receive the digits that have just been dialed. In telephone switching systems, a single supervisory pulse. On a digital connection such as a T-1 circuit, a wink is signaled by a brief change in the A and B signaling bits. On an analog line, a wink is signaled by a change in polarity (electrical + and -) on the line.

WORD SPOTTING
In speech recognition over the phone, word spotting means looking for a particular phrase or word in spoken text and ignoring everything else. For example, if the word to spot was "brown," then it wouldn't matter if you said "I

want the brown one," or "how about something in brown?" In short, word spotting is the process whereby specific words are recognized under specific speaking conditions (i.e. natural, unconstrained speech). It can also refer to the ability to ignore extraneous sounds during continuous word recognition.

WORKSTATION
In the telecom industry, a workstation is a computer and a telephone on a desk and both attached to a telecom outlet on the wall. The computer industry tends to refer to workstations as high-speed personal computers, such as Sun workstations, which are used for high-powered processing tasks like CAD/CAM, engineering, etc. A common PC -- like the one you find on my desk -- is not usually considered a workstation. The term workstation is vague.

X-WINDOWS
The UNIX equivalent of Windows. What Windows is to MS-DOS, X-Windows is to Unix. A network-based windowing system that provides a program interface for graphic window displays. X-Windows permits graphics produced on one networked station to be displayed on another. Almost all UNIX graphical interfaces, including Motif and OpenLook, are based on X-Windows. X-Windows is a networked window system developed and specified by the MIT X Consortium. Members of the X Consortium include IBM, DEC, Hewlett-Packard and Sun Microsystems. Sun Microsystems has been contracted by the MIT X Consortium to implement PEX (PHIGS Extensions to X), which will be the standard networking protocol for sending PHIGS (Programmers Hierarchical Graphics System) graphics commands through X-Windows. Some people spell it X-Windows and some spell it X Windows.

X.25
From its beginning as an international standards recommendation from ITU-T, the term X.25 has come to represent a common reference point by which mainframe computers, word processors, mini-computers, VDUs, microcomputers and a wide variety of specialized terminal equipment from many manufacturers can be made to work together over a type of data communications network called a packet switched network. On a packet switched data network (private or public), the data to be transmitted is cut up into blocks. Each block has a header with the network address of the sender and that of the destination. As the block enters the network, the number of bits in the block are put through some mathematical functions (an algorithm) to produce a check sum.

The check sum is attached as a "trailer" to the packet as it enters the network. Packets may travel different routes through the network. But, ultimately, the packets are routed by the network to the node where the destination computer or terminal is located. At the destination, the packet is disassembled. The bits are put through the same algorithm, and if the digits computed are the same as the ones attached as the trailer, there are no detected errors. An ACK, or acknowledgement, is then sent to the transmitting end. If the check sum does not match, a NAK, or Negative Acknowledgement is sent back, and the packet is retransmitted. In this manner, high speed, low error rate information can be transmitted around the country using shared telecommunications circuits on public or private data networks.

X.25 is the protocol providing devices with direct connection to a packet switched network. These devices are typically larger computers, mainframes, minicomputers, etc. Word processors, personal computers, workstations, dumb terminals, etc. do not support the X.25 packet switching protocols unless they are connected to the network via PADs -- Packet
Assembler/Disassemblers. A PAD converts between the protocol used by the smaller device and the X.25 protocol. This conversion is performed on both outgoing (from the network) and incoming data (to the network), so the transmission looks transparent to the terminal.

X.25 NETWORK
Any network that implements the internationally accepted ITU-T standard governing the operation of packet-switching networks. The X.25 standard describes a switched communications service where call setup times are relatively fast. The standard also defines how data streams are to be assembled into packets, controlled, routed, and protected as they cross the network.

X.400
X.400 is an international standard which enables disparate electronic mail systems to exchange messages. Although each e-mail system may operate internally with its own, proprietary set of protocols, the X.400 protocol acts as a translating software making communication between the electronic mail systems possible. The result is that users can now reach beyond people on their same e-mail system to the universe of users of interconnected systems. One problem with e-mail sent between X.400 networks is that the sender's name is not sent. (I kid you not.) This was one element of the protocol the committees forgot! If your message crosses an X.400 network, remember to sign your name. The X.400 standard itself is an overview which is broken down under subsequent numbers:

X.500
The ITU-T international standard designation for a directory standard that permits applications such as electronic mail to access information which can either be central or distributed. The X.500 standard for directory services provides the means to consolidate e-mail directory information through central servers situated at strategic points throughout the network. These X.500 servers then exchange directory information so each server can keep all its local mail directory information current. With X.500, any e-mail user, whether on OpenVMS, Macintosh, DOS, or UNIX workstations, can be listed in a central directory that can be accessed using an X.500-compatible user agent.

ZIP TONE
Short burst of dial tone to an ACD agent headset indicating a call is being connected to the agent console.